T0362109

CRITICAL EXCITATION METHODS IN EARTHQUAKE ENGINEERING

CRITICAL EXCITATION METHODS IN EARTHQUAKE ENGINEERING

SECOND EDITION

IZURU TAKEWAKI

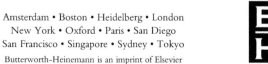

Amsterdam • Boston • Heidelberg • London
New York • Oxford • Paris • San Diego
San Francisco • Singapore • Sydney • Tokyo
Butterworth-Heinemann is an imprint of Elsevier

Butterworth-Heinemann is an imprint of Elsevier
Linacre House, Jordan Hill, Oxford OX2 8DP, UK
The Boulevard, Langford Lane, Kidlington, Oxford OX5 1GB, UK
225 Wyman Street, Waltham, MA 02451, USA

Second edition 2013

Copyright © 2013 Elsevier Ltd. All rights reserved.

No part of this publication may be reproduced, stored in a retrieval system or transmitted in
any form or by any means electronic, mechanical, photocopying, recording or otherwise
without the prior written permission of the publisher.

Permissions may be sought directly from Elsevier's Science & Technology Rights
Department in Oxford, UK: phone (+44) (0) 1865 843830; fax (+44) (0) 1865 853333;
email: permissions@elsevier.com. Alternatively you can submit your request online by
visiting the Elsevier web site at http://elsevier.com/locate/permissions, and selecting
"Obtaining permission to use Elsevier material."

Notice
No responsibility is assumed by the publisher for any injury and/or damage to persons
or property as a matter of products liability, negligence or otherwise, or from any use or
operation of any methods, products, instructions or ideas contained in the material herein.
Because of rapid advances in the medical sciences, in particular, independent verification
of diagnoses and drug dosages should be made.

British Library Cataloguing in Publication Data
A catalogue record for this book is available from the British Library.

Library of Congress Cataloging-in-Publication Data
A catalog record for this book is available from the Library of Congress.

ISBN: 978-0-08-099436-9

For information on all Elsevier publications
visit our web site at books.elsevier.com

Printed and bound by CPI Group (UK) Ltd, Croydon, CR0 4YY

Working together
to grow libraries in
developing countries

www.elsevier.com • www.bookaid.org

CONTENTS

3 Critical Excitation for Nonproportionally Damped Structural Systems · 51

4 Critical Excitation for Acceleration Response · 75

5 Critical Excitation for Elastic-Plastic Response · 97

PREFACE TO THE FIRST EDITION

There are a variety of buildings in a city. Each building has its own natural period and its original structural properties. When an earthquake occurs, a variety of ground motions are induced in the city. The combination of the building natural period with the predominant period of the induced ground motion may lead to disastrous phenomena in the city. Many past earthquake observations demonstrated such phenomena. Once a big earthquake occurs, some building codes are upgraded. However, it is true that this repetition never resolves all the issues and new damage problems occur even recently. In order to overcome this problem, a new paradigm has to be posed. To the author's knowledge, the concept of "critical excitation" and the structural design based on this concept can become one of such new paradigms.

It is believed that earthquake has a bound on its magnitude. In other words, the earthquake energy radiated from the fault has a bound. The problem is to find the most unfavorable ground motion for a building or a group of buildings (see Fig. 0.1).

A ground motion displacement spectrum or acceleration spectrum has been proposed at the rock surface depending on the seismic moment, distance from the fault, etc. (Fig. 0.2). Such spectrum may have uncertainties. One possibility or approach is to specify the acceleration or velocity power and allow the variability of the spectrum.

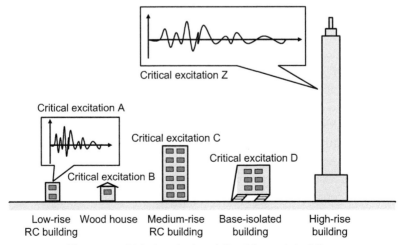

Figure 0.1 Critical excitation defined for each building.

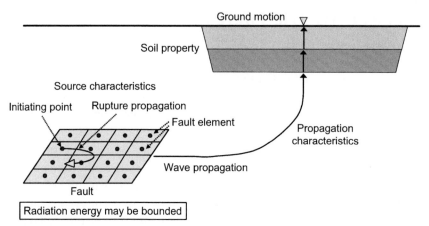

Figure 0.2 Earthquake ground motion depending on fault rupture mechanism, wave propagation and surface ground amplification, etc.

The problem of ground motion variability is very important and tough. Code-specified design ground motions are usually constructed by taking into account the knowledge from the past observation and the probabilistic insights. However, uncertainties in the occurrence of earthquakes (or ground motions), the fault rupture mechanisms, the wave propagation mechanisms, the ground properties, etc. cause much difficulty in defining reasonable design ground motions especially for important buildings in which severe damage or collapse has to be avoided absolutely (Singh 1984; Anderson and Bertero 1987; Geller et al. 1997; Takewaki 2002; Stein 2003).

A long-period ground motion has been observed in Japan recently. This type of ground motion is told to cause a large seismic demand to such structures as high-rise buildings, base-isolated buildings, oil tanks, etc. This large seismic demand results from the resonance between the long-period ground motion and the long natural period of these constructed facilities.

A significance of critical excitation is supported by its broad perspective. There are two classes of buildings in a city (see Fig. 0.3). One is the important buildings which play an important role during disastrous earthquakes. The other one is ordinary buildings. The former one should not have damage during earthquake and the latter one may be damaged partially especially for critical excitation larger than code-specified design earthquakes. The concept of critical excitation may enable structural designers to make ordinary buildings more seismic resistant.

The most critical issue in the seismic-resistant design is the resonance. The promising approaches are to shift the natural period of the building

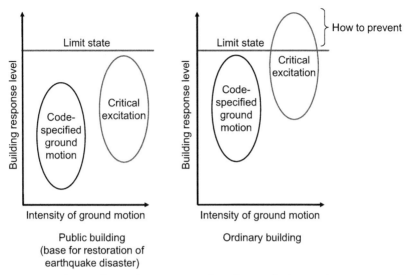

Figure 0.3 Relation of critical excitation with code-specified ground motion in public building and ordinary building.

through seismic control and to add damping in the building. However it is also true that the seismic control is under development and more sufficient time is necessary to respond to uncertain ground motions. The author hopes that this book will help the development of new seismic-resistant design methods of buildings for such unpredicted or unpredictable ground motions.

The author's research was greatly motivated by the papers by Drenick (1970) and Shinozuka (1970). The author communicated with Prof. Drenick (2002) and was informed that the work by Prof. Drenick was motivated by his communication with Japanese researchers in late 1960s. The author would like to express his appreciation to Profs. Drenick and Shinozuka.

Izuru Takewaki
Kyoto, 2006

REFERENCES

Anderson, J.C., Bertero, V.V., 1987. Uncertainties in establishing design earthquakes. J. Struct. Eng. ASCE, 113 (8), 1709–1724.
Drenick, R.F., 1970. Model-free design of aseismic structures. J. Eng. Mech. Div. ASCE, 96 (EM4), 483–493.
Drenick, R.F., 2002. Private communication.

Geller, R.J., Jackson, D.D., Kagan, Y.Y., Mulargia, F., 1997. Earthquakes cannot be predicted. Science 275, 1616.

Shinozuka, M., 1970. Maximum structural response to seismic excitations. J. Eng. Mech.
Div. ASCE, 96 (EM5), 729–738.

Singh, J.P., 1984. Characteristics of near-field ground motion and their importance in
building design. ATC-10-1 Critical aspects of earthquake ground motion and building
damage potential, ATC, 23–42.

Stein, R.S., 2003. Earthquake conversations. Sci. Am. 288 (1), 72–79.

Takewaki, I., 2002. Critical excitation method for robust design: A review. J. Struct. Eng.
ASCE, 128 (5), 665–672.

PREFACE TO THE SECOND EDITION

The largest earthquake event in the world since the first edition of this book was published in 2007 may be the March 11, 2011 event off the Pacific coast of Tohoku, Japan. Three major observations were made during that great earthquake (Takewaki et al. 2011). The first one was a devastating, giant tsunami following the earthquake, the second one was an accident at the Fukushima No.1 nuclear power plant and the last one was the occurrence of long-period ground motions that were resonant with super high-rise buildings in mega cities in Japan.

The author was convinced during and immediately after the earthquake that the critical excitation method is absolutely necessary for enhancing the earthquake resilience of building structures and engineering systems. Actually, Dr. Rudolf Drenick regarded nuclear power plant problems and super high-rise building problems as major objectives of the critical excitation method that he introduced about three decades ago. The author also took those two concerns into account before the occurrence of the 2011 Japan earthquake (see the following figure from Takewaki 2008). The author hopes that more resilient building structures and engineering systems will be designed using the advanced critical excitation method.

In the second edition, the critical excitation problem for multi-component input ground motions and that for elastic-plastic structures in

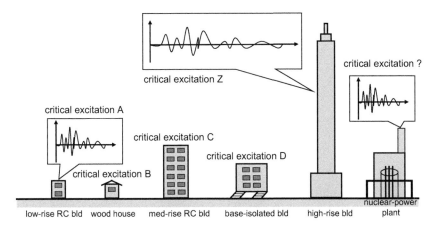

Critical excitation for nuclear power plant and high-rise buildings (Takewaki 2008). RC, reinforced concrete. (See Figure 15.1 for more information.)

a more direct way are incorporated and discussed in more depth. Finally, the problem of earthquake resilience of super high-rise buildings is discussed from broader viewpoints.

The author owes great thanks to Dr. Abbas Moustafa, Minia University, Egypt and Dr. Kohei Fujita, Kyoto University, Japan for their contributions to these themes. This second edition would have never been possible without their efforts.

Izuru Takewaki
Kyoto, 2013

REFERENCES

Takewaki, I., 2008. Critical excitation methods for important structures, invited as a Semi-Plenary Speaker. EURODYN 2008, July 7–9, Southampton, England.

Takewaki, I., Murakami, S., Fujita, K., Yoshitomi, S., Tsuji, M., 2011. The 2011 off the Pacific coast of Tohoku earthquake and response of high-rise buildings under long-period ground motions. Soil Dynamics and Earthquake Engineering 31 (11), 1511–1528.

PERMISSION DETAILS

Permission has been obtained for the following figures, tables and equations that appear in this book:

Figures 12.15–12.25, Tables 12.1, 12.2 and Equations (12.36)–(12.44) are from *The Structural Design of Tall and Special Buildings*, 20(6), K. Yamamoto, K. Fujita and I. Takewaki, Instantaneous earthquake input energy and sensitivity in base-isolated building, 631–648, 2011, with permission from Wiley and Blackwell.

Materials from Elsevier's publications:

Part of Chapter 13 has been published in 'Fujita, K., Yoshitomi, S., Tsuji, M. and Takewaki, I. (2008). Critical cross-correlation function of horizontal and vertical ground motions for uplift of rigid block, *Engineering Structures*, 30(5), 1199–1213' and 'Fujita, K. and Takewaki, I. (2010). Critical correlation of bi-directional horizontal ground motions, *Engineering Structures*, 32(1), 261–272'.

CHAPTER ONE

Overview of Seismic Critical Excitation Method

Contents

1.1. WHAT IS CRITICAL EXCITATION?

It is natural to imagine that a ground motion input resonant to the natural frequency of the structure is a critical excitation. In order to discuss this issue in detail, consider a linear elastic, viscously damped, single–degree–of–freedom (SDOF) system as shown in Fig. 1.1. Let m, k, c denote mass, stiffness and viscous damping coefficient of the SDOF system. The time derivative will be denoted by over–dot in this book. The system is subjected

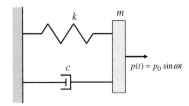

Figure 1.1 *Single-degree-of-freedom (SDOF) system subjected to external harmonic force $p(t) = p_0 \sin \omega t$.*

© 2013 Elsevier Ltd.
All rights reserved.

to an external harmonic force $p(t) = p_0 \sin \omega t$. The equation of motion of this system may be described as

$$m\ddot{u}(t) + c\dot{u}(t) + ku(t) = p_0 \sin \omega t \tag{1.1}$$

By dividing both sides by m, Eq. (1) leads to

$$\ddot{u}(t) + 2h\Omega\dot{u}(t) + \Omega^2 u(t) = (p_0/m) \sin \omega t \tag{1.2}$$

where $\Omega^2 = k/m$, $2h\Omega = c/m$. Ω and h are the undamped natural circular frequency and the critical damping ratio.

Consider first the nonresonant case, i.e. $\omega \neq \Omega$. The general solution of Eq. (1.1) can be expressed by the sum of the complementary solution of Eq. (1.1) and the particular solution of Eq. (1.1).

$$u(t) = u_c(t) + u_p(t) \tag{1.3}$$

The complementary solution is the free-vibration solution and is given by

$$u_c(t) = e^{-h\Omega t}(A \cos \Omega_D t + B \sin \Omega_D t) \tag{1.4}$$

where $\Omega_D = \sqrt{1 - h^2}\,\Omega$. On the other hand, the particular solution may be described by

$$u_p(t) = C \sin \omega t + D \cos \omega t \tag{1.5}$$

The undetermined coefficients C and D in Eq. (1.5) can be obtained by substituting Eq. (1.5) into Eq. (1.2) and comparing the coefficients on sine and cosine terms. The expressions can be found in standard textbooks. On the other hand, the undetermined coefficients A and B in Eq. (1.4) can be obtained from the initial conditions $u(0)$ and $\dot{u}(0)$.

Consider next the resonant case, i.e. $\omega = \Omega$. The solution corresponding to the initial conditions $u(0) = \dot{u}(0) = 0$ can then be written by

$$u(t) = \frac{p_0}{k}\frac{1}{2h}[e^{-h\Omega t}(\cos \Omega_D t + \frac{h}{\sqrt{1 - h^2}} \sin \Omega_D t) - \cos \Omega t] \tag{1.6}$$

Fig. 1.2 shows examples of Eq. (1.6) for several damping ratios.

Consider the undamped and resonant case, i.e. $h = 0$ and $\omega = \Omega$. As before, the general solution of Eq. (1.2) can be expressed by the sum of the complementary solution and the particular solution.

$$u(t) = u_c(t) + u_p(t) \tag{1.7}$$

The complementary solution is the free-vibration solution and is given by

$$u_c(t) = A \cos \Omega t + B \sin \Omega t \tag{1.8}$$

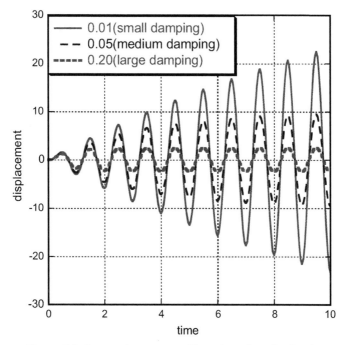

Figure 1.2 *Resonant response with various damping levels.*

On the other hand, the particular solution may be described by

$$u_p(t) = Ct \cos \Omega t \tag{1.9}$$

The final solution corresponding to the initial conditions $u(0) = \dot{u}(0) = 0$ can then be written by

$$u(t) = \frac{p_0}{2k} \left(\sin \Omega t - \Omega t \cos \Omega t \right) \tag{1.10}$$

Fig. 1.3 shows an example of Eq. (1.10) for a special frequency Ω.

1.2. ORIGIN OF CRITICAL EXCITATION METHOD (DRENICK'S APPROACH)

Newton's second law of motion may be described by

$$\frac{d}{dt}(m\dot{u}) = p \tag{1.11}$$

If the mass remains constant, the equation is reduced to

$$p = m\ddot{u} \tag{1.12}$$

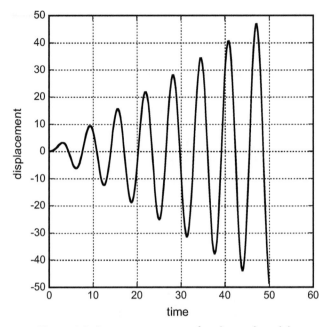

Figure 1.3 *Resonant response of undamped model.*

Consider the integration of Eq. (1.12) from time t_1 through t_2.

$$\int_{t_1}^{t_2} p\,dt = m(\dot{u}_2 - \dot{u}_1) = m\Delta\dot{u} \tag{1.13}$$

where $\dot{u}_1 = \dot{u}(t_1)$ and $\dot{u}_2 = \dot{u}(t_2)$. Assume here a unit impulse applied to a mass at rest

$$\int_{t_1}^{t_2} p\,dt = 1 \tag{1.14}$$

Then the change of velocity may be described as

$$\Delta\dot{u} = \dot{u} - 0 = \frac{1}{m} \tag{1.15}$$

Consider next a linear elastic, viscously damped SDOF system subjected to a base acceleration $\ddot{u}_g(t)$ as shown in Fig. 1.4. The equation of motion may be expressed by

$$m\ddot{u}(t) + c\dot{u}(t) + ku(t) = -m\ddot{u}_g(t) \tag{1.16}$$

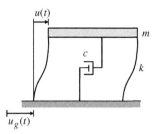

Figure 1.4 *Linear elastic, viscously damped SDOF system subjected to base motion.*

By dividing both sides by m, Eq. (1.16) leads to

$$\ddot{u}(t) + 2h\Omega\dot{u}(t) + \Omega^2 u(t) = -\ddot{u}_g(t) \tag{1.17}$$

where $\Omega^2 = k/m$, $2h\Omega = c/m$. The unit impulse response function can then be derived from Eqs. (1.4) and (1.15) as the free vibration response of the system at rest subjected to the unit impulse.

$$g(t) = H_e(t)\frac{1}{m\Omega_D}e^{-h\Omega t}\sin\Omega_D t \tag{1.18}$$

where $H_e(t)$ is the Heaviside step function. The displacement response of the system subjected to a base acceleration $\ddot{u}_g(t)$ may be obtained as the convolution.

$$u(t) = \int_0^t \{-m\ddot{u}_g(\tau)\}g(t-\tau)d\tau \tag{1.19}$$

The term $\{-m\ddot{u}_g(\tau)\}d\tau$ indicates the impulse during $d\tau$. It is interesting to note that the relative velocity and absolute acceleration can be expressed as follows with the use of the unit impulse response function.

$$\dot{u}(t) = \int_0^t \{-m\ddot{u}_g(\tau)\}\dot{g}(t-\tau)d\tau \tag{1.20}$$

$$\ddot{u}_g(t) + \ddot{u}(t) = \int_0^t \{-m\ddot{u}_g(\tau)\}\ddot{g}(t-\tau)d\tau \tag{1.21}$$

Interested readers may conduct the proof as an exercise.

The theory due to Drenick (1970) will be shown next. Consider the modified SDOF system with $\Omega = 1$. The displacement response of the system may be expressed by

$$u(t) = \int_{-\infty}^{\infty} \{-\ddot{u}_g(\tau)\}g^*(t-\tau)d\tau \qquad (1.22)$$

where

$$g^*(t) = H_e(t)\frac{1}{\sqrt{1-h^2}}e^{-ht}\sin\sqrt{1-h^2}t \qquad (1.23)$$

Consider the following constraint on the input acceleration.

$$\int_{-\infty}^{\infty} \ddot{u}_g(t)^2 dt \leq M^2 \text{ (similar to Arias Intensity)} \qquad (1.24)$$

Let us introduce the quantity N^2 by

$$\int_{-\infty}^{\infty} g^*(t)^2 dt = N^2 \qquad (1.25)$$

From Schwarz inequality,

$$|u(t)|^2 = \left[\int_{-\infty}^{\infty} \{-\ddot{u}_g(\tau)\}g^*(t-\tau)d\tau\right]^2 \leq \int_{-\infty}^{\infty} \ddot{u}_g(\tau)^2 d\tau$$

$$\times \int_{-\infty}^{\infty} g^*(t-\tau)^2 d\tau \leq M^2 N^2 \qquad (1.26)$$

Therefore

$$|u(t)| \leq MN \qquad (1.27)$$

Because the right-hand side is time-independent,

$$\max_t |u(t)| \leq MN \qquad (1.28)$$

It can be shown that the equality holds for the input.

$$\ddot{u}_g(\tau) = -\frac{M}{N}g^*(t-\tau) \qquad (1.29)$$

This is the "mirror image" of the impulse response function (see Fig. 1.5(a)). This result can also be derived from the variational approach (Drenick 1970). When $t = 0$, the equality holds certainly

$$u(0) = \int_{-\infty}^{\infty} \frac{M}{N} g^*(t - \tau)^2 d\tau = MN \qquad (1.30)$$

Fig. 1.5(b) shows the critical excitation derived by Drenick (1970) and the corresponding displacement response.

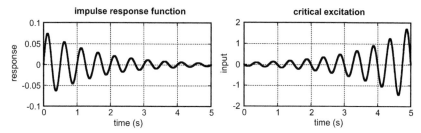

Figure 1.5(a) *Impulse response function and mirror image critical excitation.*

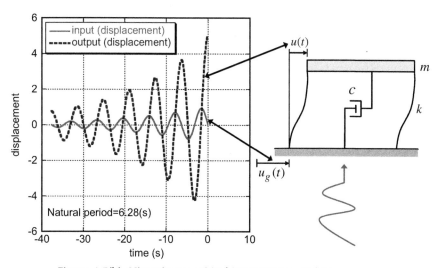

Figure 1.5(b) *Mirror image critical input motion and its response.*

1.3. SHINOZUKA'S APPROACH

Consider again a linear elastic, viscously damped SDOF system as shown in Fig. 1.4. Let $\ddot{U}_g(\omega)$ denote the Fourier transform of the ground acceleration $\ddot{u}_g(t)$. This fact can be described as

$$\ddot{U}_g(\omega) = \int_{-\infty}^{\infty} \ddot{u}_g(t) e^{-i\omega t} dt \qquad (1.31)$$

$$\ddot{u}_g(t) = \frac{1}{2\pi} \int_{-\infty}^{\infty} \ddot{U}_g(\omega) e^{i\omega t} d\omega \qquad (1.32)$$

where i denotes the imaginary unit. Parseval's theorem provides the constraint on input acceleration in the frequency domain.

$$\int_{-\infty}^{\infty} \ddot{u}_g(t)^2 dt = \frac{1}{2\pi} \int_{-\infty}^{\infty} \left| \ddot{U}_g(\omega) \right|^2 d\omega \leq M^2 \qquad (1.33)$$

The Fourier transform $U(\omega)$ of the displacement response $u(t)$ can be expressed in terms of transfer function $H(\omega)$.

$$U(\omega) = H(\omega) \ddot{U}_g(\omega) \qquad (1.34)$$

where

$$H(\omega) = U(\omega)/\ddot{U}_g(\omega) = -m/(-\omega^2 m + i\omega c + k) \qquad (1.35)$$

Since $g^*(t)$ and $H(\omega)$ are the Fourier transform's pair in case of $\Omega = 1$, Parseval's theorem provides the following relation.

$$\int_{-\infty}^{\infty} g^*(t)^2 dt = \frac{1}{2\pi} \int_{-\infty}^{\infty} \left| H(\omega) \right|^2 d\omega = N^2 \qquad (1.36)$$

From the fact that $U(\omega)$ is the Fourier transform of the displacement response $u(t)$, the following relation may be drawn.

$$u(t) = \frac{1}{2\pi} \int_{-\infty}^{\infty} H(\omega) \ddot{U}_g(\omega) e^{i\omega t} d\omega \qquad (1.37)$$

From Schwarz inequality and Eqs. (1.33), (1.36), the following relation can be derived.

$$|u(t)| = \left| \frac{1}{2\pi} \int_{-\infty}^{\infty} H(\omega)\ddot{U}_g(\omega)e^{i\omega t}d\omega \right|$$

$$\leq \frac{1}{2\pi} \int_{-\infty}^{\infty} |H(\omega)||\ddot{U}_g(\omega)|d\omega \tag{1.38}$$

$$\leq \left[\frac{1}{2\pi} \int_{-\infty}^{\infty} |\ddot{U}_g(\omega)|^2 d\omega \right]^{1/2} \left[\frac{1}{2\pi} \int_{-\infty}^{\infty} |H(\omega)|^2 d\omega \right]^{1/2}$$

$$\leq MN$$

In some practical situations, useful information on $\ddot{U}_g(\omega)$ may be available. Such additional information would provide a better estimation of the maximum response (Shinozuka 1970a, b).

If $|\ddot{U}_g(\omega)|$ has an envelope $U_e(\omega)$, i.e.

$$|\ddot{U}_g(\omega)| \leq U_e(\omega), \tag{1.39}$$

then a narrower response bound can be derived. From Eqs. (1.38) and (1.39),

$$|u(t)| \leq \frac{1}{2\pi} \int_{-\infty}^{\infty} |H(\omega)||\ddot{U}_g(\omega)|d\omega \leq \frac{1}{2\pi} \int_{-\infty}^{\infty} |H(\omega)|U_e(\omega)d\omega \equiv I_e \tag{1.40}$$

Because the right-hand side of Eq. (1.40) is time-independent,

$$\max_t |u(t)| \leq \frac{1}{2\pi} \int_{-\infty}^{\infty} |H(\omega)|U_e(\omega)d\omega = I_e \tag{1.41}$$

It may be observed that I_e can be a narrower bound than MN.

1.4. HISTORICAL SKETCH IN EARLY STAGE

As stated in Section 1.2, the method of critical excitation was proposed by Drenick (1970) for linear elastic, viscously damped SDOF systems in order to take into account inherent uncertainties in ground motions. This method is aimed at finding the excitation producing the maximum response from a class of allowable inputs. The method was outlined in the preceding sections. With the help of the Cauchy-Schwarz inequality, Drenick (1970) showed that the critical excitation for a linear elastic, viscously damped SDOF system is its impulse response function reversed in time, i.e. mirror image excitation. This implies that the critical envelope function for linear elastic, viscously damped SDOF systems in deterministic problems can be given by an increasing exponential function and the critical excitation has to be defined from the time of minus infinity. This result may be somewhat unrealistic and of only theoretical significance. Despite this, Drenick's paper (1970) is pioneering.

Drenick (1977a) pointed out later that the combination of probabilistic approaches with worst-case analyses should be employed to make the seismic resistant design robust. He claimed that the data used in the calculation of failure probabilities, usually very small numbers, in the seismic reliability analysis are scarce and reliable prediction of the failure probability is difficult only by the conventional reliability analysis, which requires the tail shapes of probability density functions of disturbances. Practical application of critical excitation methods has then been proposed extensively.

It was pointed out that the critical response by Drenick's model (1970) is conservative. To resolve this point, Shinozuka (1970a, b) discussed the same critical excitation problem in the frequency domain. He proved that, if an envelope function of Fourier amplitude spectra can be specified, a nearer upper bound of the maximum response can be obtained. The method was also outlined in the preceding section. Iyengar (1970) and Yang and Heer (1971) formulated another theory to define an envelope function of input accelerations in the time domain.

An idea similar to that due to Drenick (1970) was proposed by Papoulis (1967, 1970) independently in the field of signal analysis and circuit theory.

1.5. VARIOUS MEASURES OF CRITICALITY

Various quantities have been chosen and proposed as an objective function to be maximized in critical excitation problems.

Ahmadi (1979) posed another critical excitation problem including the response acceleration as the objective function to be maximized. He demonstrated that a rectangular wave in time domain is the critical one and recommended the introduction of another constraint in order to make the solution more realistic.

Westermo (1985) considered the following input energy during T divided by the mass m as the objective function in a new critical excitation problem.

$$E_I = \int_0^T (-\ddot{u}_g)\dot{u}\,dt \tag{1.42}$$

He also imposed a constraint on the time integral of squared input acceleration. He introduced a variational approach and demonstrated that the critical input acceleration is proportional to the response velocity. His solution is not necessarily complete and explicit because the response velocity is actually a function of the excitation to be obtained. He pointed out that the critical input acceleration includes the solution by Drenick (1970). The damage of structures may be another measure of criticality. The corresponding problems have been tackled by some researchers.

Takewaki (2004b, 2005) treated the earthquake input energy as the objective function in a new critical excitation problem. It has been shown that the formulation of the earthquake input energy in the frequency domain is essential for solving the critical excitation problem and deriving a bound on the earthquake input energy for a class of ground motions. The criticality has been expressed in terms of degree of concentration of input motion components on the maximum portion of the characteristic function defining the earthquake input energy. It should be pointed out that no mathematical programming technique is required in the solution procedure. The constancy of earthquake input energy with respect to natural period and damping ratio has been discussed. It has been shown that the constancy of earthquake input energy is directly related to the uniformity of "the Fourier amplitude spectrum" of ground motion acceleration, not the uniformity of the velocity response spectrum. The bounds under acceleration and velocity constraints (time integral of the squared base acceleration and time integral of the squared base velocity) have been clarified through numerical examinations for recorded ground motions to be meaningful in the short and intermediate/long natural period ranges, respectively.

Srinivasan et al. (1991) extended the basic approach due to Drenick (1970) to multi-degree-of-freedom (MDOF) models. They used a variational formulation and selected a quantity in terms of multiple responses as the objective function. They demonstrated that the relation among the critical displacement, velocity and acceleration responses is similar to the well-known relation among the displacement, velocity and acceleration response spectra. Similar treatment for MDOF models has been proposed by the present author in critical excitation problems for input energy.

1.6. SUBCRITICAL EXCITATION

It was suggested that the critical excitation introduced by Drenick (1970) is conservative compared to the recorded ground motions. To resolve this problem, Drenick, Wang and their colleagues proposed a concept of "subcritical excitation" (Drenick 1973; Wang et al. 1976; Wang and Drenick 1977; Wang et al. 1978; Drenick and Yun 1979; Wang and Yun 1979; Abdelrahman et al. 1979; Bedrosian et al. 1980; Wang and Philippacopoulos 1980; Drenick et al. 1980; Drenick et al. 1984). They expressed an allowable set of input accelerations as a "linear combination of recorded ground motions." Note that the site and earthquake occurrence properties of those recorded ground motions are similar. They chose several response quantities as the measure for criticality and compared the response to the subcritical excitation with those to recorded earthquake ground motions as the basis functions. They demonstrated that the conservatism of the subcritical excitations can be improved.

Abdelrahman et al. (1979) extended the idea of subcritical excitation to the method in the frequency domain. An allowable set of Fourier spectra of accelerograms has been expressed as a linear combination of Fourier spectra of recorded accelerograms. They pointed out clearly that the frequency-domain approach is more efficient than the time-domain approach.

An optimization technique was used by Pirasteh et al. (1988) in one of the subcritical excitation problems. They superimposed accelerograms recorded at similar sites to construct the candidate accelerograms, then used optimization and approximation techniques in order to find the most critical accelerogram. The most critical accelerogram was defined as the one that satisfies the constraints on peaks, Fourier spectra, intensities, growth rates and maximizes the damage index in the structure. The damage index has been defined as cumulative inelastic energy dissipation or sum of interstory drifts.

It should be remarked that the concept of subcritical excitation is based on the assumption that the critical one can be obtained from an ensemble of basis motions and that the basis motions are complete and reliable. However, for example, the record at SCT1 during Mexico Michoacan Earthquake (1985) and that at Kobe University during Hyogoken-Nanbu Earthquake

(1995) indicate that ground motions unpredictable from the past knowledge can be observed and inclusion of such ground motions is inevitable in proper and reliable implementation of subcritical excitation methods.

1.7. STOCHASTIC EXCITATION

The concept of critical excitation was extended to probabilistic problems by Iyengar and Manohar (1985, 1987); Iyengar (1989); Srinivasan et al. (1992); Manohar and Sarkar (1995); Sarkar and Manohar (1996, 1998) and Takewaki (2000a–d, 2001a–c). The papers due to Iyengar and Manohar (1985, 1987) may be the first to discuss probabilistic critical excitation methods. They used a stationary model of input ground acceleration in the paper (Iyengar and Manohar 1985) and utilized a nonstationary model of ground accelerations expressed as $\ddot{u}_g(t) = c(t)w(t)$ in the paper (Iyengar and Manohar 1987). $c(t)$ is a deterministic envelope function and $w(t)$ is a stochastic function representing a stationary random Gaussian process with zero mean.

The auto-correlation function of $w(t)$ can be expressed as

$$R_w(t_1, t_2) = E[w(t_1)w(t_2)] = \int_{-\infty}^{\infty} S_w(\omega)e^{i\omega(t_1 - t_2)}d\omega \qquad (1.43)$$

$S_w(\omega)$ is the power spectral density (PSD) function of the stochastic function $w(t)$. The PSD function of $\ddot{u}_g(t)$ may then be expressed as

$$S_g(t; \omega) = c(t)^2 S_w(\omega) \qquad (1.44)$$

Consider again a linear elastic viscously damped SDOF model. The auto-correlation function of the relative displacement of the SDOF model can be expressed as

$$R_D(t_1, t_2) = \int_{-\infty}^{\infty} \int_{-\infty}^{\infty} \int_{-\infty}^{\infty} c(\tau_1)h(t_1 - \tau_1)c(\tau_2)$$

$$\times h(t_2 - \tau_2)S_w(\omega)e^{i\omega(\tau_1 - \tau_2)}d\omega d\tau_1 d\tau_2 \qquad (1.45)$$

where $h(t)$ is the impulse response function. Substitution of $t_1 = t_2 = t$ in Eq. (1.45) leads to the mean-square relative displacement of the SDOF model.

$$\sigma_D(t)^2 = \int_{-\infty}^{\infty} \{A_C(t; \omega)^2 + A_S(t; \omega)^2\}S_w(\omega)d\omega \qquad (1.46)$$

In Eq. (1.46), the following quantities are used.

$$A_C(t; \omega) = \int_0^t c(\tau)h(t - \tau) \cos \omega\tau d\tau \qquad (1.47)$$

$$A_S(t; \omega) = \int_0^t c(\tau)h(t - \tau) \sin \omega\tau d\tau \qquad (1.48)$$

A critical excitation problem was discussed by Iyengar and Manohar (1987) to maximize $f = \max_t \sigma_D(t)^2$ subject to the constraint on $\int_{-\infty}^{\infty} S_w(\omega)d\omega$ which is equivalent to the constraint on $E[\int_0^T \ddot{u}_g^2 dt]$ ($E[\cdot]$: ensemble mean) for a given envelope function $c(t)$. Iyengar and Manohar (1987) expressed the square root of the PSD function of the excitation in terms of linear combination of orthonormal functions and obtained their coefficients through eigenvalue analysis. Srinivasan et al. (1992), Manohar and Sarkar (1995) and Sarkar and Manohar (1996, 1998) imposed a bound on the total average energy and solved linear or nonlinear programming problems. Srinivasan et al. (1992) used a nonstationary filtered shot noise model for expressing input motions. Manohar and Sarkar (1995) and Sarkar and Manohar (1996, 1998) further set a bound on the average rate of zero crossings to avoid the excessive concentration of wave components at the resonant frequency in the PSD function.

In contrast to the bound on the average rate of zero crossings, Takewaki (2000c, d, 2001a–c) introduced a new constraint on the intensity $\sup S_w(\omega)$ of the PSD function $S_w(\omega)$ of $w(t)$ in addition to the constraint on the power (integral $\int_{-\infty}^{\infty} S_w(\omega)d\omega$ of PSD function) and developed a simpler critical excitation method for both stationary and nonstationary inputs. In the case of nonstationary excitations, the following double maximization procedure must be treated.

$$\max_{S_w(\omega)} \max_t \{f(t; S_w(\omega))\} \qquad (1.49)$$

where f represents an objective function, e.g. a mean-square response. The procedure (1.49) requires determination of the time when the probabilistic index f attains its maximum under each input prescribed by $S_w(\omega)$. This procedure is quite time consuming. Takewaki (2000c, d, 2001a–c) devised a unique procedure based on the order interchange of the double maximization procedure (see Fig. 1.6), i.e.

$$\max_t \max_{S_w(\omega)} \{f(t; S_w(\omega))\} \qquad (1.50)$$

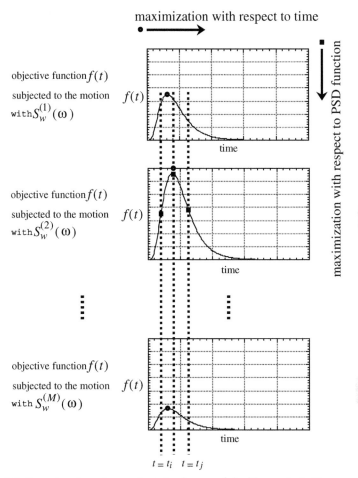

Figure 1.6 *Procedure based on order interchange of double maximization procedure.*

Takewaki (2000c, d, 2001a–c) suggested that the first maximization procedure for $S_w(\omega)$ can be performed very efficiently by utilizing the method for stationary inputs (Takewaki 2000a, b) and the second maximization procedure for time can be conducted systematically by changing the time sequentially. It was suggested that this method can be applied not only to SDOF models, but also to MDOF models if an appropriate objective function can be introduced.

1.8. CONVEX MODELS

A convex model is defined mathematically as a set of functions. Each function is a realization of an uncertain event. Several interesting convex models were proposed by Ben–Haim and Elishakoff (1990), Ben–Haim et al. (1996), Pantelides and Tzan (1996), Tzan and Pantelides (1996a) and Baratta et al. (1998) for ground motion modeling which can be constructed versatilely depending on the level of prior information available. Examples are: A local energy-bound convex model, an integral energy-bound convex model, an envelope-bound convex model, a Fourier-envelope convex model and a response-spectrum-envelope convex model (Ben–Haim et al. 1996). One of the merits of the convex models is the capability of prediction of the maximum or extreme response of structures to unknown inputs of which the appropriate probabilistic description is difficult. In addition, unlike the other methods, such as the subcritical excitation and stochastic excitation, another advantageous feature of the convex model comes from the fact that it can handle MDOF systems with the same ease as SDOF systems. The smart combination of probabilistic and convex-model approaches appears to be promising (Drenick 1977a). It is not the objective of this book to provide a detailed explanation of the convex models. Readers interested in the convex models should refer to Ben–Haim and Elishakoff (1990), Ben–Haim et al. (1996), Pantelides and Tzan (1996), Tzan and Pantelides (1996a) and Baratta et al. (1998).

1.9. NONLINEAR OR ELASTIC-PLASTIC SDOF SYSTEM

Critical excitation problems for autonomous nonlinear systems (e.g. Duffing oscillator) were considered by Iyengar (1972). By using the Schwarz inequality, he derived a response upper bound similar to that by Drenick (1970). He treated both deterministic and probabilistic inputs. Drenick and Park (1975) provided interesting and important comments on the paper due to Iyengar (1972).

An idea was proposed by Drenick (1977b) to use an equivalent linearization technique in finding a critical excitation for nonlinear systems. However, he did not discuss the applicability of the concept to actual and practical problems and his concept or scenario is restricted to deterministic equivalent linearization problems.

Westermo (1985) tackled critical excitation problems for nonlinear hysteretic and nonhysteretic systems by adopting the input energy given by Eq. (1.42) as the objective function. His approach is limited, but pioneering. He limited the class of critical excitations to periodic ones. He suggested several interesting points inherent in the critical excitation problems for nonlinear systems.

A deterministic equivalent linearization technique was used by Philippacopoulos (1980) and Philippacopoulos and Wang (1984) in critical excitation problems of nonlinear SDOF hysteretic systems. They derived critical inelastic response spectra and compared them with inelastic response spectra for recorded motions.

Takewaki (2001d, 2002) developed a new type of probabilistic critical excitation method for SDOF elastic-plastic structures. For simplicity, he treated a stationary random acceleration input $\ddot{u}_g$ of which the PSD function can be described by $S_g(\omega)$. The power $\int_{-\infty}^{\infty} S_g(\omega)d\omega$ and the intensity $\sup S_g(\omega)$ of the excitations were fixed and the critical excitation was found under these constraints. While transfer functions and unit impulse response functions can be defined and used in linear elastic structures only, such analytical expressions cannot be used in elastic-plastic structures. This situation leads to difficulty in finding a critical excitation for elastic–plastic structures. To resolve such difficulty, a statistical equivalent linearization technique has been introduced. The shape of the critical PSD function has been limited to a rectangular function attaining its upper bound in a certain frequency range. The central frequency of the rectangular PSD function has been treated as a principal parameter and changed in finding the critical PSD function. The critical excitations were obtained for two examples and compared with the corresponding recorded earthquake ground motions. Takewaki (2001d, 2002) pointed out that the central frequency of the critical rectangular PSD function is resonant to the equivalent natural frequency of the elastic–plastic SDOF system (see Chapter 5). This fact corresponds with the result by Westermo (1985).

1.10. ELASTIC-PLASTIC MDOF SYSTEM

Several interesting approaches for MDOF systems were proposed as natural extensions of the method for SDOF systems. Philippacopoulos (1980) and Philippacopoulos and Wang (1984) took full advantage of a deterministic equivalent linearization technique in critical excitation problems of nonlinear MDOF hysteretic systems. They proposed a conceptual scenario for the nonlinear MDOF hysteretic systems. However, application of the technique to practical problems is not shown in detail. For example, it is not clear for what excitation the equivalent stiffnesses and damping coefficients should be defined.

Takewaki (2001e) extended the critical excitation method for elastic-plastic SDOF models to MDOF models on deformable ground by employing a statistical equivalent linearization method for MDOF models. The linearization method was used to simulate the response of the original elastic-plastic hysteretic model. As in SDOF models, the power $\int_{-\infty}^{\infty} S_g(\omega)d\omega$ and intensity $\sup S_g(\omega)$ of the excitations are constrained. It is assumed that the shape of the critical PSD function is a rectangular one attaining its upper bound in a certain frequency range. In contrast to SDOF models, various quantities can be employed in MDOF models as the objective function to be maximized. The sum of standard deviations of story ductilities along the height has been chosen as the objective function to define the critical excitation. Note that a solution procedure similar to that for SDOF models has been used. It adopts a procedure of regarding the central frequency of the rectangular PSD function as a principal parameter for finding the critical one. The simulation results by elastic-plastic time-history response analysis disclosed that the proposed critical excitation method is reliable in the models for which the validity of the statistical equivalent linearization method is guaranteed. It was suggested that the critical response representation in terms of nonexceedance probabilities can be an appropriate candidate for expressing the criticality of recorded ground motions (see Chapter 5).

1.11. CRITICAL ENVELOPE FUNCTION

A new class of critical excitation problems may be formulated for identifying critical envelope functions for nonstationary random input (Takewaki 2004a). The nonstationary ground motion is assumed to be $\ddot{u}_g(t) = c(t)w(t)$ which is the product of a deterministic envelope function $c(t)$ and another probabilistic function $w(t)$ representing the frequency content. The former envelope function can be determined in such a way that the mean-square drift of an SDOF model attains its maximum under the constraint $E[\int_0^T \ddot{u}_g^2 dt] = \overline{C}$ on mean total energy and that $\int_{-\infty}^{\infty} S_w(\omega)d\omega = \overline{S}_w$ on power of $w(t)$ ($S_w(\omega)$ is also given). By use of the constraint on power of $w(t)$, the constraint on mean total energy can be reduced to

$$\int_0^T c(t)^2 dt = \overline{C}/\overline{S}_w \qquad (1.51)$$

A double maximization procedure for time and the envelope function is included in the critical excitation problem. The key for reaching the critical envelope function is the order interchange in the double maximization procedure. The Cauchy-Schwarz inequality can be used for obtaining an upper bound of the mean-square drift can also be derived by the use of. It can be shown that the technique is systematic and the upper bound of the response can bound the exact response efficiently within a reasonable accuracy. It can also be demonstrated that, while an increasing exponential function is the critical one in the deterministic problem tackled by Drenick (1970), the super-imposed envelope function of the envelope function of the critical excitation can be a function similar to an increasing exponential function in the probabilistic problem (see Chapter 6).

1.12. ROBUST STRUCTURAL DESIGN

In the previous sections, model parameters of structural systems were given and critical excitations were determined for the given structural system. A more interesting but difficult problem is to determine structural model parameters $\mathbf{k}$ simultaneously with respect to some proper design objectives, e.g. minimizing $f(\mathbf{k}; S_g(\omega))$. It should be remarked that the critical excitation depends on the structural model parameters. Consider a critical excitation problem for structural models subjected to a stationary excitation of which the input PSD function is denoted by $S_g(\omega)$. This problem may be expressed as (see Fig. 1.7)

$$\min_{\mathbf{k}} \max_{S_g(\omega)} \{f(\mathbf{k}; S_g(\omega))\} \qquad (1.52)$$

Since the critical excitation is defined and determined for each set of structural model parameters, this design problem is complex and highly nonlinear with respect to design variables. The following critical-excitation based design problem was considered by Takewaki (2001f) for n-story shear building structures (k_i = story stiffness in the i-th story and $\mathbf{k} = \{k_i\}$) subjected to stationary random inputs. While the integral of the PSD function and the amplitude of the PSD function are constrained for the input, the total cost of the structure as expressed by the total quantity of structural materials is constrained for the structure.

[Problem]
Find the set $\tilde{\mathbf{k}}$ of stiffnesses and the PSD function $\tilde{S}_g(\omega)$

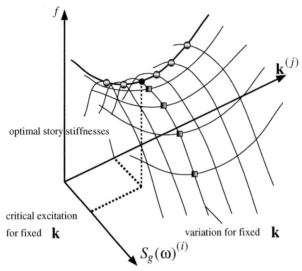

Figure 1.7 *Schematic diagram of performance upgrading based on concept of system-dependent critical excitation.*

$$\text{such that} \min_{\mathbf{k}} \ \max_{S_g(\omega)} \{ f(\mathbf{k}; S_g(\omega)) \} \tag{1.53}$$

$$\text{subject to} \int_{-\infty}^{\infty} S_g(\omega) d\omega \leq \overline{S} \tag{1.54a}$$

$$\text{to sup } S_g(\omega) \leq \overline{s} \tag{1.54b}$$

$$\text{to} \sum_{i=1}^{n} k_i = \overline{K} \tag{1.54c}$$

$$\text{and to } k_i > 0 \ (i = 1, \cdots, n) \tag{1.54d}$$

Takewaki (2001f) derived the optimality conditions for this problem via the Lagrange multiplier method and devised a solution technique based on the optimality criteria approach. It was suggested that the former theories (Takewaki 2000a–d, 2001a–c) for the critical excitation problems for given structural parameters can be utilized effectively in this new type of adaptive design problem. The key is to define a new function $\hat{f}(\mathbf{k}; \tilde{S}_g(\omega; \mathbf{k}))$ and minimize that function $\hat{f}(\mathbf{k}; \tilde{S}_g(\omega; \mathbf{k}))$ with respect to $\mathbf{k}$.

Another interesting approach was presented by Tzan and Pantelides (1996b) to find more robust designs for building structures. The optimal cross-sectional areas of a structure are found to minimize the structural volume subject to floor drift and member stress constraints in the presence of uncertainties in seismic excitation.

1.13. CRITICAL EXCITATION METHOD IN EARTHQUAKE-RESISTANT DESIGN

Earthquake inputs are uncertain even with the present knowledge and it does not appear easy to predict forthcoming events precisely both in time-history and frequency contents (Anderson and Bertero 1987; PEER Center et al. 2000). For example, recent near-field ground motions (Northridge 1994, Kobe 1995, Turkey 1999 and Chi-Chi, Taiwan 1999) and the Mexico Michoacan motion 1985 have some peculiar characteristics unpredictable before their occurrence. It is also true that the civil, mechanical and aerospace engineering structures are often required to be designed for disturbances including inherent uncertainties due mainly to their "low rate of occurrence." Worst-case analysis combined with proper information based on reliable physical data is expected to play an important role in avoiding difficulties induced by such uncertainties. Approaches based on the concept of "critical excitation" seem to be promising.

Just as the investigation on limitation states of structures plays an important role in the specification of allowable response and performance levels of structures during disturbances, the clarification of critical excitations for a given structure or a group of structures appears to provide structural designers with useful information in determining excitation parameters in a reasonable way.

A significance of critical excitation is supported by its broad perspective. In general, there are two classes of buildings in a city. One is the important buildings, which play an important role during disastrous earthquakes. The other is ordinary buildings. The former should not have damage during earthquakes and the latter may be damaged partially especially by critical excitations larger than code-specified design earthquakes (see Fig. 1.8). The concept of critical excitation may enable structural designers to make ordinary buildings more seismic-resistant.

In the case where influential active faults are known in the design stage of a structure (especially an important structure), the effects by these active faults should be taken into account in the structural design through the concept of critical excitation. If influential active faults are not necessarily known in advance, virtual or scenario faults with an appropriate *energy* may

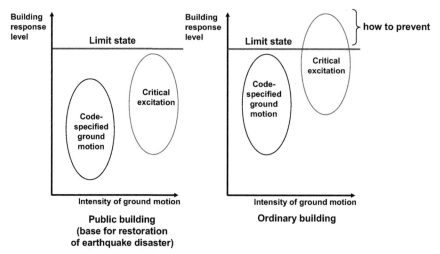

Figure 1.8 *Relation of critical excitation with code-specified ground motion in public building and ordinary building.*

be defined, especially in the design of important and socially influential structures. The combination of worst-case analysis (Takewaki 2004b, 2005) with appropriate specification of energy levels (Boore 1983) derived from the analysis of various factors, e.g. fault rupture mechanism and earthquake occurrence probability, appears to lead to the construction of a more robust and reliable seismic resistant design method (see Fig. 1.9). The appropriate setting of energy levels or information used in the worst-case analysis is important and research on this subject should be conducted more extensively.

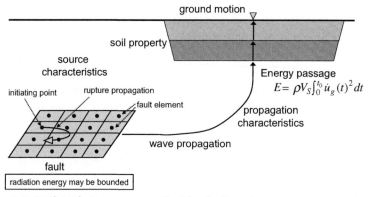

Figure 1.9 *Earthquake energy prescribed by fault rupture, wave propagation and surface soil amplification.*

Critical excitation problems for fully nonstationary excitations (see, for examples, Conte and Peng (1997), Fang and Sun (1997)) and critical excitation problems for elasto–plastic responses under those excitations are challenging problems.

As for response combination, Menun and Der Kiureghian (2000a, b) discussed the envelopes for seismic response vectors. The normal stress in a structural member under combined loading of the axial force and the bending moment may be one example. This problem is related to the interval analysis and its further development is desirable.

REFERENCES

Abdelrahman, A.M., Yun, C.B., Wang, P.C., 1979. Subcritical excitation and dynamic response of structures in frequency domain. Comput. and Struct. 10 (5), 761–771.

Ahmadi, G., 1979. On the application of the critical excitation method to aseismic design. J. Struct. Mech. 7 (1), 55–63.

Anderson, J., Bertero, V.V., 1987. Uncertainties in establishing design earthquakes. J. Struct. Eng. 113 (8), 1709–1724.

Baratta, A., Elishakoff, I., Zuccaro, G., Shinozuka, M., 1998. A generalization of the Drenick-Shinozuka model for bounds on the seismic response of a Single-Degree-Of-Freedom system. Earthquake Engrg. and Struct. Dyn. 27 (5), 423–437.

Bedrosian, B., Barbela, M., Drenick, R.F., Tsirk, A., 1980. Critical excitation method for calculating earthquake effects on nuclear plant structures: An assessment study. NUREG/CR-1673, RD. U.S. Nuclear Regulatory Commission, Burns and Roe, Inc, Oradell, N.J.

Ben-Haim, Y., Elishakoff, I., 1990. Convex Models of Uncertainty in Applied Mechanics. Elsevier, Amsterdam.

Ben-Haim, Y., Chen, G., Soong, T., 1996. Maximum structural response using convex models. J. Engrg. Mech. 122 (4), 325–333.

Boore, D.M., 1983. Stochastic simulation of high-frequency ground motions based on seismological models of the radiated spectra. Bulletin of the Seismological Society of America 73 (6A), 1865–1894.

Conte, J.P., Peng, B.F., 1997. Fully nonstationary analytical earthquake ground-motion model. J. Eng. Mech. 123 (1), 15–24.

Drenick, R.F., 1970. Model-free design of aseismic structures. J. Engrg. Mech. Div. 96 (4), 483–493.

Drenick, R.F., 1973. Aseismic design by way of critical excitation. J. Engrg. Mech. Div. 99 (4), 649–667.

Drenick, R.F., 1977a. On a class of non-robust problems in stochastic dynamics. In: Clarkson, B.L. (Ed.), Stochastic Problems in Dynamics. Pitman, London, pp. 237–255.

Drenick, R.F., 1977b. The critical excitation of nonlinear systems. J. Appl. Mech. 44 (2), 333–336.

Drenick, R.F., Park, C.B., 1975. Comments on "Worst inputs and a bound on the highest peak statistics of a class of non-linear systems." J. Sound and Vibration 41 (1), 129–130.

Drenick, R.F., Yun, C.B., 1979. Reliability of seismic resistance predictions. J. Struct. Div. 105 (10), 1879–1891.

Drenick, R.F., Wang, P.C., Yun, C.B., Philippacopoulos, A.J., 1980. Critical seismic response of nuclear reactors. J. Nuclear Engrg. and Design 59, 425–435.

Drenick, R.F., Novmestky, F., Bagchi, G., 1984. Critical excitation of structures. In: Wind and Seismic Effects, Proc. of the 12th Joint UJNR Panel Conference. NBS Special Publication, pp. 133–142.

Fang, T., Sun, M., 1997. A unified approach to two types of evolutionary random response problems in engineering. Archive of Appl. Mech. 67 (7), 496–506.

Iyengar, R.N., 1970. Matched Inputs. Report No.47, Series J. Center for Applied Stochastics, Purdue University, W. Lafayette, Ind.

Iyengar, R.N., 1972. Worst inputs and a bound on the highest peak statistics of a class of non-linear systems. J. Sound and Vibration 25 (1), 29–37.

Iyengar, R.N., 1989. Critical seismic excitation for structures. In: Proc. of 5th ICOSSAR. San Francisco, ASCE Publications, New York.

Iyengar, R.N., Manohar, C.S., 1985. System dependent critical stochastic seismic excitations. In: M15/6, Proc. of the 8th Int. Conf. on SMiRT. Belgium, Brussels.

Iyengar, R., Manohar, C., 1987. Nonstationary random critical seismic excitations. J. Eng. Mech. 113 (4), 529–541.

Manohar, C.S., Sarkar, A., 1995. Critical earthquake input power spectral density function models for engineering structures. Earthquake Engrg. Struct. Dyn. 24 (12), 1549–1566.

Menun, C., Kiureghian, A., 2000a. Envelopes for seismic response vectors: I Theory. J. Struct. Eng. 126 (4), 467–473.

Menun, C., Kiureghian, A., 2000b. Envelopes for seismic response vectors: II Application. J. Struct. Eng. 126 (4), 474–481.

Pantelides, C.P., Tzan, S.R., 1996. Convex model for seismic design of structures: I analysis. Earthquake Engrg. Struct. Dyn. 25 (9), 927–944.

Papoulis, A., 1967. Limits on bandlimited signals. Proc. of the IEEE 55 (10), 1677–1686.

Papoulis, A., 1970. Maximum response with input energy constraints and the matched filter principle. IEEE Trans. on Circuit Theory 17 (2), 175–182.

PEER Center, ATC, Japan Ministry of Education, Science, Sports, and Culture, US-NSF, 2000. Effects of near-field earthquake shaking. In: Proc. of 5th ICOSSAR. San Francisco, March 20–21, 2000.

Philippacopoulos, A.J., 1980. Critical Excitations for Linear and Nonlinear Structural Systems. Ph.D. Dissertation, Polytechnic Institute of New York.

Philippacopoulos, A., Wang, P., 1984. Seismic inputs for nonlinear structures. J. Eng. Mech. 110 (5), 828–836.

Pirasteh, A.A., Cherry, J.L., Balling, R.J., 1988. The use of optimization to construct critical accelerograms for given structures and sites. Earthquake Engrg. Struct. Dyn. 16 (4), 597–613.

Sarkar, A., Manohar, C.S., 1996. Critical cross power spectral density functions and the highest response of multi-supported structures subjected to multi-component earthquake excitations. Earthq. Engrg. Struct. Dyn. 25, 303–315.

Sarkar, A., Manohar, C.S., 1998. Critical seismic vector random excitations for multiply supported structures. J. Sound and Vibration 212 (3), 525–546.

Shinozuka, M., 1970a. Maximum structural response to seismic excitations. J. Engrg. Mech. Div. 96 (5), 729–738.

Shinozuka, M., 1970b. Maximum structural response to earthquake accelerations Chapter 5. In: Lind, N.C. (Ed.), Structural Reliability and Codified Design. University of Waterloo, Waterloo (CA), pp. 73–85.

Srinivasan, M., Ellingwood, B., Corotis, R., 1991. Critical base excitations of structural systems. J. Eng. Mech. 117 (6), 1403–1422.

Srinivasan, M., Corotis, R., Ellingwood, B., 1992. Generation of critical stochastic earthquakes. Earthquake Engrg. Struct. Dyn. 21 (4), 275–288.

Takewaki, I., 2000a. Optimal damper placement for critical excitation. Probabilistic Engrg. Mech. 15 (4), 317–325.

Takewaki, I., 2000b. Effective damper placement for critical excitation. Confronting Urban Earthquakes, Report of fundamental research on the mitigation of urban disasters caused by near-field earthquakes for Grant in Aid of Scientific Research on Priority Areas, Ministry of Education, Science, Sports and Culture (Japan), 558–561.

Takewaki, I., 2000c. A new probabilistic critical excitation method. J. Struct. Constr. Engrg. (Transactions of AIJ) 533, 69–74 (in Japanese).

Takewaki, I., 2000d. A nonstationary random critical excitation method for MDOF linear structural models. J. Struct. Constr. Engrg. (Transactions of AIJ) 536, 71–77 (in Japanese).

Takewaki, I., 2001a. A new method for non-stationary random critical excitation. Earthquake Engrg. Struct. Dyn. 30 (4), 519–535.

Takewaki, I., 2001b. Nonstationary random critical excitation for nonproportionally damped structural systems. Comput. Meth. Appl. Mech. Engrg. 190 (31), 3927–3943.

Takewaki, I., 2001c. Nonstationary random critical excitation for acceleration response. J. Engrg. Mech. 127 (6), 544–556.

Takewaki, I., 2001d. Critical excitation for MDOF elastic-plastic structures via statistical equivalent linearization. J. Struct. Engrg. B. 47B, 187–194 (in Japanese).

Takewaki, I., 2001e. Probabilistic critical excitation for MDOF elastic-plastic structures on compliant ground. Earthq. Engrg. Struct. Dyn. 30 (9), 1345–1360.

Takewaki, I., 2001f. Maximum global performance design for variable critical excitations. J. Struct. Constr. Engrg. (Transactions of AIJ) 539, 63–69 (in Japanese).

Takewaki, I., 2002. Critical excitation for elastic–plastic structures via statistical equivalent linearization. Probabilistic Engrg. Mech. 17 (1), 73–84.

Takewaki, I., 2004a. Critical envelope functions for non-stationary random earthquake input. Computers & Structures 82 (20–21), 1671–1683.

Takewaki, I., 2004b. Bound of earthquake input energy. J. Struct. Eng. 130 (9), 1289–1297.

Takewaki, I., 2005. Bound of earthquake input energy to soil–structure interaction systems. Soil Dynamics and Earthquake Engineering 25 (7–10), 741–752.

Tzan, S.R., Pantelides, C.P., 1996a. Convex models for impulsive response of structures. J. Eng. Mech. 122 (6), 521–529.

Tzan, S.-R., Pantelides, C.P., 1996b. Convex model for seismic design of structures—II: design of conventional and active structures. Earthquake Engrg. Struct. Dyn. 25 (9), 945–963.

Wang, P.C., Drenick, R.F., 1977. Critical seismic excitation and response of structures. In: Proc. of 6WCEE, vol. II. K.A. Rastogi for Sarita Prakashan, Meerut, India, 1040–1045.

Wang, P.C., Philippacopoulos, A.J., 1980. Critical seismic assessment of life-line structures. Proc. of 7WCEE vol. 8, 257–264.

Wang, P.C., Yun, C.B., 1979. Site-dependent critical design spectra. Earthquake Engrg. Struct. Dyn. 7 (6), 569–578.

Wang, P.C., Wang, W., Drenick, R., Vellozzi, J., 1976. Critical excitation and response of free standing chimneys. In: Proc. of the Int. Symposium on Earthquake Struct. Engrg, vol. I. Mo., St. Louis. Aug., 269–284.

Wang, P.C., Wang, W.Y.L., Drenick, R.F., 1978. Seismic assessment of high-rise buildings. J. Engrg. Mech. Div. 104 (2), 441–456.

Westermo, B.D., 1985. The critical excitation and response of simple dynamic systems. J. Sound and Vibration 100 (2), 233–242.

Yang, J.N., Heer, E., 1971. Maximum dynamic response and proof testing. J. Engrg. Mech. Div. 97 (4), 1307–1313.

CHAPTER TWO

Critical Excitation for Stationary and Nonstationary Random Inputs

Contents

2.1. INTRODUCTION

In this chapter, critical excitation methods for stationary and nonstationary random inputs are discussed. It is natural to assume that earthquake ground motions are samples or realizations of a nonstationary random process. Therefore critical excitation methods for nonstationary random inputs may be desirable for constructing and developing realistic earthquake–resistant design methods. However, it may also be relevant to develop the critical excitation methods for stationary random inputs as the basis for further development for nonstationary random inputs. First of all, critical excitation methods for stationary random inputs are discussed and some fundamental and important results are derived. These results play a significant role in the critical excitation methods explained in this book. Secondly, critical excitation methods for nonstationary random inputs are developed based on the theory for stationary random inputs.

© 2013 Elsevier Ltd.
All rights reserved.

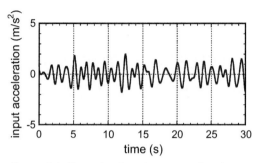

Figure 2.1 *Example of stationary random input.*

Fig. 2.1 shows a sample of a stationary random process. The mean value and the standard deviation of this process are constant every time. The phase of this process is assumed to be uniformly random.

 ## 2.2. STATIONARY INPUT TO SINGLE-DEGREE-OF-FREEDOM (SDOF) MODEL

Consider an SDOF model of mass m, viscous damping coefficient c and stiffness k as shown in Fig. 2.2. When this model is subjected to the base motion $u_g(t)$ as a stationary Gaussian random process with zero mean, the equation of motion may be described by

$$m\ddot{u}(t) + c\dot{u}(t) + ku(t) = -m\ddot{u}_g(t) \tag{2.1}$$

Fourier transformation of Eq. (2.1) leads to

$$(-\omega^2 m + i\omega c + k)U(\omega) = -m\ddot{U}_g(\omega) \tag{2.2}$$

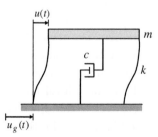

Figure 2.2 *SDOF model subjected to horizontal ground motion.*

where $U(\omega)$ and $\ddot{U}_g(\omega)$ are Fourier transforms of $u(t)$ and $\ddot{u}_g(t)$, respectively. The transfer function may then be derived as

$$H(\omega) = U(\omega)/\ddot{U}_g(\omega) = -m/(-\omega^2 m + i\omega c + k) \tag{2.3}$$

Let $S_g(\omega)$ denote the power spectral density (PSD) function of $\ddot{u}_g(t)$. The mean square response of the structural deformation $D(t) = u(t)$ may be expressed by

$$\sigma_D^2 = \int_{-\infty}^{\infty} |H(\omega)|^2 S_g(\omega) d\omega \tag{2.4}$$

$$f = \sigma_D^2 = \int_{-\infty}^{\infty} |H(\omega)|^2 S_g(\omega) d\omega = \int_{-\infty}^{\infty} F(\omega) S_g(\omega) d\omega \tag{2.5}$$

where

$$F(\omega) = |H(\omega)|^2 \tag{2.6}$$

The critical excitation problem for stationary random inputs may be stated as follows.

[Problem CESS]
Given floor mass, story stiffness and structural viscous damping, find the critical PSD function $\widetilde{S}_g(\omega)$ to maximize f defined by Eq. (2.5) subject to

$$\int_{-\infty}^{\infty} S_g(\omega) d\omega \leq \overline{S} \ (\overline{S}; \text{given power limit}) \tag{2.7}$$

$$\sup S_g(\omega) \leq \bar{s} \ (\bar{s}; \text{given PSD amplitude limit}) \tag{2.8}$$

Equation (2.7) limits the power of the excitation and Eq. (2.8) is introduced to keep the present excitation model physically realistic. It is well known that a PSD function, a Fourier amplitude spectrum and an undamped velocity response spectrum of an earthquake have an approximate relationship. If the time duration of the earthquake is fixed, the PSD function corresponds to the Fourier amplitude spectrum and almost corresponds to the undamped velocity response spectrum. Therefore the present limitation on the peak of the PSD function approximately indicates the specification of a bound on the undamped velocity response spectrum.

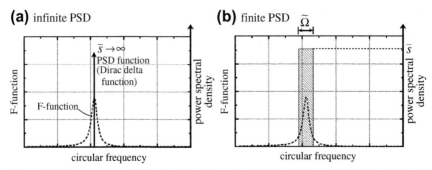

Figure 2.3 *Power spectral density function of critical excitation: (a) infinite PSD, (b) finite PSD.*

The solution to the above-mentioned problem can be obtained in a simple manner. Fig. 2.3 shows an example of the function $F(\omega) = |H(\omega)|^2$. The critical PSD function is a rectangular function overlapped around the natural frequency of the structural model.

In case of the infinite PSD amplitude limit, i.e. $\bar{s} \to \infty$, $\tilde{S}_g(\omega)$ is reduced to the Dirac delta function (see Fig. 2.3(a)) and the value f takes

$$f = \bar{S}F(\omega_M) \tag{2.9}$$

where ω_M is characterized by

$$F(\omega_M) = \max_{\omega} F(\omega) \tag{2.10}$$

This implies that the critical excitation is almost resonant to the fundamental natural frequency of the structural model.

When $\bar{s}$ is finite, $\tilde{S}_g(\omega)$ turns out to be a constant $\bar{s}$ in a finite interval $\tilde{\Omega} = \bar{S}/\bar{s}$ (see Fig. 2.3(b)). This input is called hereafter "the input with a rectangular PSD function." The optimization procedure is very simple because of the positive definiteness of the functions $F(\omega)$ and $S_g(\omega)$ in Eq. (2.5) and it is sufficient to find the finite interval $\tilde{\Omega}$ which can be searched for by decreasing a horizontal line in the figure of the function $F(\omega)$ until the interval length attains $\bar{S}/\bar{s}$ and finding their intersections (see Fig. 2.4).

2.3. STATIONARY INPUT TO MULTI-DEGREE-OF-FREEDOM (MDOF) MODEL

Consider an n-story shear building model, as shown in Fig. 2.5, subjected to the base acceleration $\ddot{u}_g(t)$ which is regarded as a stationary Gaussian random

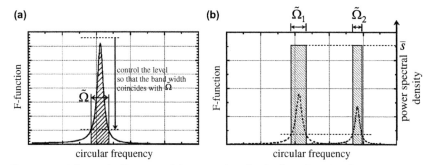

Figure 2.4 *Schematic diagram of the procedure for finding the critical excitation with a rectangular PSD function: (a) single case, (b) multiple isolated case.*

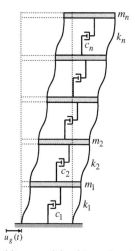

Figure 2.5 *n-story shear building model subjected to horizontal base acceleration.*

process with zero mean. $\mathbf{M}$, $\mathbf{C}$, $\mathbf{K}$, $\mathbf{r} = \{1 \cdots 1\}^T$ are the system mass, viscous damping, stiffness matrices and the influence coefficient vector, respectively. Equations of motion of this model in the frequency domain may be written as

$$(-\omega^2 \mathbf{M} + i\omega \mathbf{C} + \mathbf{K})\mathbf{U}(\omega) = -\mathbf{M}\mathbf{r}\ddot{U}_g(\omega) \tag{2.11}$$

$\mathbf{U}(\omega)$ and $\ddot{U}_g(\omega)$ denote the Fourier transforms of the floor displacements $\mathbf{u}(t)$ and the Fourier transform of the input acceleration $\ddot{u}_g(t)$, respectively. Eq. (2.11) can be simplified to the following compact form.

$$\mathbf{A}\mathbf{U}(\omega) = \mathbf{B}\ddot{U}_g(\omega) \tag{2.12}$$

where

$$\mathbf{A} = \left(-\omega^2\mathbf{M} + i\omega\mathbf{C} + \mathbf{K}\right) \text{(tridiagonal matrix)} \tag{2.13a}$$

$$\mathbf{B} = -\mathbf{Mr} \tag{2.13b}$$

Let $d_i(t)$ denote the interstory drift in the i-th story. Define the set $\mathbf{d}(t) = \{d_i(t)\}$ and their Fourier transforms $\mathbf{D}(\omega) = \{D_i(\omega)\}$. $\mathbf{D}(\omega)$ can be expressed in terms of $\mathbf{U}(\omega)$ by

$$\mathbf{D}(\omega) = \mathbf{TU}(\omega) \tag{2.14}$$

$\mathbf{T}$ is a constant matrix consisting of 1 (diagonal components), -1 and 0. Substitution of $\mathbf{U}(\omega)$ in Eq. (2.12) into Eq. (2.14) leads to

$$\mathbf{D}(\omega) = \mathbf{TA}^{-1}\mathbf{B}\ddot{U}_g(\omega) \tag{2.15}$$

Eq. (2.15) can be expressed simply as

$$\mathbf{D}(\omega) = \mathbf{H}_D(\omega)\ddot{U}_g(\omega) \tag{2.16}$$

In Eq. (2.16), $\mathbf{H}_D(\omega) = \{H_{D_i}(\omega)\}$ are the transfer functions of interstory drifts to the input acceleration and are described as

$$\mathbf{H}_D(\omega) = \mathbf{TA}^{-1}\mathbf{B} \tag{2.17}$$

Since $\mathbf{A}$ is a tridiagonal matrix, its inverse can be obtained in closed form.

Let $S_g(\omega)$ denote the PSD function of $\ddot{u}_g(t)$. According to the random vibration theory, the mean-square response of the i-th interstory drift can be computed from

$$\sigma_{D_i}^2 = \int_{-\infty}^{\infty} |H_{D_i}(\omega)|^2 S_g(\omega)d\omega = \int_{-\infty}^{\infty} H_{D_i}(\omega)H_{D_i}^*(\omega)S_g(\omega)d\omega \tag{2.18}$$

where $(\ )^*$ indicates the complex conjugate.

The sum of the mean squares of the interstory drifts can be expressed by

$$f = \sum_{i=1}^{n} \sigma_{D_i}^2 = \int_{-\infty}^{\infty} F(\omega)S_g(\omega)d\omega \tag{2.19}$$

where

$$F(\omega) = \sum_{i=1}^{n} |H_{D_i}(\omega)|^2 = \sum_{i=1}^{n} H_{D_i}(\omega)H_{D_i}^*(\omega) \tag{2.20}$$

The problem of critical excitation for stationary inputs may be described as:

[Problem CESM]

Given floor masses, story stiffnesses and story viscous dampings, find the critical PSD function $\tilde{S}_g(\omega)$ maximizing f defined by Eq. (2.19) subject to

$$\int_{-\infty}^{\infty} S_g(\omega)d\omega \leq \overline{S} \ (\overline{S}; \text{given value of power limit}) \tag{2.21}$$

$$\sup S_g(\omega) \leq \overline{s} \ (\overline{s}; \text{given value of PSD amplitude limit}) \tag{2.22}$$

Fig. 2.6 shows examples of $F(\omega)$ for 2-DOF models and Fig. 2.7 presents the variation of the function f with respect to $1/\overline{s}$ for various damping ratios.

Almost the same solution procedure as for an SDOF model can also be applied to this problem. In the case where $\overline{s} \to \infty$, it is known that $\tilde{S}_g(\omega)$ is reduced to the Dirac delta function (see Fig. 2.3(a)) and the value f can be expressed by

$$f = \overline{S}F(\omega_M) \tag{2.23}$$

where the frequency ω_M is characterized by

$$F(\omega_M) = \max_\omega F(\omega) \tag{2.24}$$

This means that the frequency content of the critical excitation is almost resonant to the fundamental natural frequency of the structural model.

In the case where $\overline{s}$ is finite, $\tilde{S}_g(\omega)$ is found to be a constant $\overline{s}$ in a finite interval $\tilde{\Omega} = \overline{S}/\overline{s}$ (see Fig. 2.3(b)). This input will be called "the input with a rectangular PSD function." The optimization procedure is simple because of

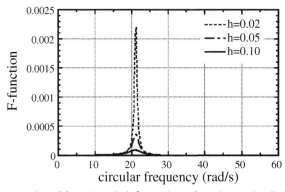

Figure 2.6 *Examples of function F(ω) for various damping ratios (2-DOF model).*

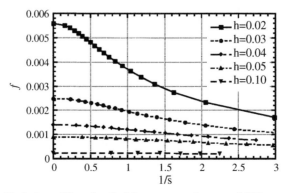

Figure 2.7 *Variation of function f with respect to inverse of PSD amplitude* $1/\bar{s}$.

the positive definiteness of the functions $F(\omega)$ and $S_g(\omega)$ in Eq. (2.19). It is sufficient to find the finite interval $\tilde{\Omega}$ which can be determined by decreasing a horizontal line in the figure of the function $F(\omega)$ until the interval length attains $\overline{S}/\overline{s}$ and finding their intersections (see Fig. 2.4(a)). When higher-mode effects are significant in MDOF systems, the critical PSD function will result in a multiple isolated rectangular PSD function (see Fig. 2.4(b)).

2.4. CONSERVATIVENESS OF BOUNDS

The level of conservativeness of the explained critical excitation is examined through comparison with the results for recorded earthquakes. A representative four recorded earthquake ground motions are taken into account, i.e. El Centro NS 1940, Taft EW 1952, Hyogoken-Nanbu, Kobe University NS 1995 and Hyogoken-Nanbu, JMA Kobe NS 1995. Let us consider an SDOF elastic model with a damping ratio $= 0.02$. The interstory drift is chosen as the response parameter to be compared.

Fig. 2.8 illustrates the PSD functions in a relaxed sense (approximate treatment for nonstationary motions using the Fourier transform and time duration) for these four ground motions. It can be observed that a sharp peak appears around the period of about 1.2(s) in Hyogoken-Nanbu, Kobe University NS 1995.

The solid line in Fig. 2.9 shows the standard deviation of the interstory drift for each recorded ground motion. The value to the critical excitation is also plotted to the undamped natural period of the SDOF model. The PSD functions in Fig. 2.8 have been used in Eq. (2.8) to estimate approximately the standard deviation of the interstory drift. Note that the coefficient $\sqrt{2}$ is multiplied in order to take into account the nonstationarity of ground

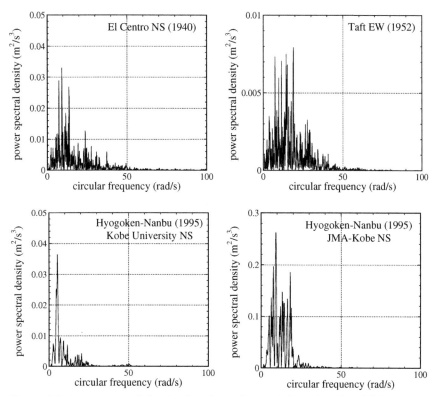

Figure 2.8 *Power spectral density functions of recorded earthquakes (El Centro NS 1940; Taft EW 1952; Hyogoken-Nanbu, Kobe University NS 1995; Hyogoken-Nanbu, JMA Kobe NS 1995).*

motions. One-third of the displacement response spectrum for each ground motion is also plotted in Fig. 2.9 (broken line). The coefficient "three" approximately represents the so-called peak factor. In Fig. 2.9, the area of the PSD function and the peak value of the PSD function have been computed for each recorded ground motion. These values are specified by $\overline{S} = 0.278(\text{m}^2/\text{s}^4)$, $\overline{s} = 0.0330(\text{m}^2/\text{s}^3)$ for El Centro NS 1940, $\overline{S} = 0.0901(\text{m}^2/\text{s}^4)$, $\overline{s} = 0.00792(\text{m}^2/\text{s}^3)$ for Taft EW 1952, $\overline{S} = 0.185(\text{m}^2/\text{s}^4)$, $\overline{s} = 0.0364(\text{m}^2/\text{s}^3)$ for Hyogoken-Nanbu, Kobe University NS 1995 and $\overline{S} = 2.55(\text{m}^2/\text{s}^4)$, $\overline{s} = 0.262(\text{m}^2/\text{s}^3)$ for Hyogoken-Nanbu, JMA Kobe NS 1995. It can be observed that, while the level of conservativeness is about 2 or 3 in the natural period range of interest in El Centro NS 1940 and Taft EW 1952, a closer coincidence can be seen around the natural period of 1.2(s) in Hyogoken-Nanbu, Kobe University NS 1995. This indicates that Hyogoken-Nanbu, Kobe University NS 1995

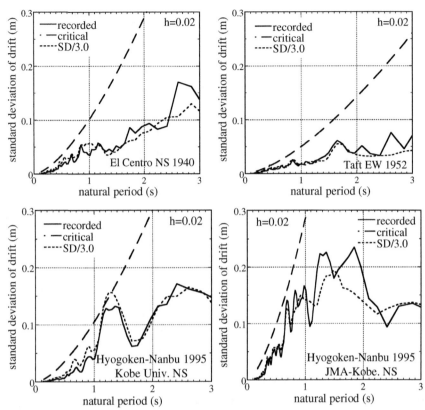

Figure 2.9 *Standard deviation of interstory drift of an SDOF model (damping ratio = 0.02) subjected to recorded earthquakes, that relate to the present critical excitation and one-third of the displacement response spectrum.*

has a predominant period around 1.2(s) and the resonant property of this ground motion can be represented by the explained critical excitation.

2.5. NONSTATIONARY INPUT TO SDOF MODEL

The key idea for stationary inputs can be used for nonstationary inputs. In this section, it is assumed that the input base acceleration can be described by the following uniformly modulated nonstationary random process.

$$\ddot{u}_g(t) = c(t)w(t) \qquad (2.25)$$

In Eq. (2.25), $c(t)$ is a given deterministic envelope function and $w(t)$ is a stationary Gaussian process with zero mean to be determined. More complex nonuniformly modulated nonstationary models have been proposed

(Conte and Peng 1997; Fang and Sun 1997). Advanced nonstationary critical excitation methods for such complex models may be interesting.

$S_w(\omega)$ denotes the PSD function of $w(t)$. In this case the PSD function of $\ddot{u}_g(t)$ can be expressed by $S_g(t; \omega) = c(t)^2 S_w(\omega)$. Let us consider an SDOF model of the natural circular frequency ω_1 and the damping ratio h. The mean-square deformation of the SDOF model can then be expressed by

$$
\sigma_x(t)^2 = \int\limits_{-\infty}^{\infty} \left[\int\limits_0^t c(\tau_1)g(t - \tau_1)e^{i\omega\tau_1}\mathrm{d}\tau_1 \right]\left[\int\limits_0^t c(\tau_2)g(t - \tau_2)e^{-i\omega\tau_2}\mathrm{d}\tau_2 \right] S_w(\omega)\mathrm{d}\omega
$$

$$
= \int\limits_{-\infty}^{\infty} \{A_C(t; \omega)^2 + A_S(t; \omega)^2\} S_w(\omega)\mathrm{d}\omega
$$

$$
= \int\limits_{-\infty}^{\infty} H(t; \omega) S_w(\omega)\mathrm{d}\omega
$$

$$
(2.26)
$$

In Eq. (2.26), the function $g(t) = H_e(t)(1/\omega_{1d})e^{-h\omega_1 t} \sin \omega_{1d}t$ is the well-known unit impulse response function. $H_e(t)$ is the Heaviside step function and $\omega_{1d} = \sqrt{1 - h^2}\omega_1$. The functions $A_C(t; \omega)$, $A_S(t; \omega)$ are defined by

$$
A_C(t; \omega) = \int\limits_0^t c(\tau)g(t - \tau) \cos \omega\tau\mathrm{d}\tau \qquad (2.27a)
$$

$$
A_S(t; \omega) = \int\limits_0^t c(\tau)g(t - \tau) \sin \omega\tau\mathrm{d}\tau \qquad (2.27b)
$$

$A_C(t; \omega)$ and $A_S(t; \omega)$ indicate the displacement response of an SDOF model to the amplitude modulated cosine function $c(\tau) \cos \omega\tau$ and to the amplitude modulated sine function $c(\tau) \sin \omega\tau$, respectively. The detailed expressions of $A_C(t; \omega)$ and $A_S(t; \omega)$ for a specific envelope function $c(t)$ are shown in the Appendix. It is important to note that the function $H(t; \omega)$ includes the effects of the envelope function $c(t)$ and zero initial conditions and its frequency content is time-dependent.

The envelope function $c(t)$ is assumed here to be given. The problem of critical excitation may be stated as:

[Problem CENSS]
Given the natural frequency and damping ratio of an SDOF model and the excitation envelope function $c(t)$, find the critical PSD function $\tilde{S}_w(\omega)$ of $w(t)$ to maximize the

specific mean square deformation $\sigma_x(t^*)^2$ *(t^* is the time when the maximum mean square deformation to $S_w(\omega)$ is attained) subject to the excitation power limit*

$$\int_{-\infty}^{\infty} S_w(\omega)d\omega \leq \overline{S}_w \quad (\overline{S}_w : \text{ given power limit}) \tag{2.28}$$

and to the PSD amplitude limit.

$$\sup S_w(\omega) \leq \overline{s}_w \quad (\overline{s}_w : \text{ given PSD amplitude limit}) \tag{2.29}$$

This problem consists of the double maximization procedures described by

$$\max_{S_w(\omega)} \max_t \{f(t; S_w(\omega)) \equiv \sigma_x(t; S_w(\omega))^2\}$$

The first maximization is implemented with respect to time for a given PSD function $S_w(\omega)$ (see Fig. 2.10) and the second maximization is conducted with respect to the PSD function $S_w(\omega)$. In the first maximization, the time t^* causing the maximum mean square deformation must be obtained for each PSD function. This original problem is complex and needs a lot of computation. To overcome this problem, a smart procedure based on the interchange of the order of the maximization procedures is introduced. The procedure can be expressed by

$$\max_t \max_{S_w(\omega)} \{f(t; S_w(\omega)) \equiv \sigma_x(t; S_w(\omega))^2\}$$

The first maximization process with respect to the PSD function for a given time can be implemented effectively and efficiently by using the critical excitation method for stationary inputs. The second maximization with respect to time can be conducted by comparing the values at various times directly.

The algorithm may be summarized as:

(i) Compute the transfer function $H(t_i; \omega)$ in Eq. (2.26) for a specific time $t = t_i$.

(ii) Find the critical PSD function at time $t = t_i$ as the rectangular PSD function (the procedure used for stationary inputs can be applied).

(iii) Compute $\sigma_x(t_i)^2$ using Eq. (2.26) to the rectangular PSD function obtained in step (ii).

(iv) Repeat (i)–(iii) for various times and find $\sigma_x(t_m)^2 = \max \sigma_x(t_i)^2$.

(v) The PSD function for $t = t_m$ is determined as the critical one.

It is important to note that the present algorithm based on the interchange of the order of the double maximization procedures is applicable to more complex nonuniformly modulated nonstationary excitation models. In that

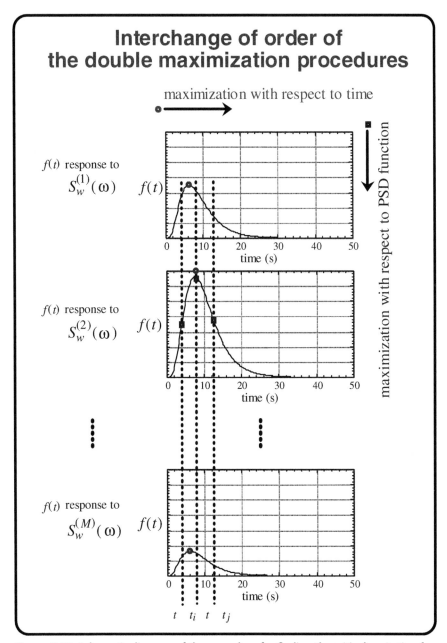

Figure 2.10 *Schematic diagram of the procedure for finding the critical excitation for nonstationary random inputs (order interchange of double maximization procedure).*

case, the expression of Eq. (2.26) must be modified and a new critical excitation problem must be stated.

2.6. NONSTATIONARY INPUT TO MDOF MODEL

Let us consider an n-story shear building model with proportional damping subjected to a nonstationary random base acceleration $\ddot{u}_g$ which can be described by the uniformly modulated nonstationary random process of Eq. (2.25). The parameters ω_j and h_j denote the j-th undamped natural circular frequency and the j-th damping ratio. The equations of motion can be written as

$$\mathbf{M\ddot{u}}(t) + \mathbf{C\dot{u}}(t) + \mathbf{Ku}(t) = -\mathbf{Mr}\ddot{u}_g(t) \tag{2.30}$$

The quantities in Eq. (2.30) have been defined in Section 2.3. Introduce the following coordinate transformation from physical coordinates $\mathbf{u}(t)$ to normal coordinates $\mathbf{q}(t) = \{q_j\}$ through the modal matrix $\mathbf{\Phi}$.

$$\mathbf{u}(t) = \mathbf{\Phi q}(t) \tag{2.31}$$

Substitution of Eq. (2.31) into Eq. (2.30) provides n independent equations.

$$\ddot{q}_j + 2h_j\omega_j\dot{q}_j + \omega_j^2 q_j = -\Gamma_j\ddot{u}_g \quad (j = 1, \cdots, n) \tag{2.32}$$

The mean-square response of the k-th interstory drift can then be expressed by

$$\sigma_{D_k}(t)^2 = \int_{-\infty}^{\infty} \left\{ \sum_{j=1}^{n} \Gamma_j\left(\phi_k^{(j)} - \phi_{k-1}^{(j)}\right) \int_0^t c(\tau_1)g_j(t - \tau_1)e^{i\omega\tau_1}d\tau_1 \right\} \cdot$$

$$\left\{ \sum_{j=1}^{n} \Gamma_j\left(\phi_k^{(j)} - \phi_{k-1}^{(j)}\right) \int_0^t c(\tau_2)g_j(t - \tau_2)e^{-i\omega\tau_2}d\tau_2 \right\} S_w(\omega)d\omega$$

$$= \int_{-\infty}^{\infty} \left\{ \sum_{j=1}^{n} \Gamma_j\left(\phi_k^{(j)} - \phi_{k-1}^{(j)}\right)[A_{Cj}(t; \omega) + iA_{Sj}(t; \omega)] \right\} \cdot$$

$$\left\{ \sum_{j=1}^{n} \Gamma_j\left(\phi_k^{(j)} - \phi_{k-1}^{(j)}\right)[A_{Cj}(t; \omega) - iA_{Sj}(t; \omega)] \right\} S_w(\omega)d\omega$$

$$= \int_{-\infty}^{\infty} \left[\left\{ \sum_{j=1}^{n} \Gamma_j\left(\phi_k^{(j)} - \phi_{k-1}^{(j)}\right)A_{Cj}(t; \omega) \right\}^2 \right.$$

$$\left. + \left\{ \sum_{j=1}^{n} \Gamma_j\left(\phi_k^{(j)} - \phi_{k-1}^{(j)}\right)A_{Sj}(t; \omega) \right\}^2 \right] S_w(\omega)d\omega \tag{2.33}$$

Γ_j is the j-th participation factor, i.e. $\Gamma_j = \boldsymbol{\phi}^{(j)^T}\mathbf{Mr}/\boldsymbol{\phi}^{(j)^T}\mathbf{M}\boldsymbol{\phi}^{(j)}$, and $\phi_k^{(j)}$ is the k-th component in the j-th eigenvector $\boldsymbol{\phi}^{(j)}$. $A_{Cj}(t;\omega)$ and $A_{Sj}(t;\omega)$ in Eq. (2.33) are defined by

$$A_{Cj}(t;\omega) = \int_0^t g_j(t-\tau)\{c(\tau)\cos\omega\tau\}d\tau \qquad (2.34a)$$

$$A_{Sj}(t;\omega) = \int_0^t g_j(t-\tau)\{c(\tau)\sin\omega\tau\}d\tau \qquad (2.34b)$$

The function $g_j(t)$ in Eqs. (2.33) and (2.34) indicates the impulse response function with ω_j, h_j as the undamped natural circular frequency and damping ratio. $A_{Cj}(t;\omega)$ and $A_{Sj}(t;\omega)$ for a specific envelope function $c(t)$ can be derived by regarding ω_1 and h for SDOF models in Eqs. (2.27a, b) as ω_j and h_j.

The sum of the time-dependent mean-square interstory drifts can be written as

$$f(t) = \sum_{k=1}^n \sigma_{D_k}(t)^2 = \int_{-\infty}^{\infty} H_M(t;\omega)S_w(\omega)d\omega \qquad (2.35)$$

where

$$H_M(t;\omega) = \sum_{k=1}^n \left[\left\{\sum_{j=1}^n \Gamma_j\left(\phi_k^{(j)} - \phi_{k-1}^{(j)}\right)A_{Cj}(t;\omega)\right\}^2 \right.$$
$$\left. + \left\{\sum_{j=1}^n \Gamma_j\left(\phi_k^{(j)} - \phi_{k-1}^{(j)}\right)A_{Sj}(t;\omega)\right\}^2\right] \qquad (2.36)$$

The problem of critical excitation may be stated as:

[Problem CENSM]

Given floor masses, story stiffnesses and structural viscous damping of a shear building model and the excitation envelope function $c(t)$, find the critical PSD function $\tilde{S}_w(\omega)$ of $w(t)$ to maximize the specific function $f(t^*)$ (t^* is the time when the maximum value of $f(t)$ to $S_w(\omega)$ is attained) subject to the excitation power limit

$$\int_{-\infty}^{\infty} S_w(\omega)d\omega \leq \overline{S}_w \qquad (2.37)$$

and to the PSD amplitude limit.

$$\sup S_w(\omega) \leq \bar{s}_w \tag{2.38}$$

It may be possible to adopt another quantity, e.g. the top-floor acceleration (see Chapter 3), as the criticality measure $f(t)$. The algorithm explained for the SDOF model subjected to nonstationary inputs may be used directly by regarding $H_M(t; \omega)$ in Eq. (2.35) as $H(t; \omega)$ in Eq. (2.26).

2.7. NUMERICAL EXAMPLES FOR SDOF MODEL

The following envelope function is used (see Fig. 2.11).

$$c(t) = e^{-\alpha t} - e^{-\beta t} \tag{2.39}$$

This type of envelope function was introduced in the 1960s in the field of earthquake engineering. The parameters $\alpha = 0.13, \beta = 0.45$ are chosen so as to have a peak around the time of 4(s) and a duration of about 30(s). The functions $H(t; \omega)$ of the model with $T_1 = 0.5(s), h = 0.02$ (undamped natural period and damping ratio) are plotted in Fig. 2.12 at every two seconds. It can be seen that, while $H(t; \omega)$ indicates a rather wide-band frequency content around the time of 4 and 6(s), it indicates a narrower frequency content at later times. It can also be found that the function $H(t; \omega)$ has a conspicuous peak around the natural frequency of the model and its amplitude attains the maximum around the time of 8(s). Fig. 2.13 illustrates the time history of the mean-square deformation to the corresponding critical excitation for various values of the PSD amplitude $\bar{s}_w$. The power of $w(t)$ of the critical excitation has been chosen as $\bar{S}_w = 1.51(m^2/s^4)$. This implies that it has the same value as that of El Centro NS 1940 at the time of 3.9(s) when the acceleration amplitude attains its maximum.

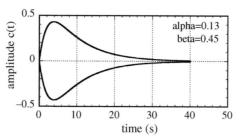

Figure 2.11 *Envelope function* c(t).

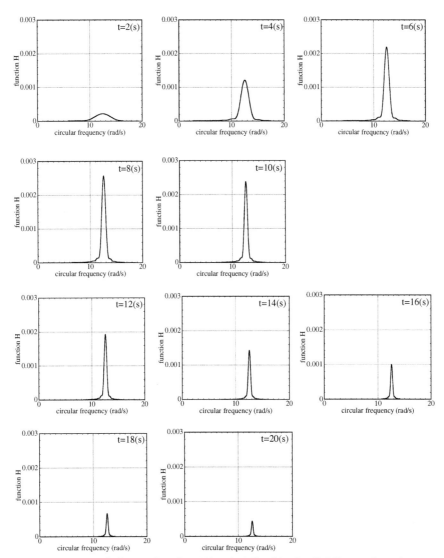

Figure 2.12 *Evolutionary functions* $H(t; \omega)$ *defined by Eq. (2.26) at various times.*

In view of the evolutionary transfer functions $H(t; \omega)$ in Fig. 2.12, a rectangular PSD function with the natural frequency of the SDOF model as the central point in the frequency range turns out to be a good and acceptable approximation of the PSD function of the critical excitation.

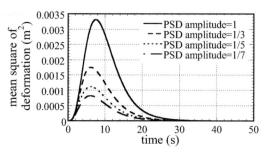

Figure 2.13 *Time-dependent mean-square deformation of the SDOF model for various PSD amplitudes.*

2.8. NUMERICAL EXAMPLES FOR MDOF MODEL

Let us consider a 6-story shear building model. The floor masses are chosen as $m_i = 32 \times 10^3 (\text{kg})(i = 1, \cdots, 6)$ and the story stiffnesses are $k_i = 3.76 \times 10^7 (\text{N/m})(i = 1, \cdots, 6)$. The fundamental natural period of the model is found to be 0.760(s). The viscous damping matrix of the shear building model has been given so as to be proportional to the stiffness matrix. The lowest-mode damping ratio is specified as 0.02. The same envelope function as in an SDOF model has been adopted. Fig. 2.14 illustrates the evolutionary transfer functions $H_M(t; \omega)$, defined by Eq. (2.36), of the 6-story shear building model at 2 second intervals. It can be seen that the evolutionary transfer functions $H_M(t; \omega)$ of the 6-story MDOF model have a similar tendency to those of the SDOF model shown in Fig. 2.12 except the peak frequency value (the fundamental natural frequency). Fig. 2.15 presents the time history of $f(t)$ defined by Eq. (2.35) for various values of the PSD amplitude. The power of the critical excitation has the same value as that for the SDOF model. It can be understood that, while the time giving the peak value differs from that for the SDOF model, the tendency is nearly the same as in the SDOF model.

Two time-history samples of the critical excitations are shown in Fig. 2.16. Both have the same PSD power $\overline{S}_w = 1.51(\text{m}^2/\text{s}^4)$. The upper one indicates a sample of the critical excitation with the PSD amplitude of 1.0 (m^2/s^3) and the lower one shows that with the PSD amplitude of 1/7 (m^2/s^3). The uniformly random phase is assumed here. It can be observed that, while the upper one has a narrow-band property, the lower one has a wide-band characteristic. These characteristics represent well the PSD function properties of the two critical excitations.

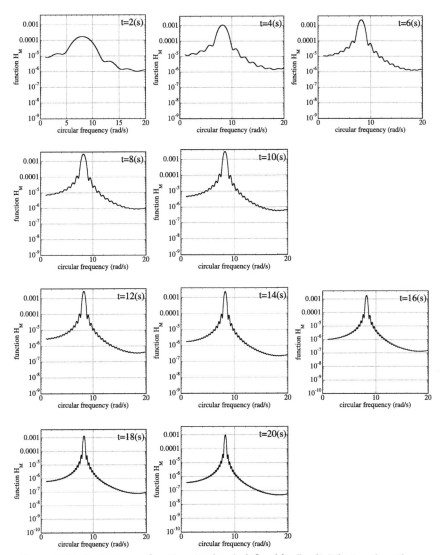

Figure 2.14 *Evolutionary functions* $H_M(t; \omega)$ *defined by Eq. (2.36) at various times.*

2.9. CONCLUSIONS

The conclusions may be summarized as follows:

(1) A critical excitation method for stationary random inputs can be developed by introducing a stochastic response index as the objective function to be maximized. The power and the intensity of the excitations are prescribed.

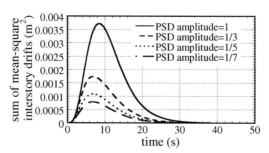

Figure 2.15 *Time-dependent function* $f(t)$ *(the sum of the mean-square interstory drifts of the MDOF model) for various PSD amplitudes.*

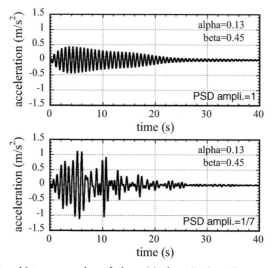

Figure 2.16 *Time-history samples of the critical excitations (upper: PSD amplitude* $= 1.0\text{m}^2\text{s}^{-3}$*, lower: PSD amplitude* $= 1/7\text{m}^2\text{s}^{-3}$*).*

(2) When the restriction on the excitation intensity does not exist, the critical PSD function turns out to be the Dirac delta function. In the existence of the restriction on the excitation intensity, the critical excitation is reduced to a rectangular PSD function with the maximum intensity limit.

(3) The level of conservativeness of the derived critical excitation for stationary random inputs is about 2 or 3 in recorded ground motions without conspicuous predominant frequency. On the other hand, it is close to unity in ground motions with a conspicuous predominant frequency. The resonant characteristic of such ground motion can be represented suitably by the explained critical excitation.

(4) The key idea for stationary random inputs can be used in SDOF models under nonstationary random inputs which can be described by a uniformly modulated excitation model. The interchange of the order of the double maximization procedures is taken full advantage of with respect to time and to the PSD function. A similar algorithm can also be devised for MDOF proportionally damped models.

It is extremely difficult even in the advanced computational environment to consider all the possible design earthquakes in the structural design stage (PEER Center et al. 2000). It is expected that the critical excitation approach can take into account uncertainties in establishing the design of buildings for earthquakes. Only the elastic response has been considered in this chapter for a simple explanation of a new concept. However, the consideration of elastic-plastic responses in the seismic-resistant design of structures may be important. The present critical excitation method could be extended to nonlinear problems by using equivalent linearization techniques (Drenick 1977). The critical excitation methods for elastic-plastic responses will be discussed in Chapter 5.

APPENDIX FUNCTIONS A_C $(t;\omega)$, A_S $(t;\omega)$ FOR A SPECIFIC ENVELOPE FUNCTION

$A_C(t;\omega)$ in Eq. (2.27a) can be expressed by substituting the specific envelope function shown in Eq. (2.39) into Eq. (2.27a).

$$A_C(t;\omega) = \frac{1}{\omega_{1d}}e^{-h\omega_1 t}\left[\sin \omega_{1d}t \int_0^t \left\{e^{(h\omega_1-\alpha)\tau} - e^{(h\omega_1-\beta)\tau}\right\}\right.$$

$$\times \frac{1}{2}\{\cos (\omega_{1d} + \omega)\tau + \cos (\omega_{1d} - \omega)\tau\}d\tau$$

$$- \cos \omega_{1d}t \int_0^t \left\{e^{(h\omega_1-\alpha)\tau} - e^{(h\omega_1-\beta)\tau}\right\}\frac{1}{2}\{\sin (\omega_{1d} + \omega)\tau$$

$$\left. + \sin (\omega_{1d} - \omega)\tau\}d\tau\right]$$

$$= \frac{1}{\omega_{1d}}e^{-h\omega_1 t}\{\sin \omega_{1d}tG_{CC}(t;\omega) - \cos \omega_{1d}tG_{SC}(t;\omega)\}$$

$$(A2.1)$$

where $G_{CC}(t; \omega)$ and $G_{SC}(t; \omega)$ are defined by

$$G_{CC}(t; \omega) = \int_0^t \left\{ e^{(h\omega_1 - \alpha)\tau} - e^{(h\omega_1 - \beta)\tau} \right\} \frac{1}{2} \{ \cos (\omega_{1d} + \omega)\tau$$

$$+ \cos (\omega_{1d} - \omega)\tau \} d\tau \qquad \text{(A2.2)}$$

$$G_{SC}(t; \omega) = \int_0^t \left\{ e^{(h\omega_1 - \alpha)\tau} - e^{(h\omega_1 - \beta)\tau} \right\} \frac{1}{2} \{ \sin (\omega_{1d} + \omega)\tau$$

$$+ \sin (\omega_{1d} - \omega)\tau \} d\tau \qquad \text{(A2.3)}$$

The double of the bracket in Eq. (A2.1) can be manipulated as

$$2\{ \sin \omega_{1d}t G_{CC}(t; \omega) - \cos \omega_{1d}t G_{SC}(t; \omega) \}$$

$$= \frac{1}{(h\omega_1 - \alpha)^2 + (\omega_{1d} + \omega)^2}$$

$$\times \left[(h\omega_1 - \alpha)e^{(h\omega_1 - \alpha)t}(-\sin \omega t) + (\omega_{1d} + \omega)e^{(h\omega_1 - \alpha)t}(\cos \omega t) \right.$$

$$\left. - (h\omega_1 - \alpha) \sin \omega_{1d}t - (\omega_{1d} + \omega) \cos \omega_{1d}t \right]$$

$$+ \frac{1}{(h\omega_1 - \alpha)^2 + (\omega_{1d} - \omega)^2}$$

$$\times \left[(h\omega_1 - \alpha)e^{(h\omega_1 - \alpha)t}(\sin \omega t) + (\omega_{1d} - \omega)e^{(h\omega_1 - \alpha)t}(\cos \omega t) \right.$$

$$\left. - (h\omega_1 - \alpha) \sin \omega_{1d}t - (\omega_{1d} - \omega) \cos \omega_{1d}t \right]$$

$$- \frac{1}{(h\omega_1 - \beta)^2 + (\omega_{1d} + \omega)^2}$$

$$\times \left[(h\omega_1 - \beta)e^{(h\omega_1 - \beta)t}(-\sin \omega t) + (\omega_{1d} + \omega)e^{(h\omega_1 - \beta)t}(\cos \omega t) \right.$$

$$\left. - (h\omega_1 - \beta) \sin \omega_{1d}t - (\omega_{1d} + \omega) \cos \omega_{1d}t \right]$$

$$- \frac{1}{(h\omega_1 - \beta)^2 + (\omega_{1d} - \omega)^2}$$

$$\times \left[(h\omega_1 - \beta)e^{(h\omega_1 - \beta)t}(\sin \omega t) + (\omega_{1d} - \omega)e^{(h\omega_1 - \beta)t}(\cos \omega t) \right.$$

$$\left. - (h\omega_1 - \beta) \sin \omega_{1d}t - (\omega_{1d} - \omega) \cos \omega_{1d}t \right]$$

$$\text{(A2.4)}$$

Similarly $A_S(t;\omega)$ in Eq. (2.27b) can be expressed by substituting the specific envelope function shown in Eq. (2.39) into Eq. (2.27b).

$$A_S(t;\omega) = \frac{1}{\omega_{1d}}e^{-h\omega_1 t}\left[\sin \omega_{1d}t \int_0^t \left\{e^{(h\omega_1-\alpha)\tau} - e^{(h\omega_1-\beta)\tau}\right\}\frac{1}{2}\{\sin (\omega_{1d}+\omega)\tau\right.$$

$$+ \sin (\omega_{1d}-\omega)\tau\}d\tau - \cos \omega_{1d}t \int_0^t \left\{e^{(h\omega_1-\alpha)\tau}\right.$$

$$\left. - e^{(h\omega_1-\beta)\tau}\right\}\frac{1}{2}\{\cos (\omega_{1d}+\omega)\tau - \cos (\omega_{1d}-\omega)\tau\}d\tau\right]$$

$$= \frac{1}{\omega_{1d}}e^{-h\omega_1 t}\{\sin \omega_{1d}tG_{CS}(t;\omega) - \cos \omega_{1d}tG_{SS}(t;\omega)\}$$

$$\text{(A2.5)}$$

where $G_{CS}(t;\omega)$ and $G_{SS}(t;\omega)$ are defined by

$$G_{CS}(t;\omega) = \int_0^t \left\{e^{(h\omega_1-\alpha)\tau} - e^{(h\omega_1-\beta)\tau}\right\}\frac{1}{2}\{\sin (\omega_{1d}+\omega)\tau$$

$$- \sin (\omega_{1d}-\omega)\tau\}d\tau$$

$$\text{(A2.6)}$$

$$G_{SS}(t;\omega) = \int_0^t \left\{e^{(h\omega_1-\alpha)\tau} - e^{(h\omega_1-\beta)\tau}\right\}\frac{1}{2}\{\cos (\omega_{1d}-\omega)\tau$$

$$- \cos (\omega_{1d}+\omega)\tau\}d\tau$$

$$\text{(A2.7)}$$

The double of the bracket in Eq. (A2.5) can be manipulated as
$$2\{\sin \omega_{1d}tG_{CS}(t;\omega) - \cos \omega_{1d}tG_{SS}(t;\omega)\}$$

$$= \frac{1}{(h\omega_1 - \alpha)^2 + (\omega_{1d} + \omega)^2}$$

$$\times \left[(h\omega_1 - \alpha)e^{(h\omega_1-\alpha)t}(\cos \omega t) - (\omega_{1d} + \omega)e^{(h\omega_1-\alpha)t}(- \sin \omega t)\right.$$

$$\left. - (h\omega_1 - \alpha) \cos \omega_{1d}t + (\omega_{1d} + \omega) \sin \omega_{1d}t\right]$$

$$+ \frac{1}{(h\omega_1 - \alpha)^2 + (\omega_{1d} - \omega)^2}$$

$$\times \left[(h\omega_1 - \alpha)e^{(h\omega_1-\alpha)t}(- \cos \omega t) + (\omega_{1d} - \omega)e^{(h\omega_1-\alpha)t}(\sin \omega t)\right.$$

$$\left. + (h\omega_1 - \alpha) \cos \omega_{1d}t - (\omega_{1d} - \omega) \sin \omega_{1d}t\right]$$

$$- \frac{1}{(h\omega_1 - \beta)^2 + (\omega_{1d} + \omega)^2}$$
$$\times \left[(h\omega_1 - \beta)e^{(h\omega_1 - \beta)t}(\cos \omega t) - (\omega_{1d} + \omega)e^{(h\omega_1 - \beta)t}(-\sin \omega t) \right.$$
$$\left. - (h\omega_1 - \beta)\cos \omega_{1d}t + (\omega_{1d} + \omega)\sin \omega_{1d}t \right]$$
$$- \frac{1}{(h\omega_1 - \beta)^2 + (\omega_{1d} - \omega)^2}$$
$$\times \left[(h\omega_1 - \beta)e^{(h\omega_1 - \beta)t}(-\cos \omega t) + (\omega_{1d} - \omega)e^{(h\omega_1 - \beta)t}(\sin \omega t) \right.$$
$$\left. + (h\omega_1 - \beta)\cos \omega_{1d}t - (\omega_{1d} - \omega)\sin \omega_{1d}t \right]$$

$$\text{(A2.8)}$$

REFERENCES

Conte, J.P., Peng, B.F., 1997. Fully nonstationary analytical earthquake ground-motion model. J. Eng. Mech. 123 (1), 15–24.

Drenick, R.F., 1977. The critical excitation of nonlinear systems. J. Applied Mech. 44 (2), 333–336.

Fang, T., Sun, M., 1997. A unified approach to two types of evolutionary random response problems in engineering. Archive of Applied Mech. 67 (7), 496–506.

PEER Center, ATC, Japan Ministry of Education, Science, Sports, and Culture, US-NSF, 2000. Effects of near-field earthquake shaking. In: Proc. of US-Japan workshop. San Francisco, March 20–21, 2000.

CHAPTER THREE

Critical Excitation for Nonproportionally Damped Structural Systems

Contents

3.1. INTRODUCTION

The purpose of this chapter is to explain a new probabilistic critical excitation method for nonstationary inputs to nonproportionally damped structural systems. In contrast to most of the conventional critical excitation methods, a stochastic response index is treated here as the objective function to be maximized as in Chapter 2. The power and the intensity of the excitations are fixed and the critical excitation is found under these restrictions. The key for finding the new nonstationary random critical excitation for nonproportionally damped structural systems is the interchange of the order of the double maximization procedures with respect to time and to the PSD function. It is shown that the spirit of the previously explained critical excitation method for stationary inputs is partially applicable to multi-degree-of-freedom structural systems with nonproportional

© 2013 Elsevier Ltd.
All rights reserved.   51

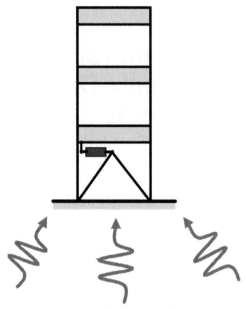

Figure 3.1 *Nonproportionally damped system subjected to nonstationary input.*

damping subjected to nonstationary random inputs. Numerical examples of a 6-DOF shear building model with nonproportional damping subjected to nonstationary inputs are presented to demonstrate the effectiveness and validity of the present method.

It is usual in the practical structural design that the mass, stiffness and damping distributions do not satisfy the condition on proportional damping. Fig. 3.1 shows a building with an added viscous damper in the 1st story. Such a structural system is a typical example of nonproportionally damped structural systems. Fig. 3.2 illustrates other examples of story-stiffness and damping distributions. Model A, B and C include additional damping in the 1st, 3rd and 6th stories, respectively, and Model D corresponds to a base-isolated building. Model PD is a proportionally damped system.

3.2. MODELING OF INPUT MOTIONS

In this chapter, the input horizontal base acceleration is assumed to be described by the following uniformly modulated nonstationary random process as in Chapter 2.

$$\ddot{u}_g(t) = c(t)w(t) \tag{3.1}$$

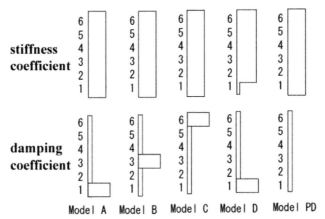

Figure 3.2 *Examples of nonproportionally damped structural system.*

where $c(t)$ is a given deterministic envelope function and $w(t)$ is a stationary Gaussian process with zero mean.

Let $S_w(\omega)$ denote the PSD function of $w(t)$. In this case the PSD function of $\ddot{u}_g$ may be expressed by $S_g(t; \omega) = c(t)^2 S_w(\omega)$.

3.3. RESPONSE OF NONPROPORTIONALLY DAMPED MODEL TO NONSTATIONARY RANDOM EXCITATION

Consider an n-story shear building model, as shown in Fig. 3.3, with a nonproportional damping subjected to a nonstationary random horizontal base acceleration $\ddot{u}_g$ which can be described by the uniformly modulated nonstationary random process of Eq. (3.1). The horizontal displacements of the floors are expressed by $\mathbf{u}(t)$. Let $\mathbf{M}, \mathbf{C}, \mathbf{K}, \mathbf{r} = \{1 \cdots 1\}^T$ denote the system mass matrix, the system viscous damping matrix, the system stiffness matrix and the influence coefficient vector, respectively. The equations of motion of the model can be expressed by

$$\mathbf{M}\ddot{\mathbf{u}}(t) + \mathbf{C}\dot{\mathbf{u}}(t) + \mathbf{K}\mathbf{u}(t) = -\mathbf{M}\mathbf{r}\ddot{u}_g(t) \tag{3.2}$$

Let $\mathbf{U}^{(j)} = \{U_k^{(j)}\}$ denote the j-th damped complex eigenvector. The time history $D_k(t)$ of the interstory drift in the k-th story may be expressed in terms of complex-type Duhamel integrals.

$$D_k(t) = 2\sum_{j=1}^{n} \mathrm{Re}\left[\left(U_k^{(j)} - U_{k-1}^{(j)}\right) \frac{\mathbf{U}^{(j)^T}\mathbf{M}\mathbf{r}}{F_j} \int_0^t \left(-\ddot{u}_g(\tau)\right) e^{\lambda_j(t-\tau)} d\tau \right]$$

$$\tag{3.3}$$

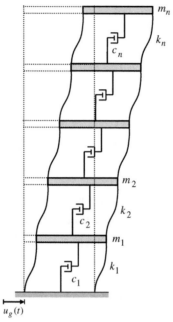

Figure 3.3 n-story shear building model subjected to horizontal base acceleration.

where

$$\lambda_j = -h_j\omega_j + \omega_{jd}i$$

$$\omega_j^2 = \frac{\overline{\mathbf{U}}^{(j)^T}\mathbf{K}\mathbf{U}^{(j)}}{\overline{\mathbf{U}}^{(j)^T}\mathbf{M}\mathbf{U}^{(j)}}$$

$$\omega_{jd} = \omega_j\sqrt{1 - h_j^2} \qquad\qquad (3.4a-e)$$

$$F_j = 2\omega_j(\sqrt{1 - h_j^2}i - h_j)\mathbf{U}^{(j)^T}\mathbf{M}\mathbf{U}^{(j)} + \mathbf{U}^{(j)^T}\mathbf{C}\mathbf{U}^{(j)}$$

$$h_j = \frac{\overline{\mathbf{U}}^{(j)^T}\mathbf{C}\mathbf{U}^{(j)}}{2\omega_j\overline{\mathbf{U}}^{(j)^T}\mathbf{M}\mathbf{U}^{(j)}}$$

$\text{Re}[\cdot]$ denotes the real part of a complex number and i denotes the imaginary unit. The over-bar indicates the complex conjugate. Let us define the following complex number.

$$\alpha_k^{(j)} \equiv 2\left(U_k^{(j)} - U_{k-1}^{(j)}\right)\frac{\mathbf{U}^{(j)^T}\mathbf{M}\mathbf{r}}{F_j} \qquad\qquad (3.5)$$

The autocorrelation function of the k-th interstory drift can be expressed as

$$
R_{D_k}(t_1, t_2) = E\left[\operatorname{Re}\left[\sum_{j=1}^{n} \alpha_k^{(j)} \int_0^{t_1} c(\tau_1) e^{(-h_j\omega_j + \omega_{jd}\mathrm{i})(t_1 - \tau_1)} w(\tau_1) d\tau_1\right]\right.
$$

$$
\left.\cdot \operatorname{Re}\left[\sum_{j=1}^{n} \alpha_k^{(j)} \int_0^{t_2} c(\tau_2) e^{(-h_j\omega_j + \omega_{jd}\mathrm{i})(t_2 - \tau_2)} w(\tau_2) d\tau_2\right]\right]
$$

$$
= \int_0^{t_2} \int_0^{t_1} \operatorname{Re}\left[\sum_{j=1}^{n} \alpha_k^{(j)} c(\tau_1) e^{(-h_j\omega_j + \omega_{jd}\mathrm{i})(t_1 - \tau_1)}\right]
$$

$$
\cdot \operatorname{Re}\left[\sum_{j=1}^{n} \alpha_k^{(j)} c(\tau_2) e^{(-h_j\omega_j + \omega_{jd}\mathrm{i})(t_2 - \tau_2)}\right] E[w(\tau_1)w(\tau_2)] d\tau_1 d\tau_2
$$

$$(3.6)$$

where $E[\cdot]$ indicates the ensemble mean. Application of the Wiener-Khintchine theorem for $w(t)$ to Eq. (3.6) provides

$$
R_{D_k}(t_1, t_2) = \int_{-\infty}^{\infty} \int_0^{t_2} \int_0^{t_1} \operatorname{Re}\left[\sum_{j=1}^{n} \left\{\alpha_k^{(j)} c(\tau_1) e^{(-h_j\omega_j + \omega_{jd}\mathrm{i})(t_1 - \tau_1)}\right\}\right]
$$

$$
\cdot \operatorname{Re}\left[\sum_{j=1}^{n} \left\{\alpha_k^{(j)} c(\tau_2) e^{(-h_j\omega_j + \omega_{jd}\mathrm{i})(t_2 - \tau_2)}\right\}\right] e^{\mathrm{i}(\tau_1 - \tau_2)\omega} S_w(\omega) d\tau_1 d\tau_2 d\omega
$$

$$
= \int_{-\infty}^{\infty} \int_0^{t_1} \operatorname{Re}\left[\sum_{j=1}^{n} \left\{\alpha_k^{(j)} c(\tau_1) e^{(-h_j\omega_j + \omega_{jd}\mathrm{i})(t_1 - \tau_1)}\right\}\right] e^{\mathrm{i}\tau_1\omega} d\tau_1
$$

$$
\cdot \int_0^{t_2} \operatorname{Re}\left[\sum_{j=1}^{n} \left\{\alpha_k^{(j)} c(\tau_2) e^{(-h_j\omega_j + \omega_{jd}\mathrm{i})(t_2 - \tau_2)}\right\}\right] e^{-\mathrm{i}\tau_2\omega} d\tau_2 S_w(\omega) d\omega
$$

$$
= \int_{-\infty}^{\infty} \int_0^{t_1} \left\{\sum_{j=1}^{n} \operatorname{Re}\left[\alpha_k^{(j)} c(\tau_1) e^{(-h_j\omega_j + \omega_{jd}\mathrm{i})(t_1 - \tau_1)}\right]\right\} e^{\mathrm{i}\tau_1\omega} d\tau_1
$$

$$
\cdot \int_0^{t_2} \left\{\sum_{j=1}^{n} \operatorname{Re}\left[\alpha_k^{(j)} c(\tau_2) e^{(-h_j\omega_j + \omega_{jd}\mathrm{i})(t_2 - \tau_2)}\right]\right\} e^{-\mathrm{i}\tau_2\omega} d\tau_2 S_w(\omega) d\omega
$$

$$(3.7)$$

The mean-square of the k-th interstory drift can be derived from Eq. (3.7) by substituting $t_1 = t_2 = t$.

$$\sigma_{D_k}(t)^2 = \int_{-\infty}^{\infty} \left[\sum_{j=1}^{n} \left\{ B_{Ck}^{(j)}(t;\omega) + iB_{Sk}^{(j)}(t;\omega) \right\} \right]$$

$$\times \left[\sum_{j=1}^{n} \left\{ B_{Ck}^{(j)}(t;\omega) - iB_{Sk}^{(j)}(t;\omega) \right\} \right] S_w(\omega) d\omega \qquad (3.8)$$

where

$$B_{Ck}^{(j)}(t;\omega) = \int_0^t \left(\mathrm{Re}\left[\alpha_k^{(j)} c(\tau) e^{(-h_j\omega_j + \omega_{jd}i)(t-\tau)} \right] \right) \cos \omega\tau d\tau \qquad (3.9a)$$

$$B_{Sk}^{(j)}(t;\omega) = \int_0^t \left(\mathrm{Re}\left[\alpha_k^{(j)} c(\tau) e^{(-h_j\omega_j + \omega_{jd}i)(t-\tau)} \right] \right) \sin \omega\tau d\tau \qquad (3.9b)$$

Rearrangement of Eq. (3.8) may be reduced to

$$\sigma_{D_k}(t)^2 = \int_{-\infty}^{\infty} \left\{ \sum_{j=1}^{n} B_{Ck}^{(j)}(t;\omega) + i\sum_{j=1}^{n} B_{Sk}^{(j)}(t;\omega) \right\}$$

$$\times \left\{ \sum_{j=1}^{n} B_{Ck}^{(j)}(t;\omega) - i\sum_{j=1}^{n} B_{Sk}^{(j)}(t;\omega) \right\} S_w(\omega) d\omega$$

$$= \int_{-\infty}^{\infty} \left[\left\{ \sum_{j=1}^{n} B_{Ck}^{(j)}(t;\omega) \right\}^2 + \left\{ \sum_{j=1}^{n} B_{Sk}^{(j)}(t;\omega) \right\}^2 \right] S_w(\omega) d\omega \qquad (3.10)$$

Let us define the following quantity.

$$f_N(t) = \sum_{k=1}^{n} \sigma_{D_k}(t)^2 = \int_{-\infty}^{\infty} H_{MN}(t;\omega) S_w(\omega) d\omega \qquad (3.11)$$

where

$$H_{MN}(t;\omega) = \sum_{k=1}^{n} \left[\left\{ \sum_{j=1}^{n} B_{Ck}^{(j)}(t;\omega) \right\}^2 + \left\{ \sum_{j=1}^{n} B_{Sk}^{(j)}(t;\omega) \right\}^2 \right] \qquad (3.12)$$

Part of the integrand in Eqs. (3.9a, b) can be reduced to

$$\text{Re}\left[\alpha_k^{(j)}c(\tau)e^{(-h_j\omega_j+\omega_{jd}i)(t-\tau)}\right]$$

$$= c(\tau)e^{-h_j\omega_j(t-\tau)}\left\{\text{Re}\left[\alpha_k^{(j)}\right]\text{Re}\left[e^{\omega_{jd}i(t-\tau)}\right]\right.$$

$$\left. -\text{Im}\left[\alpha_k^{(j)}\right]\text{Im}\left[e^{\omega_{jd}i(t-\tau)}\right]\right\}$$

$$= c(\tau)e^{-h_j\omega_j(t-\tau)}\left\{\text{Re}\left[\alpha_k^{(j)}\right]\cos\omega_{jd}(t-\tau)\right.$$

$$\left. -\text{Im}\left[\alpha_k^{(j)}\right]\sin\omega_{jd}(t-\tau)\right\} \tag{3.13}$$

where $\text{Im}[\cdot]$ denotes the imaginary part of a complex number. In the case where the structural system has proportional damping, the complex number $\alpha_k^{(j)}$ defined in Eq. (3.5) is reduced to a pure imaginary number and the real and imaginary parts of $\alpha_k^{(j)}$ are reduced to $\text{Re}\left[\alpha_k^{(j)}\right] = 0, \text{Im}\left[\alpha_k^{(j)}\right] = -\Gamma_j(\phi_k^{(j)} - \phi_{k-1}^{(j)})/\omega_{jd}$. This expression coincides with that for a structural system with proportional damping. Substitution of Eq. (3.13) into Eq. (3.9a) leads to

$$B_{Ck}^{(j)}(t;\omega) = \int_0^t c(\tau)e^{-h_j\omega_j(t-\tau)}\left\{\text{Re}\left[\alpha_k^{(j)}\right]\cos\omega_{jd}(t-\tau)\right.$$

$$\left. -\text{Im}\left[\alpha_k^{(j)}\right]\sin\omega_{jd}(t-\tau)\right\}\cos\omega\tau d\tau$$

$$= \omega_{jd}\text{Re}\left[\alpha_k^{(j)}\right]\int_0^t c(\tau)e^{-h_j\omega_j(t-\tau)}\frac{1}{\omega_{jd}}\cos\omega_{jd}(t-\tau)\cos\omega\tau d\tau$$

$$- \omega_{jd}\text{Im}\left[\alpha_k^{(j)}\right]\int_0^t c(\tau)e^{-h_j\omega_j(t-\tau)}\frac{1}{\omega_{jd}}\sin\omega_{jd}(t-\tau)\cos\omega\tau d\tau$$

$$= \omega_{jd}\text{Re}\left[\alpha_k^{(j)}\right]A_{Cj}^*(t;\omega) - \omega_{jd}\text{Im}\left[\alpha_k^{(j)}\right]A_{Cj}(t;\omega)$$

$$\tag{3.14}$$

where

$$A_{Cj}(t;\omega) = \int_0^t c(\tau)g_j(t-\tau)\cos\omega\tau d\tau$$

(3.15a)

$$= \int_0^t c(\tau)e^{-h_j\omega_j(t-\tau)}\frac{1}{\omega_{jd}}\sin\omega_{jd}(t-\tau)\cos\omega\tau d\tau$$

$$A_{Cj}^*(t;\omega) = \int_0^t c(\tau)e^{-h_j\omega_j(t-\tau)}\frac{1}{\omega_{jd}}\cos\omega_{jd}(t-\tau)\cos\omega\tau d\tau \qquad (3.15b)$$

The function $g_j(t)$ in Eq. (3.15a) is the impulse response function with ω_j, h_j as the undamped natural circular frequency and the damping ratio. $A_{Cj}(t;\omega)$ is a displacement response to an amplitude-modulated cosine acceleration input $-c(t)\cos\omega t$. Similarly, substitution of Eq. (3.13) into Eq. (3.9b) leads to

$$B_{Sk}^{(j)}(t;\omega) = \int_0^t c(\tau)e^{-h_j\omega_j(t-\tau)}\left\{\mathrm{Re}\left[\alpha_k^{(j)}\right]\cos\omega_{jd}(t-\tau)\right.$$

$$\left. - \mathrm{Im}\left[\alpha_k^{(j)}\right]\sin\omega_{jd}(t-\tau)\right\}\sin\omega\tau d\tau$$

$$= \omega_{jd}\mathrm{Re}\left[\alpha_k^{(j)}\right]\int_0^t c(\tau)e^{-h_j\omega_j(t-\tau)}\frac{1}{\omega_{jd}}\cos\omega_{jd}(t-\tau)\sin\omega\tau d\tau$$

$$- \omega_{jd}\mathrm{Im}\left[\alpha_k^{(j)}\right]\int_0^t c(\tau)e^{-h_j\omega_j(t-\tau)}\frac{1}{\omega_{jd}}\sin\omega_{jd}(t-\tau)\sin\omega\tau d\tau$$

$$= \omega_{jd}\mathrm{Re}\left[\alpha_k^{(j)}\right]A_{Sj}^*(t;\omega) - \omega_{jd}\mathrm{Im}\left[\alpha_k^{(j)}\right]A_{Sj}(t;\omega)$$

(3.16)

where

$$A_{Sj}(t;\omega) = \int_0^t c(\tau)g_j(t-\tau)\sin\omega\tau d\tau$$

(3.17a)

$$= \int_0^t c(\tau)e^{-h_j\omega_j(t-\tau)}\frac{1}{\omega_{jd}}\sin\omega_{jd}(t-\tau)\sin\omega\tau d\tau$$

$$A_{Sj}^*(t;\omega) = \int_0^t c(\tau)e^{-h_j\omega_j(t-\tau)}\frac{1}{\omega_{jd}}\cos\omega_{jd}(t-\tau)\sin\omega\tau d\tau \qquad (3.17\text{b})$$

$A_{Sj}(t;\omega)$ is a displacement response to an amplitude-modulated sine acceleration input $-c(t)\sin\omega t$. The detailed expressions of $A_{Cj}(t;\omega)$ and $A_{Sj}(t;\omega)$ in Eqs. (3.15), (3.17) and integral forms in Eqs. (3.14), (3.16) for a specific envelope function $c(t)$ given by Eq. (3.30) are shown in the Appendix.

3.4. CRITICAL EXCITATION PROBLEM

The problem of finding the critical excitation may be described as:

[Problem CENM]
Given floor masses, story stiffnesses and structural nonproportional viscous damping of a shear building model and the excitation envelope function $c(t)$, find the critical PSD function $\tilde{S}_w(\omega)$ to maximize the specific function $f_N(t^*)$ (t^* is the time when the maximum of $f_N(t)$ is attained) subject to the excitation power limit (integral of the PSD function in the frequency range)

$$\int_{-\infty}^{\infty} S_w(\omega)d\omega \leq \overline{S}_w \qquad (3.18)$$

and to the PSD amplitude limit

$$\sup S_w(\omega) \leq \overline{s}_w \qquad (3.19)$$

3.5. SOLUTION PROCEDURE

This problem consists of the double maximization procedures which may be described mathematically by

$$\max_{S_w(\omega)}\ \max_t\{f_N(t; S_w(\omega))\}$$

The first maximization is performed with respect to time for a given PSD function $S_w(\omega)$ (see Fig. 3.4) and the second maximization is done with respect to the PSD function $S_w(\omega)$. In the first maximization, the time t^* when the maximum of $f_N(t)$ is attained must be obtained for each PSD function. This original problem is complicated and needs much computation. To overcome this difficulty, a new sophisticated procedure based on

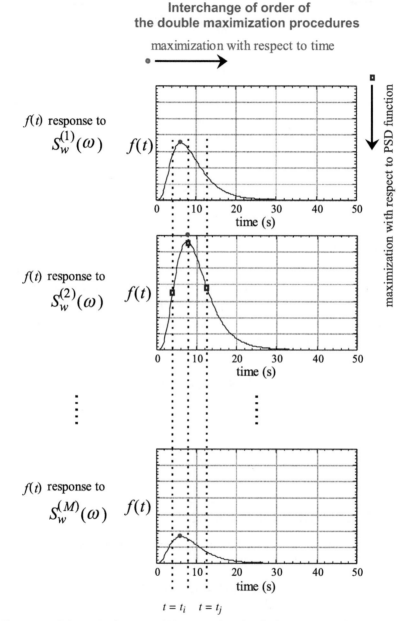

Figure 3.4 *Schematic diagram of the procedure for finding the critical excitation for nonstationary random inputs (order interchange of double maximization procedure).*

the interchange of the order of the maximization procedures is proposed. The proposed procedure can be expressed by

$$\max_{t} \max_{S_w(\omega)} \{ f_N(t; S_w(\omega)) \}$$

The first maximization process with respect to the PSD function for a given time can be pursued very efficiently (Fig. 3.4) by utilizing the critical excitation method for stationary inputs (Takewaki 2000) (see Fig. 2.3). The critical excitation obtained for a specific time in this stage has a rectangular PSD function as shown in Fig. 2.3. The second maximization procedure with respect to time can be implemented by comparing the values at various times directly (Fig. 3.4).

The proposed algorithm may be summarized as:

(i) Compute $H_{MN}(t_i; \omega)$ in Eq. (3.12) for a specific time $t = t_i$.

(ii) Find the critical PSD function at time $t = t_i$ as the rectangular PSD function (the procedure used for stationary inputs is utilized straightforwardly).

(iii) Compute $f_N(t_i)$ to the rectangular PSD function obtained in step (ii) from Eq. (3.11).

(iv) Repeat steps (i)–(iii) for various times and obtain $f_N(t_m) = \max f_N(t_i)$.

(v) The PSD function for $t = t_m$ is determined as the PSD function of the critical excitation.

It is important to note that the present algorithm based on the interchange of the order of the double maximization procedures is applicable to more sophisticated nonuniformly modulated nonstationary excitation models although the expression of Eq. (3.11) must be modified and a new critical excitation problem must be stated.

3.6. CRITICAL EXCITATION FOR ACCELERATION (PROPORTIONAL DAMPING)

An acceleration can be adopted as the objective function to be maximized in another critical excitation problem. For simplicity of expression, an n-story shear building model with proportional damping is considered in this section.

The equations of motion for this model can be expressed as Eq. (3.2) by replacing the damping matrix by the proportional damping matrix. Let ω_j and h_j denote the j-th undamped natural circular frequency and the j-th damping ratio, respectively. Let us introduce the following coordinate transformation from $\mathbf{u}(t)$ to $\mathbf{q}(t) = \{q_j\}$ through the modal matrix $\mathbf{\Phi}$.

$$\mathbf{u}(t) = \mathbf{\Phi}\mathbf{q}(t) \tag{3.20}$$

Substitution of Eq. (3.20) into Eq. (3.2) leads to the following n independent equations.

$$\ddot{q}_j + 2h_j\omega_j\dot{q}_j + \omega_j^2 q_j = -\Gamma_j\ddot{u}_g \quad (j = 1, \cdots, n) \tag{3.21}$$

where Γ_j is the j-th participation factor, i.e. $\Gamma_j = \boldsymbol{\phi}^{(j)^T}\mathbf{Mr}/\boldsymbol{\phi}^{(j)^T}\mathbf{M}\boldsymbol{\phi}^{(j)}$, and $\boldsymbol{\phi}^{(j)} = \{\phi_k^{(j)}\}$ is the j-th undamped eigenvector. Substitution of the relation $q_j = \Gamma_j q_{0j}$ into Eq. (3.21) and rearrangement of the resulting equation lead to

$$\ddot{u}_g + \ddot{q}_{0j} = -2h_j\omega_j\dot{q}_{0j} - \omega_j^2 q_{0j} \quad (j = 1, \cdots, n) \tag{3.22}$$

Recall that

$$q_{0j}(t) = \int_0^t g_j(t - \tau)\{-\ddot{u}_g(\tau)\}d\tau \tag{3.23a}$$

$$\dot{q}_{0j}(t) = \int_0^t \dot{g}_j(t - \tau)\{-\ddot{u}_g(\tau)\}d\tau \tag{3.23b}$$

where $g_j(t)$ is the impulse response function defined in Eq. (3.15) and its time derivative can be expressed as

$$\dot{g}_j(t) = -h_j\omega_j g_j(t) + e^{-h_j\omega_j t}\cos\omega_{jd}t \tag{3.24}$$

Substitution of Eqs. (3.23a, b) into Eq. (3.22) with the aid of Eq. (3.24) provides

$$\ddot{u}_g + \ddot{q}_{0j} = \omega_j^2(2h_j^2 - 1)\int_0^t g_j(t - \tau)\{-\ddot{u}_g(\tau)\}d\tau$$

$$- 2h_j\omega_j\omega_{jd}\int_0^t \frac{1}{\omega_{jd}}e^{-h_j\omega_j(t-\tau)}\cos\omega_{jd}(t - \tau)\{-\ddot{u}_g(\tau)\}d\tau$$

$$\tag{3.25}$$

With the help of the relation $\sum_{j=1}^n \Gamma_j\phi_n^{(j)} = 1$, the top-floor absolute acceleration can be expressed as

$$\ddot{u}_g + \ddot{u}_n = \ddot{u}_g + \sum_{j=1}^n \Gamma_j\phi_n^{(j)}\ddot{q}_{0j} = \sum_{j=1}^n \Gamma_j\phi_n^{(j)}(\ddot{u}_g + \ddot{q}_{0j}) \tag{3.26}$$

Following the similar process from Eq. (3.6) through Eq. (3.11) and keeping Eqs. (3.25) and (3.26) in mind, the following mean-square absolute acceleration at the top floor n to the input given by Eq. (3.1) can be derived as the objective function.

$$f^*(t) = \sigma_{A_n}(t)^2 = \int_{-\infty}^{\infty} H_M(t; \omega) S_w(\omega) d\omega \qquad (3.27)$$

where

$$H_{MA}(t; \omega) = \left[\sum_{j=1}^{n} \Gamma_j \phi_n^{(j)} D_{Cj}(t; \omega)\right]^2 + \left[\sum_{j=1}^{n} \Gamma_j \phi_n^{(j)} D_{Sj}(t; \omega)\right]^2$$

$$(3.28)$$

The parameters $D_{Cj}(t; \omega)$ and $D_{Sj}(t; \omega)$ are defined by

$$D_{Cj}(t; \omega) = \omega_j^2 (2h_j^2 - 1) A_{Cj}(t; \omega) - 2h_j \omega_j \omega_{jd} A_{Cj}^*(t; \omega) \qquad (3.29a)$$

$$D_{Sj}(t; \omega) = \omega_j^2 (2h_j^2 - 1) A_{Sj}(t; \omega) - 2h_j \omega_j \omega_{jd} A_{Sj}^*(t; \omega) \qquad (3.29b)$$

where $A_{Cj}(t; \omega)$ and $A_{Sj}(t; \omega)$ are defined in Eqs. (3.15), (3.17). Detailed expressions of $A_{Cj}(t; \omega)$ and $A_{Sj}(t; \omega)$ and the integral expressions in Eqs. (3.29a, b) for a specific function $c(t)$ given by Eq. (3.30) can be found in the Appendix.

Another critical excitation problem for acceleration can be defined by replacing the objective function of Eq. (3.11) by $f^*(t)$ given by Eq. (3.27). The solution procedure stated in Section 3.5 can be applied to this problem straightforwardly.

3.7. NUMERICAL EXAMPLES (PROPORTIONAL DAMPING)

A numerical example is shown for the model with the objective function $f_N(t)$ given by Eq. (3.11). The following envelope function of the input acceleration is used as an example (see Fig. 3.5).

$$c(t) = e^{-\alpha t} - e^{-\beta t} \qquad (3.30)$$

The parameters $\alpha = 0.13, \beta = 0.45$ are employed. The power of the stochastic part $w(t)$ in the excitation has been chosen as $\overline{S}_w = 1.51 (\mathrm{m}^2/\mathrm{s}^4)$ so that it has the same value of El Centro NS 1940 at the time $= 3.9(\mathrm{s})$ when

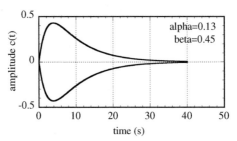

Figure 3.5 *Envelope function c(t).*

the acceleration amplitude attains its maximum. The intensity of $w(t)$ has been chosen to be $\bar{s}_w = 1.0 (\mathrm{m^2/s^3})$.

Consider a 6-story shear building model. The floor masses are assumed to be $m_i = 32 \times 10^3 (\mathrm{kg})(i = 1, \cdots, 6)$ and the story stiffnesses are $k_i = 3.76 \times 10^7 (\mathrm{N/m})(i = 1, \cdots, 6)$. Then the fundamental natural period of the model is 0.760(s). The viscous damping coefficients of the building have been given by $c_i = 3.76 \times 10^5 (\mathrm{N \cdot s/m})(i = 1, \cdots, 6)$. The viscous damping matrix of the shear building model is a proportional damping matrix in this case and the lowest-mode damping ratio is almost equal to 0.04. The evolutionary functions $H_{MN}(t; \omega)$ of this model are plotted at various times in Fig. 3.6. It can be observed that, while the evolutionary function $H_{MN}(t; \omega)$ indicates a rather wide-band frequency content around the time of 2, 4 and 6(s), it indicates a narrower frequency content afterward. This implies that the components around the fundamental natural frequency of the model decay as the time goes and only the resonant component remains afterward. It can also be found that the function $H_{MN}(t; \omega)$ has a clear peak around the natural frequency of the model and its amplitude attains the maximum around the time $= 8$(s). Fig. 3.7 shows the time history of the function $f_N(t)$ defined by Eq. (3.11), i.e. the sum of mean-square interstory drifts, to the corresponding critical excitation.

It is interesting to note that, in view of the evolutionary functions $H_{MN}(t; \omega)$ in Fig. 3.6, a rectangular PSD function with the fundamental natural frequency of the model as the central point is a good approximation of the PSD function of the critical excitation. This approximation can reduce the computational effort drastically.

3.8. NUMERICAL EXAMPLES (NONPROPORTIONAL DAMPING)

Consider next a nonproportionally damped 6-story shear building model. A numerical example is shown for the model with the objective function $f_N(t)$

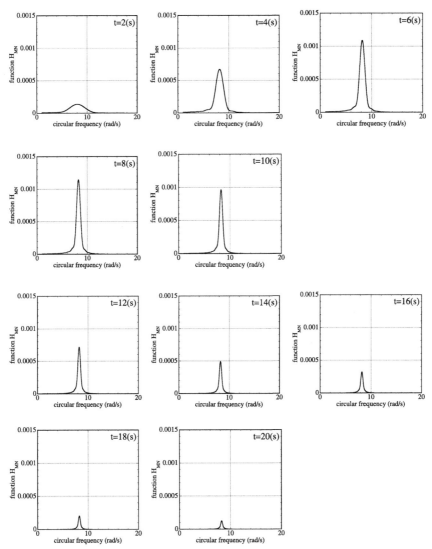

Figure 3.6 *Evolutionary functions H_{MN} (t; ω) defined by Eq. (3.12) at various times (proportional damping model).*

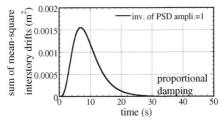

Figure 3.7 *Time-dependent function f_N (t) (sum of mean-square interstory drifts) for inverse of PSD amplitude = 1 (proportional damping model).*

given by Eq. (3.11). The floor masses and story stiffnesses are the same as the model treated in Section 3.7. The viscous damping coefficients of the building have been given by $c_1 = 3.76 \times 10^6 \ (N \cdot s/m)$, $c_i = 3.76 \times 10^5 (N \cdot s/m)$ $(i = 2, \cdots, 6)$. The viscous damping matrix of the shear building model is a nonproportional damping matrix in this case. Fig. 3.8 shows the evolutionary

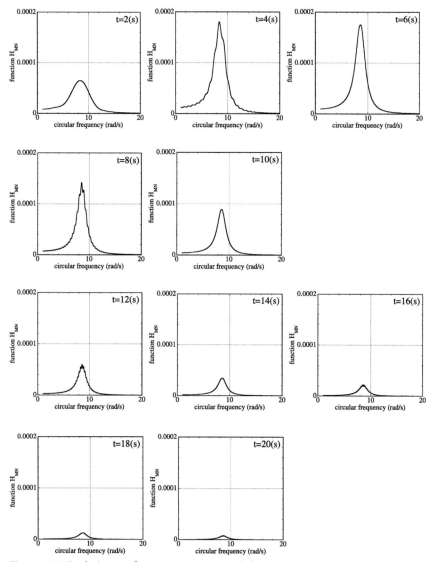

Figure 3.8 *Evolutionary functions $H_{MN}(t; \omega)$ defined by Eq. (3.12) at various times (nonproportional damping model).*

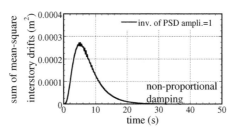

Figure 3.9 *Time-dependent function $f_N(t)$ (sum of mean-square interstory drifts) for inverse of PSD amplitude = 1 (nonproportional damping model).*

functions $H_{MN}(t;\omega)$ of the model at various times. It can be observed that the evolutionary functions $H_{MN}(t;\omega)$ of the present model have a different tendency from that of the proportionally damped model shown in Fig. 3.6. An irregular phenomenon appears almost every four seconds. Fig. 3.9 shows the time history of the function $f_N(t)$ defined by Eq. (3.11). The power and intensity of the critical excitation are the same as in Section 3.7. It can be seen that the function $f_N(t)$ indicates a nonsmooth property in contrast to that for the proportionally damped model. This property may result from the complicated inter-mode correlation due to nonproportional damping.

Fig. 3.10 shows examples of the critical excitation proposed in this chapter for four inverse PSD amplitudes 1, 3, 5, 7 ($\bar{s}_w = 1, 1/3, 1/5, 1/7 (\text{m}^2/\text{s}^3)$). Uniform random numbers have been adopted as the phase angles and the superposition of 100 cosine waves has been employed as the wave generation method. The power of $w(t)$ is $\bar{S}_w = 1.51 (\text{m}^2/\text{s}^4)$. It can be observed from Fig. 3.10 that as the frequency content of the excitation becomes wide-band ($\bar{s}_w$ decreases and the inverse PSD amplitude increases), wave components with various frequencies arise.

3.9. NUMERICAL EXAMPLES (VARIOUS TYPES OF DAMPING CONCENTRATION)

Consider again nonproportionally damped 6-story shear building models (see Fig. 3.2). A numerical example is also shown for the model with the objective function $f_N(t)$ given by Eq. (3.11). The floor masses and story stiffnesses are the same as the model treated in Sections 3.7 and 3.8. Three types of the viscous damping coefficient distribution have been treated: (i) $c_1 = 3.76 \times 10^6 (\text{N·s/m})$, $c_i = 3.76 \times 10^5 (\text{N·s/m})$ $(i \neq 1)$, (ii) $c_3 = 3.76 \times 10^6 (\text{N·s/m})$, $c_i = 3.76 \times 10^5 (\text{N·s/m})(i \neq 3)$, (iii) $c_6 = 3.76 \times$

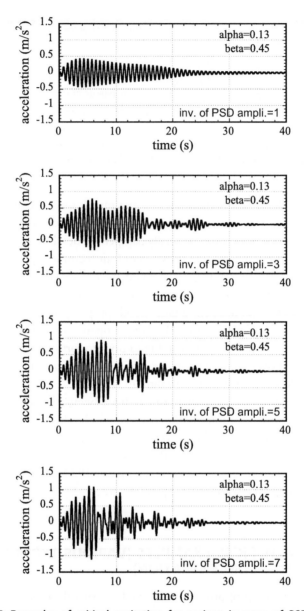

Figure 3.10 *Examples of critical excitation for various inverses of PSD amplitude (nonproportional damping model).*

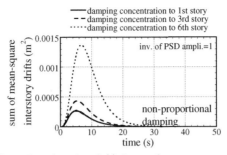

Figure 3.11 *Time-dependent function $f_N(t)$ (sum of mean-square interstory drifts) for three models with various damping concentration types (inverse of PSD amplitude = 1).*

$10^6 (\text{N} \cdot \text{s/m})$, $c_i = 3.76 \times 10^5 (\text{N} \cdot \text{s/m})(i \neq 6)$. Case (i) represents the model with the damping concentration to the 1st story, case (ii) does the model with the damping concentration to the 3rd story and case (iii) does the model with the damping concentration to the 6th story.

Fig. 3.11 shows the time histories of the function $f_N(t)$ defined by Eq. (3.11) for the aforementioned three models. The power and intensity of the critical excitation are the same as in Section 3.7. It can be observed from Fig. 3.11 that the model with the damping concentration to the 1st story is the most effective model for reducing the overall story drift index.

3.10. CONCLUSIONS

The conclusions may be summarized as follows:

(1) A new probabilistic critical excitation method can be developed for nonstationary vibrations of nonproportionally damped structural systems by restricting both the power and intensity of the input power spectral density function.

(2) It has been found that the integrand of the objective function to be maximized in the context of the critical excitation method can be expressed as the product of a positive function and the power spectral density function of a stochastic part in the input motion.

(3) The previously explained idea for stationary inputs can be utilized partially in finding the nonstationary critical excitation that can be described by a uniformly modulated excitation model. The key is the exchange of the order of the double maximization procedures with respect to time and to the PSD function.

(4) Numerical examples disclosed the time-varying characteristics of the nonstationary transfer function multiplied by the envelope function of the input motion model and demonstrated the validity of the present critical excitation method.

(5) An acceleration can be adopted as the objective function to be maximized in another critical excitation problem.

APPENDIX FUNCTIONS $A_{Cj}(t; \omega)$, $A_{Sj}(t; \omega)$ AND $A^*_{Cj}(t; \omega)$, $A^*_{Sj}(t; \omega)$ FOR A SPECIFIC FUNCTION $C(t)$

Detailed expressions are shown of the functions $A_{Cj}(t; \omega)$, $A_{Sj}(t; \omega)$ and $A^*_{Cj}(t; \omega)$, $A^*_{Sj}(t; \omega)$ for a specific function $c(t)$ given by Eq. (3.30).

The function $A_{Cj}(t; \omega)$ may be expressed as

$$
A_{Cj}(t; \omega) = \frac{1}{\omega_{jd}} e^{-h_j\omega_j t} \left[\sin \omega_{jd} t \int_0^t \left\{ e^{(h_j\omega_j-\alpha)\tau} - e^{(h_j\omega_j-\beta)\tau} \right\} \frac{1}{2} \left\{ \cos (\omega_{jd} \right. \right.
$$

$$
\left. + \omega)\tau + \cos (\omega_{jd} - \omega)\tau \right\} d\tau - \cos \omega_{jd} t \int_0^t \left\{ e^{(h_j\omega_j-\alpha)\tau} \right.
$$

$$
\left. \left. - e^{(h_j\omega_j-\beta)\tau} \right\} \frac{1}{2} \left\{ \sin (\omega_{jd} + \omega)\tau + \sin (\omega_{jd} - \omega)\tau \right\} d\tau \right]
$$

$$
= \frac{1}{\omega_{jd}} e^{-h_j\omega_j t} \left\{ \sin \omega_{jd} t G_{CC}(t; \omega) - \cos \omega_{jd} t G_{SC}(t; \omega) \right\}
$$

(A3.1)

where

$$
G_{CC}(t; \omega) = \int_0^t \left\{ e^{(h_j\omega_j-\alpha)\tau} - e^{(h_j\omega_j-\beta)\tau} \right\} \frac{1}{2} \left\{ \cos (\omega_{jd} + \omega)\tau \right.
$$

$$
\left. + \cos (\omega_{jd} - \omega)\tau \right\} d\tau
$$

(A3.2)

$$
G_{SC}(t; \omega) = \int_0^t \left\{ e^{(h_j\omega_j-\alpha)\tau} - e^{(h_j\omega_j-\beta)\tau} \right\} \frac{1}{2} \left\{ \sin (\omega_{jd} + \omega)\tau \right.
$$

$$
\left. + \sin (\omega_{jd} - \omega)\tau \right\} d\tau
$$

(A3.3)

The double of the parenthesis in Eq. (A3.1) can be reduced to

$$2\{ \sin \omega_{jd}t G_{CC}(t; \omega) - \cos \omega_{jd}t G_{SC}(t; \omega)\}$$

$$= \frac{1}{(h_j\omega_j - \alpha)^2 + (\omega_{jd} + \omega)^2} \left[\left(h_j\omega_j - \alpha\right)e^{(h_j\omega_j - \alpha)t}(- \sin \omega t) + (\omega_{jd} + \omega)e^{(h_j\omega_j - \alpha)t}(\cos \omega t) \right.$$

$$\left. -(h_j\omega_j - \alpha) \sin \omega_{jd}t - (\omega_{jd} + \omega) \cos \omega_{jd}t \right]$$

$$+ \frac{1}{(h_j\omega_j - \alpha)^2 + (\omega_{jd} - \omega)^2} \left[\left(h_j\omega_j - \alpha\right)e^{(h_j\omega_j - \alpha)t}(\sin \omega t) + (\omega_{jd} - \omega)e^{(h_j\omega_j - \alpha)t}(\cos \omega t) \right.$$

$$\left. -(h_j\omega_j - \alpha) \sin \omega_{jd}t - (\omega_{jd} - \omega) \cos \omega_{jd}t \right]$$

$$- \frac{1}{(h_j\omega_j - \beta)^2 + (\omega_{jd} + \omega)^2} \left[\left(h_j\omega_j - \beta\right)e^{(h_j\omega_j - \beta)t}(- \sin \omega t) + (\omega_{jd} + \omega)e^{(h_j\omega_j - \beta)t}(\cos \omega t) \right.$$

$$\left. -(h_j\omega_j - \beta) \sin \omega_{jd}t - (\omega_{jd} + \omega) \cos \omega_{jd}t \right]$$

$$- \frac{1}{(h_j\omega_j - \beta)^2 + (\omega_{jd} - \omega)^2} \left[\left(h_j\omega_j - \beta\right)e^{(h_j\omega_j - \beta)t}(\sin \omega t) + (\omega_{jd} - \omega)e^{(h_j\omega_j - \beta)t}(\cos \omega t) \right.$$

$$\left. -(h_j\omega_j - \beta) \sin \omega_{jd}t - (\omega_{jd} - \omega) \cos \omega_{jd}t \right]$$

$$\text{(A3.4)}$$

The function $A_{Sj}(t; \omega)$ may be expressed as

$$A_{Sj}(t; \omega) = \frac{1}{\omega_{jd}}e^{-h_j\omega_j t} \left[\sin \omega_{jd}t \int_0^t \left\{ e^{(h_j\omega_j - \alpha)\tau} - e^{(h_j\omega_j - \beta)\tau} \right\} \right.$$

$$\times \frac{1}{2} \left\{ \sin (\omega_{jd} + \omega)\tau - \sin (\omega_{jd} - \omega)\tau \right\} d\tau$$

$$- \cos \omega_{jd}t \int_0^t \left\{ e^{(h_j\omega_j - \alpha)\tau} - e^{(h_j\omega_j - \beta)\tau} \right\}$$

$$\left. \times \frac{1}{2} \left\{ \cos (\omega_{jd} - \omega)\tau - \cos (\omega_{jd} + \omega)\tau \right\} d\tau \right]$$

$$= \frac{1}{\omega_{jd}}e^{-h_j\omega_j t}\{ \sin \omega_{jd}t G_{CS}(t; \omega) - \cos \omega_{jd}t G_{SS}(t; \omega)\}$$

$$\text{(A3.5)}$$

where

$$G_{CS}(t; \omega) = \int_0^t \left\{ e^{(h_j \omega_j - \alpha)\tau} - e^{(h_j \omega_j - \beta)\tau} \right\}$$

(A3.6)

$$\times \frac{1}{2} \left\{ \sin(\omega_{jd} + \omega)\tau - \sin(\omega_{jd} - \omega)\tau \right\} d\tau$$

$$G_{SS}(t; \omega) = \int_0^t \left\{ e^{(h_j \omega_j - \alpha)\tau} - e^{(h_j \omega_j - \beta)\tau} \right\}$$

(A3.7)

$$\times \frac{1}{2} \left\{ \cos(\omega_{jd} - \omega)\tau - \cos(\omega_{jd} + \omega)\tau \right\} d\tau$$

The double of the parenthesis in Eq. (A3.5) can be reduced to

$$2 \left\{ \sin \omega_{jd} t G_{CS}(t; \omega) - \cos \omega_{jd} t G_{SS}(t; \omega) \right\}$$

$$= \frac{1}{(h_j \omega_j - \alpha)^2 + (\omega_{jd} + \omega)^2} \left[(h_j \omega_j - \alpha) e^{(h_j \omega_j - \alpha)t} (\cos \omega t) - (\omega_{jd} + \omega) e^{(h_j \omega_j - \alpha)t} (-\sin \omega t) \right.$$

$$\left. - (h_j \omega_j - \alpha) \cos \omega_{jd} t + (\omega_{jd} + \omega) \sin \omega_{jd} t \right]$$

$$+ \frac{1}{(h_j \omega_j - \alpha)^2 + (\omega_{jd} - \omega)^2} \left[(h_j \omega_j - \alpha) e^{(h_j \omega_j - \alpha)t} (-\cos \omega t) + (\omega_{jd} - \omega) e^{(h_j \omega_j - \alpha)t} (\sin \omega t) \right.$$

$$\left. + (h_j \omega_j - \alpha) \cos \omega_{jd} t - (\omega_{jd} - \omega) \sin \omega_{jd} t \right]$$

$$- \frac{1}{(h_j \omega_j - \beta)^2 + (\omega_{jd} + \omega)^2} \left[(h_j \omega_j - \beta) e^{(h_j \omega_j - \beta)t} (\cos \omega t) - (\omega_{jd} + \omega) e^{(h_j \omega_j - \beta)t} (-\sin \omega t) \right.$$

$$\left. - (h_j \omega_j - \beta) \cos \omega_{jd} t + (\omega_{jd} + \omega) \sin \omega_{jd} t \right]$$

$$- \frac{1}{(h_j \omega_j - \beta)^2 + (\omega_{jd} - \omega)^2} \left[(h_j \omega_j - \beta) e^{(h_j \omega_j - \beta)t} (-\cos \omega t) + (\omega_{jd} - \omega) e^{(h_j \omega_j - \beta)t} (\sin \omega t) \right.$$

$$\left. + (h_j \omega_j - \beta) \cos \omega_{jd} t - (\omega_{jd} - \omega) \sin \omega_{jd} t \right]$$

(A3.8)

The function $A^*_{Cj}(t; \omega)$ can be expressed as

$$A^*_{Cj}(t; \omega) = \frac{1}{\omega_{jd}} e^{-h_j \omega_j t} \{\cos \omega_{jd} t G_{CC}(t; \omega) + \sin \omega_{jd} t G_{SC}(t; \omega)\}$$

$$(A3.9)$$

The double of the parenthesis in Eq. (A3.9) can be reduced to

$$2\{\cos \omega_{jd} t G_{CC}(t; \omega) + \sin \omega_{jd} t G_{SC}(t; \omega)\}$$

$$= \frac{1}{(h_j \omega_j - \alpha)^2 + (\omega_{jd} + \omega)^2} \Big[(h_j \omega_j - \alpha) e^{(h_j \omega_j - \alpha) t}(\cos \omega t)$$

$$+ (\omega_{jd} + \omega) e^{(h_j \omega_j - \alpha) t}(\sin \omega t) - (h_j \omega_j - \alpha) \cos \omega_{jd} t$$

$$+ (\omega_{jd} + \omega) \sin \omega_{jd} t \Big] + \frac{1}{(h_j \omega_j - \alpha)^2 + (\omega_{jd} - \omega)^2}$$

$$\times \Big[(h_j \omega_j - \alpha) e^{(h_j \omega_j - \alpha) t}(\cos \omega t)$$

$$+ (\omega_{jd} - \omega) e^{(h_j \omega_j - \alpha) t}(-\sin \omega t) - (h_j \omega_j - \alpha) \cos \omega_{jd} t$$

$$+ (\omega_{jd} - \omega) \sin \omega_{jd} t \Big] - \frac{1}{(h_j \omega_j - \beta)^2 + (\omega_{jd} + \omega)^2}$$

$$\times \Big[(h_j \omega_j - \beta) e^{(h_j \omega_j - \beta) t}(\cos \omega t) + (\omega_{jd} + \omega) e^{(h_j \omega_j - \beta) t}(\sin \omega t)$$

$$- (h_j \omega_j - \beta) \cos \omega_{jd} t + (\omega_{jd} + \omega) \sin \omega_{jd} t \Big]$$

$$- \frac{1}{(h_j \omega_j - \beta)^2 + (\omega_{jd} - \omega)^2} \Big[(h_j \omega_j - \beta) e^{(h_j \omega_j - \beta) t}(\cos \omega t)$$

$$+ (\omega_{jd} - \omega) e^{(h_j \omega_j - \beta) t}(-\sin \omega t) \Big] \qquad (A3.10)$$

The function $A^*_{Sj}(t; \omega)$ can be expressed as

$$A^*_{Sj}(t; \omega) = \frac{1}{\omega_{jd}} e^{-h_j \omega_j t} \{\cos \omega_{jd} t G_{CS}(t; \omega) + \sin \omega_{jd} t G_{SS}(t; \omega)\}$$

$$(A3.11)$$

The double of the parenthesis in Eq. (A3.11) can be reduced to

$$2\{\cos \omega_{jd}t\, G_{CS}(t;\omega) + \sin \omega_{jd}t\, G_{SS}(t;\omega)\}$$

$$= \frac{1}{(h_j\omega_j - \alpha)^2 + (\omega_{jd} + \omega)^2}\left[(h_j\omega_j - \alpha)e^{(h_j\omega_j - \alpha)t}(\sin \omega t) \right.$$

$$+ (\omega_{jd} + \omega)e^{(h_j\omega_j - \alpha)t}(-\cos \omega t)$$

$$\left. + (h_j\omega_j - \alpha)\sin \omega_{jd}t + (\omega_{jd} + \omega)\cos \omega_{jd}t\right]$$

$$+ \frac{1}{(h_j\omega_j - \alpha)^2 + (\omega_{jd} - \omega)^2}\left[(h_j\omega_j - \alpha)e^{(h_j\omega_j - \alpha)t}\{\sin (2\omega_{jd} - \omega)t\} \right.$$

$$+ (\omega_{jd} - \omega)e^{(h_j\omega_j - \alpha)t}\{-\cos (2\omega_{jd} - \omega)t\}$$

$$\left. - (h_j\omega_j - \alpha)\sin \omega_{jd}t + (\omega_{jd} - \omega)\cos \omega_{jd}t\right]$$

$$- \frac{1}{(h_j\omega_j - \beta)^2 + (\omega_{jd} + \omega)^2}\left[(h_j\omega_j - \beta)e^{(h_j\omega_j - \beta)t}(\sin \omega t) \right.$$

$$+ (\omega_{jd} + \omega)e^{(h_j\omega_j - \beta)t}(-\cos \omega t)$$

$$\left. - (h_j\omega_j - \beta)\sin \omega_{jd}t + (\omega_{jd} + \omega)\cos \omega_{jd}t\right]$$

$$- \frac{1}{(h_j\omega_j - \beta)^2 + (\omega_{jd} - \omega)^2}\left[(h_j\omega_j - \beta)e^{(h_j\omega_j - \beta)t}\{\sin (2\omega_{jd} - \omega)t\} \right.$$

$$+ (\omega_{jd} - \omega)e^{(h_j\omega_j - \beta)t}\{-\cos (2\omega_{jd} - \omega)t\}$$

$$\left. - (h_j\omega_j - \beta)\sin \omega_{jd}t + (\omega_{jd} - \omega)\cos \omega_{jd}t\right]$$

$$\text{(A3.12)}$$

REFERENCES

Takewaki, I., 2000. Optimal damper placement for critical excitation. Probabilistic Engineering Mechanics 15 (4), 317–325.

CHAPTER FOUR

Critical Excitation for Acceleration Response

Contents

4.1. INTRODUCTION

The purpose of this chapter is to explain and discuss a probabilistic critical excitation method for *acceleration* responses of nonproportionally damped structural systems to nonstationary inputs. Recently, acceleration responses have started to be considered important from the viewpoint of the protection and maintenance of functionality in buildings (see Fig. 4.1). It is therefore natural and desirable to develop critical excitation methods for acceleration.

In contrast to most of the conventional critical excitation methods, a stochastic acceleration response at a point is treated as the objective function to be maximized. The power and the intensity of the excitations are fixed, and the critical excitation is found under these restrictions. As in Chapters 2 and 3, the key for finding the new nonstationary random critical excitation for nonproportionally damped structural systems is the order interchange in the double maximization procedure with respect to time and to the power spectral density (PSD) function. It is shown that the

© 2013 Elsevier Ltd.
All rights reserved.

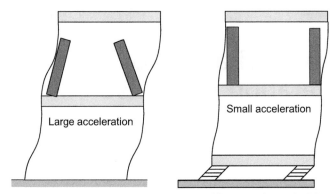

Figure 4.1 *Protection and maintenance of functionality in building.*

spirit of the previously explained critical excitation method for stationary inputs (Takewaki 2000a) is partially applicable to multi-degree-of-freedom structural systems with nonproportional damping subjected to nonstationary random inputs. Numerical examples are presented to demonstrate the effectiveness and validity of the present method.

4.2. MODELING OF INPUT MOTIONS

The one-directional horizontal base acceleration is assumed here to be described by the following uniformly modulated nonstationary random process as in Chapters 2 and 3:

$$\ddot{u}_g(t) = c(t)w(t) \tag{4.1}$$

The function $c(t)$ is a given deterministic envelope function and the function $w(t)$ represents a stationary Gaussian process with zero mean for the critical one to be found. Differentiation with respect to time is denoted by an over-dot.

Let $S_w(\omega)$ denote the PSD function of the stochastic function $w(t)$. The time-variant PSD function of $\ddot{u}_g$ may then be expressed by $S_g(t; \omega) = c(t)^2 S_w(\omega)$.

4.3. ACCELERATION RESPONSE OF NONPROPORTIONALLY DAMPED MODEL TO NONSTATIONARY RANDOM INPUT

Let us consider an *n*-story shear building model, as shown in Fig. 4.2, with nonproportional damping and subjected to a nonstationary base

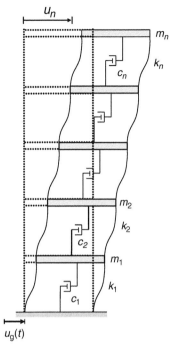

Figure 4.2 *n-story shear building model subjected to one-dimensional horizontal base motion.*

acceleration $\ddot{u}_\mathrm{g}$. Assume that this base acceleration can be described by the uniformly modulated nonstationary random process of Eq. (4.1). The horizontal displacements of the floors relative to the base are denoted by $\mathbf{u}$ (t). Let $\mathbf{M}$, $\mathbf{C}$, $\mathbf{K}$, $\mathbf{r} = \{1 \cdot \cdot \cdot 1\}^\mathrm{T}$ denote the system mass, viscous damping and stiffness matrices and the influence coefficient vector of the input, respectively. The equations of motion of the shear building model can be expressed by

$$\mathbf{M}\ddot{\mathbf{u}}(t) + \mathbf{C}\dot{\mathbf{u}}(t) + \mathbf{K}\mathbf{u}(t) = -\mathbf{M}\mathbf{r}\ddot{u}_g(t) \tag{4.2}$$

Let $\mathbf{U}^{(j)} = \left\{ U_k^{(j)} \right\}$ denote the j-th complex eigenvector of damped mode. The displacement $u_n(t)$ of the top floor may be expressed in terms of complex-type Duhamel integrals:

$$u_n(t) = 2 \sum_{j=1}^{n} \mathrm{Re}\left[U_n^{(j)} \frac{\mathbf{U}^{(j)^\mathrm{T}} \mathbf{Mr}}{F_j} \int_0^t (-\ddot{u}_\mathrm{g}(\tau)) e^{\lambda_j(t-\tau)} \mathrm{d}\tau \right] \tag{4.3}$$

where the following quantities are used:

$$\lambda_j = -h_j\omega_j + \omega_{jd}\mathrm{i}$$

$$\omega_j^2 = \frac{\overline{\mathbf{U}}^{(j)\mathrm{T}}\mathbf{K}\mathbf{U}^{(j)}}{\overline{\mathbf{U}}^{(j)\mathrm{T}}\mathbf{M}\mathbf{U}^{(j)}}$$

$$\omega_{jd} = \omega_j\sqrt{1 - h_j^2} \tag{4.4a--e}$$

$$F_j = 2\omega_j(\sqrt{1 - h_j^2}\,\mathrm{i} - h_j)\ \mathbf{U}^{(j)\mathrm{T}}\mathbf{M}\mathbf{U}^{(j)} + \mathbf{U}^{(j)\mathrm{T}}\mathbf{C}\mathbf{U}^{(j)}$$

$$h_j = \frac{\overline{\mathbf{U}}^{(j)\mathrm{T}}\mathbf{C}\mathbf{U}^{(j)}}{2\omega_j\overline{\mathbf{U}}^{(j)\mathrm{T}}\mathbf{M}\mathbf{U}^{(j)}}$$

Re[·] indicates the real part of a complex number and i denotes the imaginary unit. The over-bar indicates the complex conjugate. Eq. (4.3) may also be written as

$$u_n(t) = \sum_{j=1}^{n}\mathrm{Re}[\beta_j U_n^{(j)}q_j(t)] \tag{4.5}$$

where the following quantities are used:

$$\beta_j = 2\omega_{jd}\mathbf{U}^{(j)\mathrm{T}}\mathbf{Mri}/F_j$$

$$q_j(t) = q_{Rj}(t) - \mathrm{i}q_{Ij}(t)$$

$$q_{Rj}(t) = q_{0j}(t)$$

$$q_{0j}(t) = \frac{1}{\omega_{jd}}\int_0^t \{-\ddot{u}_\mathrm{g}(\tau)\}e^{-h_j\omega_j(t-\tau)}\sin\omega_{jd}(t - \tau)\mathrm{d}\tau \tag{4.6a--f}$$

$$q_{Ij}(t) = q_{0j}^*(t) = \{\dot{q}_{0j}(t) + h_j\omega_j q_{0j}(t)\}/\omega_{jd}$$

$$q_{0j}^*(t) = \frac{1}{\omega_{jd}}\int_0^t \{-\ddot{u}_\mathrm{g}(\tau)\}e^{-h_j\omega_j(t-\tau)}\cos\omega_{jd}(t - \tau)\mathrm{d}\tau$$

By using Eq. (4.5), the top-floor absolute acceleration may be expressed in terms of complex modal coordinates:

$$\ddot{u}_g + \ddot{u}_n = \ddot{u}_g + \sum_{j=1}^{n} \text{Re}\left[\beta_j U_n^{(j)} \ddot{\bar{q}}_j(t)\right] \tag{4.7}$$

where the complex modal coordinate $\ddot{\bar{q}}_j(t)$ is defined by

$$\ddot{\bar{q}}_j(t) = \ddot{q}_{Rj}(t) - i\ddot{q}_{Ij}(t)$$

$$= \left\{ (2h_j^2 - 1)\omega_j^2 q_{0j}(t) - 2h_j\omega_j\omega_{jd}q_{0j}^*(t) - \ddot{u}_g(t) \right\}$$

$$-i\left\{ 2h_j\omega_j\omega_{jd}q_{0j}(t) + (2h_j^2 - 1)\omega_j^2 q_{0j}^*(t) + \frac{h_j\omega_j}{\omega_{jd}}\ddot{u}_g(t) - \frac{\dddot{u}_g(t)}{\omega_{jd}} \right\} \tag{4.8}$$

Note that Eq. (4.6e), the equation of motion for a single-degree-of-freedom (SDOF) model in terms of modal coordinates, and the following relation have been used in Eq. (4.8) including $u_g(t)$:

$$\dot{q}_{Ij}(t) = -h_j\omega_j q_{Ij}(t) - \omega_{jd}q_{0j}(t) - \frac{\ddot{u}_g(t)}{\omega_{jd}} \tag{4.9}$$

Substitution of Eq. (4.8) into (4.7) yields

$$\ddot{u}_g + \ddot{u}_n = \ddot{u}_g + \sum_{j=1}^{n} \text{Re}\left[\beta_j U_n^{(j)}\right]\left\{ \left(2h_j^2 - 1\right)\omega_j^2 q_{0j}(t) - 2h_j\omega_j\omega_{jd}q_{0j}^*(t) - \ddot{u}_g(t) \right\}$$

$$+ \sum_{j=1}^{n} \text{Im}\left[\beta_j U_n^{(j)}\right]\left\{ 2h_j\omega_j\omega_{jd}q_{0j}(t) + \left(2h_j^2 - 1\right)\omega_j^2 q_{0j}^*(t) \right.$$

$$\left. + \frac{h_j\omega_j}{\omega_{jd}}\ddot{u}_g(t) - \frac{\dddot{u}_g(t)}{\omega_{jd}} \right\} \tag{4.10}$$

In Eq. (4.10), $\text{Im}[\cdot]$ indicates the imaginary part of a complex number. After some manipulation, Eq. (4.10) may be reduced to the following compact form:

$$\ddot{u}_g + \ddot{u}_n = A\ddot{u}_g + B\dddot{u}_g(t) + \sum_{j=1}^{n} a_j q_{0j}(t) + \sum_{j=1}^{n} b_j q_{0j}^*(t) \tag{4.11}$$

where the following quantities are used:

$$A = 1 - \sum_{j=1}^{n} \mathrm{Re}\left[\beta_j U_n^{(j)}\right] + \sum_{j=1}^{n} \mathrm{Im}\left[\beta_j U_n^{(j)}\right] \frac{h_j \omega_j}{\omega_{jd}}$$

$$B = -\sum_{j=1}^{n} \mathrm{Im}\left[\beta_j U_n^{(j)}\right] / \omega_{jd}$$

$$a_j = \mathrm{Re}\left[\beta_j U_n^{(j)}\right] \left(2h_j^2 - 1\right)\omega_j^2 + \mathrm{Im}\left[\beta_j U_n^{(j)}\right] 2h_j \omega_j \omega_{jd}$$

$$b_j = -\mathrm{Re}\left[\beta_j U_n^{(j)}\right] 2h_j \omega_j \omega_{jd} + \mathrm{Im}\left[\beta_j U_n^{(j)}\right] \left(2h_j^2 - 1\right)\omega_j^2$$

(4.12a–d)

The mean-square top-floor absolute acceleration may then be obtained as

$$E[\{\ddot{u}_g(t) + \ddot{u}_n(t)\}^2] = E\left[\left\{A\ddot{u}_g + B\ddot{u}_g(t) + \sum_{j=1}^{n} a_j q_{0j}(t) + \sum_{j=1}^{n} b_j q_{0j}^*(t)\right\}^2\right]$$

(4.13)

Note that the following can be derived by constructing auto-correlation and cross-correlation functions and using the Wiener–Khintchine theorem for $w(t)$:

$$E\left[\left\{\sum_{j=1}^{n} a_j q_{0j}(t)\right\}^2\right] = \int_{-\infty}^{\infty} \left[\left\{\sum_{j=1}^{n} a_j A_{Cj}(t; \omega)\right\}^2\right.$$

$$\left. + \left\{\sum_{j=1}^{n} a_j A_{S_j}(t; \omega)\right\}^2\right] S_w(\omega) d\omega$$

$$E\left[\left\{\sum_{j=1}^{n} b_j q_{0j}^*(t)\right\}^2\right] = \int_{-\infty}^{\infty} \left[\left\{\sum_{j=1}^{n} b_j A_{Cj}^*(t; \omega)\right\}^2\right.$$

$$\left. + \left\{\sum_{j=1}^{n} b_j A_{S_j}^*(t; \omega)\right\}^2\right] S_w(\omega) d\omega$$

$$E\left[\left\{\sum_{j=1}^{n} a_j q_{0j}(t)\right\}\left\{\sum_{j=1}^{n} b_j q_{0j}^*(t;\omega)\right\}\right]$$

$$= \int_{-\infty}^{\infty}\left[\left\{\sum_{j=1}^{n} a_j A_{Cj}(t;\omega)\right\}\left\{\sum_{j=1}^{n} b_j A_{Cj}^*(t;\omega)\right\} + \left\{\sum_{j=1}^{n} a_j A_{Sj}(t;\omega)\right\}\right.$$

$$\left. \times \left\{\sum_{j=1}^{n} b_j A_{Sj}^*(t;\omega)\right\}\right] S_w(\omega)d\omega$$

$$E[\ddot{u}_g(t)^2] = \int_{-\infty}^{\infty}\left[c(t)^2\right]S_w(\omega)d\omega$$

$$E[\dddot{u}_g(t)^2] = \int_{-\infty}^{\infty}\left[\omega^2 c(t)^2 + \dot{c}(t)^2\right]S_w(\omega)d\omega$$

$$E[\ddot{u}_g(t)\dddot{u}_g(t)] = \int_{-\infty}^{\infty}[c(t)\dot{c}(t)]S_w(\omega)d\omega$$

$$E[q_{0j}(t)\ddot{u}_g(t)] = \int_{-\infty}^{\infty}[\{A_{Cj}(t;\omega)\cos\omega t + A_{Sj}(t;\omega)\sin\omega t\}\{-c(t)\}]S_w(\omega)d\omega$$

$$E[q_{0j}^*(t)\ddot{u}_g(t)] = \int_{-\infty}^{\infty}[\{A_{Cj}^*(t;\omega)\cos\omega t + A_{Sj}^*(t;\omega)\sin\omega t\}\{-c(t)\}]S_w(\omega)d\omega$$

$$E[q_{0j}(t)\dddot{u}_g(t)] = \int_{-\infty}^{\infty}[\{A_{Cj}(t;\omega)\cos\omega t + A_{Sj}(t;\omega)\sin\omega t\}\{-\dot{c}(t)\}$$

$$+\{-A_{Cj}(t;\omega)\sin\omega t + A_{Sj}(t;\omega)\cos\omega t\}\{-\omega c(t)\}]S_w(\omega)d\omega$$

$$E[q_{0j}^*(t)\dddot{u}_g(t)] = \int_{-\infty}^{\infty}[\{A_{Cj}^*(t;\omega)\cos\omega t + A_{Sj}^*(t;\omega)\sin\omega t\}\{-\dot{c}(t)\}$$

$$+\{-A_{Cj}^*(t;\omega)\sin\omega t + A_{Sj}^*(t;\omega)\cos\omega t\}\{-\omega c(t)\}]S_w(\omega)d\omega$$

$$(4.14a-j)$$

where $E[\cdot]$ denotes the ensemble mean and the followings are introduced:

$$A_{Cj}(t;\omega) = \frac{1}{\omega_{jd}} \int_0^t c(\tau) e^{-h_j\omega_j(t-\tau)} \sin\omega_{jd}(t-\tau)\cos\omega\tau \; d\tau$$

$$A_{Sj}(t;\omega) = \frac{1}{\omega_{jd}} \int_0^t c(\tau) e^{-h_j\omega_j(t-\tau)} \sin\omega_{jd}(t-\tau)\sin\omega\tau \; d\tau$$

$$A^*_{Cj}(t;\omega) = \frac{1}{\omega_{jd}} \int_0^t c(\tau) e^{-h_j\omega_j(t-\tau)} \cos\omega_{jd}(t-\tau)\cos\omega\tau \; d\tau$$

$$A^*_{Sj}(t;\omega) = \frac{1}{\omega_{jd}} \int_0^t c(\tau) e^{-h_j\omega_j(t-\tau)} \cos\omega_{jd}(t-\tau)\sin\omega\tau \; d\tau$$

$$(4.15a-d)$$

In Eqs. (4.14a–j), together with the properties on the auto-correlation function $R_w(\tau)$ and the corresponding PSD function $S_w(\omega)$ of $w(t)$ that $(S_w(\omega), R_w(\tau))$, $(i\omega S_w(\omega), R'_w(\tau))$ and $(-\omega^2 S_w(\omega), R''_w(\tau))$ constitute the Fourier transformation pairs, the following relations have been used:

$$E[w(t)w(t+\tau)] = R_w(\tau)$$

$$E[w(t)\dot{w}(t+\tau)] = R'_w(\tau)$$

$$E[\dot{w}(t)w(t+\tau)] = -R'_w(\tau)$$

$$(4.16a-d)$$

$$E[\dot{w}(t)\dot{w}(t+\tau)] = -R''_w(\tau)$$

where $(\;)'$ indicates differentiation with respect to time lag.

The mean-square top-floor absolute acceleration can then be described as

$$f_{NA}(t) = \sigma_{A_n}(t)^2 = \int_{-\infty}^{\infty} H_{A_n}(t;\omega) S_w(\omega) d\omega \qquad (4.17)$$

In Eq. (4.17), $H_{A_n}(t; \omega)$ is defined by

$$
H_{A_n}(t; \omega) = \left\{ \sum_{j=1}^{n} a_j A_{Cj}(t; w) + \sum_{j=1}^{n} b_j A_{Cj}^{*}(t; w) - Ac(t) \cos \omega t \right.
$$

$$
\left. -B\dot{c}(t) \cos \omega t + Bc(t)\omega \sin \omega t \right\}^{2}
$$

$$
+ \left\{ \sum_{j=1}^{n} a_j A_{Sj}(t; w) + \sum_{j=1}^{n} b_j A_{Sj}^{*}(t; w) - Ac(t) \sin \omega t \right.
$$

$$
\left. -B\dot{c}(t) \sin \omega t + Bc(t)\omega \cos \omega t \right\}^{2} \tag{4.18}
$$

$H_{A_n}(t; \omega)$ will be called an evolutionary transfer function later in accordance with Iyengar and Manohar (1987) (it may be appropriate to call it a time-varying frequency response function). It should be remarked that the integrand in Eq. (4.17) to be maximized in the critical excitation problem can be expressed as the product of a positive function (sum of the squared numbers) and the positive PSD function of a stochastic part in the input motion expressed by Eq. (4.1).

When considering proportional damping, A and B defined in Eqs. (4.12a, b) are reduced to zero. Then the expression of Eq. (4.18) coincides completely with that for a proportionally damped structural system (Takewaki 2000b).

$A_{Cj}(t; \omega)$, $A_{Sj}(t; \omega)$, $A_{Cj}^{*}(t; \omega)$ and $A_{Sj}^{*}(t; \omega)$ in Eqs. (4.15a–d) for a specific envelope function c (t) given by Eq. (4.21) are shown in Chapter 3.

4.4. CRITICAL EXCITATION PROBLEM

The problem of critical excitation for acceleration may be described as follows.

[Problem CENMA]

Given floor masses, story stiffnesses and nonproportional viscous damping coefficients of a shear building model and the excitation envelope function c(t), find the critical PSD function $\tilde{S}_w(\omega)$ maximizing the specific function f_{NA} (t^*) (t^* is the time when *the maximum value of f_{NA} (t) to $S_w(\omega)$ is attained) subject to the excitation power limit constraint (integral of the PSD function in the frequency range)*

$$\int_{-\infty}^{\infty} S_w(\omega)d\omega \leq \overline{S}_w \qquad (4.19)$$

and to the PSD amplitude limit constraint

$$\sup S_w(\omega) \leq \overline{s}_w \qquad (4.20)$$

4.5. SOLUTION PROCEDURE

The aforementioned problem includes the double maximization procedures described mathematically by

$$\max_{S_w(\omega)} \max_{t} \{f_{NA}(t; S_w(\omega))\}$$

The first maximization is implemented with respect to time for a given PSD function $S_w(\omega)$ (see Fig. 4.3) and the second maximization is conducted with respect to the PSD function $S_w(\omega)$. In the first maximization, the time t^*causing the maximum value of f_{NA} (t) has to be obtained for each PSD function. This original problem is time consuming. To remedy this, a smart procedure based on the interchange of the order of the maximization procedures is explained here. The procedure can be described by

$$\max_{t} \max_{S_w(\omega)} \{f_{NA}(t; S_w(\omega))\}$$

The first maximization with respect to the PSD function for a given time can be performed effectively and efficiently by utilizing the critical excitation method for stationary inputs (Takewaki 2000a) (see Fig. 4.3). In case of an infinite amplitude limit $\overline{s}_w$, the critical PSD function is found to be the Dirac delta function. On the other hand, when $\overline{s}_w$ is finite, the critical PSD function is reduced to a constant $\overline{s}_w$ in a finite interval $\tilde{\Omega} = \overline{S}_w/(2\overline{s}_w)$. The intervals, $\tilde{\Omega}_1$, $\tilde{\Omega}_2$, ..., constituting $\tilde{\Omega}$ can be found by changing the level in the diagram of $H_{A_n}(t_i; \omega)$ (see Fig. 4.4). The critical excitation obtained for a specific time up to this stage turns out to be a rectangular PSD function as shown in Fig. 4.4. The second maximization with respect to time can be implemented sequentially by comparing the values at various times directly.

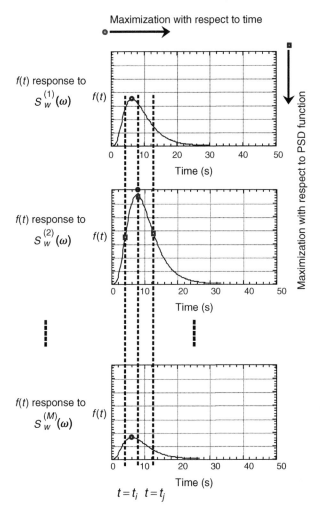

Figure 4.3 *Schematic diagram of the procedure for finding the critical excitation for nonstationary random inputs (order interchange of double maximization procedure).*

The solution algorithm may be summarized as:

(i) Compute the evolutionary function $H_{A_n}(t_i; \omega)$ in Eq. (4.18) for a specific time $t = t_i$.

(ii) Determine the critical PSD function at time $t = t_i$ as the rectangular PSD function (the procedure devised for stationary inputs is used directly).

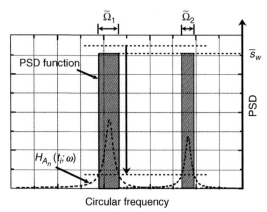

Figure 4.4 *Isolated PSD function of critical excitation (critical excitation for finite PSD amplitude $\bar{s}_w$).*

(iii) Compute $f_{NA}(t_i)$ from Eq. (4.17) to the rectangular PSD function determined in step (ii).

(iv) Repeat steps (i)–(iii) for sequential times and obtain $f_{NA}(t_m) = \max f_{NA}(t_i)$.

(v) The PSD function derived for $t = t_m$ is determined as the true PSD function of the critical excitation.

It is important to note that the global optimality is guaranteed in this algorithm. The global optimality in the maximization with respect to the shape of PSD functions is supported by the property of a single-valued function of $H_{A_n}(t_i; \omega)$ and that with respect to time is guaranteed by a sequential search algorithm in time. It is also meaningful to note that the present algorithm based on the interchange of the order of the double maximization procedures is applicable to more complex nonuniformly modulated nonstationary excitation models although the expression of Eq. (4.17) must be modified and a new critical excitation problem must be stated.

4.6. NUMERICAL EXAMPLES

Five different combinations of the stiffness distribution and the damping distribution are taken into account for 6-story shear building models (see Fig. 4.5). The purpose of the following sections is to present versatile characteristics of evolutionary transfer functions $H_{A_n}(t; \omega)$ of the

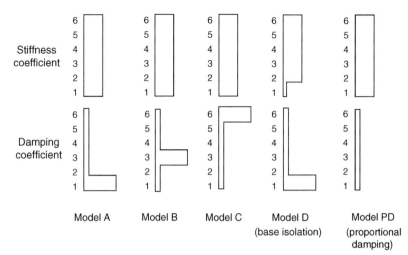

Figure 4.5 *Five 6-story shear building models that have different combinations of stiffness distribution and damping distribution.*

models with various stiffness and damping distributions. An example of critical PSD functions is also shown. Once the characteristics of $H_{A_n}(t; \omega)$ are made clear, it is straightforward and simple to find the critical excitation for those evolutionary transfer functions $H_{A_n}(t; \omega)$ by taking full advantage of the procedure shown in Section 4.5.

The following envelope function of the horizontal input acceleration is adopted as an example (see Fig. 4.6):

$$c(t) = e^{-\alpha t} - e^{-\beta t} \tag{4.21}$$

The parameters $\alpha = 0.13$, and $\beta = 0.45$ are used. The power of the stochastic part $w(t)$ in the excitation has been determined as

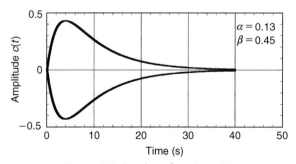

Figure 4.6 *Envelope function c (t).*

$\bar{S}_w = 1.51$ m^2/s^4. It is intended that the power at the time $= 3.9$ s (when the acceleration amplitude attains its maximum) has the same value as that of the time-averaging PSD function (Baratta et al. 1998) of El Centro NS 1940. The intensity of $w(t)$ has been assumed as $\bar{s}_w = 1.0$ m^2/s^3.

4.7. MODEL WITH NONPROPORTIONAL DAMPING-1

Three numerical examples are demonstrated for the 6-story shear building models with various nonproportional dampings (Models A–C in Fig. 4.5). Note that damping is concentrated at the 1st story in Model A, at the 3rd story in Model B and at the 6th story in Model C.

The floor masses and the story stiffnesses are given by $m_i = 32 \times 10^3$ kg ($i = 1,\ldots, 6$) and $k_i = 3.76 \times 10^7$ N/m ($i = 1,\ldots, 6$). The viscous damping coefficients of the building are given by $c_1 = 3.76 \times 10^6$ N·s/m, $c_i = 3.76 \times 10^5$ N·s/m ($i \neq 1$) for Model A; $c_3 = 3.76 \times 10^6$ N· s/m, $c_i = 3.76 \times 10^5$ N· s/m ($i \neq 3$) for Model B and $c_i = 3.76 \times 10^5$ N· s/m ($i \neq 6$), $c_6 = 3.76 \times 10^6$ N· s/m for Model C. The lowest three undamped natural circular frequencies of Models A–C are $\omega_1 = 8.71$ rad/s, $\omega_2 = 27.7$ rad/s and $\omega_3 = 44.4$ rad/s.

Figs. 4.7–4.9 show the evolutionary transfer functions $H_{A_n}(t;\ \omega)$ of these models plotted at every 2 s. It can be seen from these figures that, while the function $H_{A_n}(t;\ \omega)$ exhibits a similar tendency in Models A and B, it indicates somewhat different characteristics in Model C. This implies that the function $H_{A_n}(t;\ \omega)$ can be influenced greatly by the position of concentrated damping. Especially, the magnitude is small in Model C. This means that the damping installation in upper stories is effective in reducing the acceleration. Note that, while the damping installation in lower stories is effective in the reduction of interstory drifts, the damping installation in upper stories is effective in reducing the acceleration. It can also be found that the function $H_{A_n}(t;\ \omega)$ has a rather clear peak at ω^* between the second and third natural frequencies of the model in Models A and B and the peak values around the fundamental natural frequency and around ω^* are comparable in Model C.

Fig. 4.10 indicates the time history of the mean-square top-floor absolute acceleration in Model A subjected to the critical excitation. The effect of the evolutionary function $H_{A_n}(t;\ \omega)$ greater than about 100 rad/s has been disregarded in the computation of the critical excitation. It can be seen that the peak value occurs around $t = 4$s.

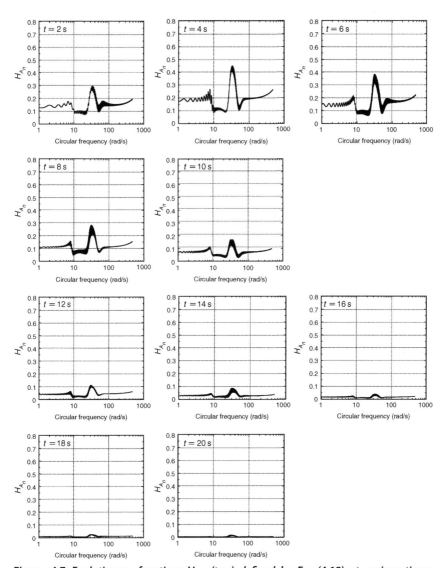

Figure 4.7 *Evolutionary functions* H_{A_n} *(t; ω) defined by Eq. (4.18) at various times (Model A).*

4.8. MODEL WITH NONPROPORTIONAL DAMPING-2

Consider another nonproportionally damped shear building model of 6-stories expressed as Model D in Fig. 4.5 (base-isolated

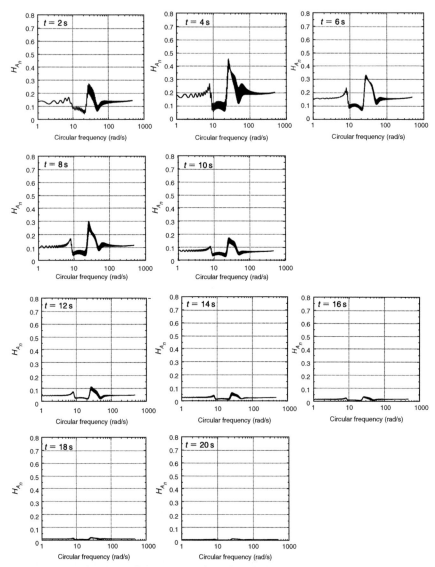

Figure 4.8 *Evolutionary functions* $H_{A_n}(t; \omega)$ *defined by Eq. (4.18) at various times (Model B).*

model). The floor masses are the same as those of the model treated in Section 4.7. The story stiffnesses are specified here by $k_1 = 7.52 \times 10^5$ N/m, $k_i = 3.76 \times 10^7$ N/m ($i = 2,..., 6$) to give a small 1st-story stiffness. On the other hand, the viscous damping coefficients of the

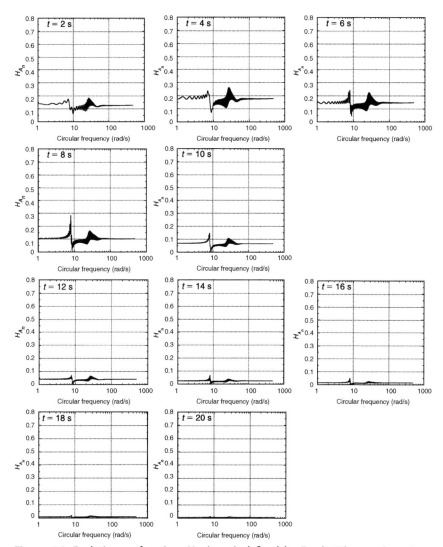

Figure 4.9 *Evolutionary functions* $H_{A_n}(t; \omega)$ *defined by Eq. (4.18) at various times (Model C).*

building are specified by $c_1 = 3.76 \times 10^6$ N · s/m, $c_i = 3.76 \times 10^5$ N · s/m $(i = 2,\ldots, 6)$. Fig. 4.11 presents the evolutionary functions $H_{A_n}(t; \omega)$ of the model at every 2 s. It can be seen that the functions $H_{A_n}(t; \omega)$ of the present model have a similar tendency to those of Models A and B shown in Figs. 4.7 and 4.8.

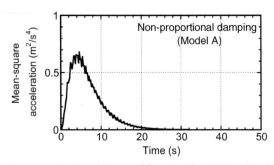

Figure 4.10 *Time history of function f_{NA} (t) (Model A).*

4.9. MODEL WITH PROPORTIONAL DAMPING

Consider finally a proportionally damped shear building model of 6-story expressed as Model PD in Fig. 4.5. The floor masses are the same as those of the model treated in Section 4.7. The story stiffnesses and viscous damping coefficients are given by $k_i = 3.76 \times 10^7$ N/m$(i = 1,\ldots, 6)$, $c_i = 3.76 \times 10^5$ N $\cdot$ s/m$(i = 1,\ldots, 6)$. The lowest three damped natural circular frequencies of this model are found to be $\omega_{1d} = 8.27$ rad/s, $\omega_{2d} = 24.3$ rad/s and $\omega_{3d} = 39.0$ rad/s.

The evolutionary functions $H_{A_n}(t; \omega)$ of Model PD at every 2 s are shown in Fig. 4.12. It can be understood that, while the peak around the second natural frequency is dominant at $t = 2$ and 4 s, the peak values around the fundamental and second natural frequencies are comparable after $t = 6$ s. Fig. 4.13 illustrates the time history of the mean-square top-floor absolute acceleration of Model PD subjected to the corresponding critical excitation. It can be seen that the peak value occurs around the time $t = 4$ s. The critical PSD function for Model PD is shown in Fig. 4.14. Three isolated rectangular PSD functions are found. Isolated rectangles not necessarily centered at the natural frequencies of the system may not be physically understandable. However, the procedure explained here is based on an exact mathematical treatment and can reveal new features of the random critical excitation for acceleration.

4.10. CONCLUSIONS

The conclusions may be summarized as follows:
(1) Acceleration is considered to be important from the viewpoint of the protection and maintenance of functionality in buildings. A probabilistic

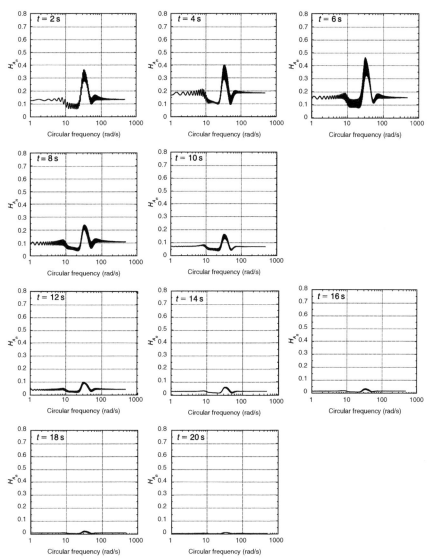

Figure 4.11 *Evolutionary functions* $H_{A_n}(t; \omega)$ *defined by Eq. (4.18) at various times (Model D).*

critical excitation method can be developed for nonstationary *acceleration* responses of nonproportionally damped structural systems by constraining both the power and the intensity of the PSD function of the nonstationary input.

(2) The integrand of the objective function to be maximized in the critical excitation problem can be expressed as the product of a positive

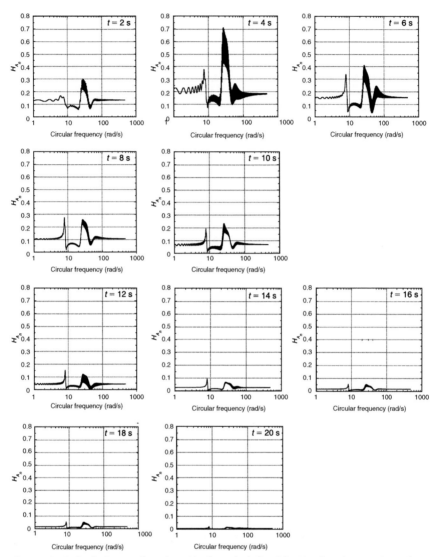

Figure 4.12 *Evolutionary functions* $H_{A_n}(t; \omega)$ *defined by Eq. (4.18) at various times (Model PD).*

function (sum of the squared numbers) and the positive PSD function of a stochastic part in the input.

(3) The idea for stationary random inputs can be used partially in obtaining the nonstationary random critical excitation which can be described by a uniformly modulated excitation model. The key idea is the order

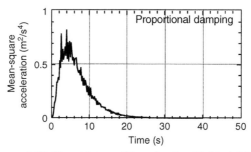

Figure 4.13 *Time history of function f$_{NA}$ (t) (Model PD).*

interchange in the double maximization procedures with respect to time and to the shape of the PSD function.

(4) It can be observed from numerical examples that there exist peculiar time-varying characteristics of the generalized nonstationary transfer function multiplied by the envelope function of the input motion model. The effectiveness of the present solver in the critical excitation problem has also been demonstrated numerically.

(5) While the damping installation in lower stories is effective in general for the purpose of the reduction of interstory drifts, the damping installation in upper stories is effective in reducing the acceleration.

Items (2) and (3) are novel in particular in the development of probabilistic critical excitation methods for acceleration. The validity of selection of the

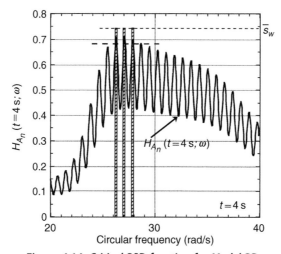

Figure 4.14 *Critical PSD function for Model PD.*

power $\overline{S}_w$ and the intensity $\overline{s}_w$ should further be investigated from the viewpoint of occurrence possibility and fault rupture mechanism of physical events.

APPENDIX FUNCTIONS $A_{Cj}(t; \omega)$, $A_{Sj}(t; \omega)$, $A_{Cj}^{*}(t; \omega)$, $A_{Sj}^{*}(t; \omega)$ FOR A SPECIFIC $\dot{C}(t)$

These are shown in Chapter 3.

REFERENCES

Baratta, A., Elishakoff, I., Zuccaro, G., Shinozuka, M., 1998. A generalization of the Drenick–Shinozuka model for bounds on the seismic response of a Single-Degree-of-Freedom system. Earthquake Eng. Struct. Dyn. 27 (5), 423–437.

Iyengar, R.N., Manohar, C.S., 1987. Non-stationary random critical seismic excitations. J. Eng. Mech. 113 (4), 529–541.

Takewaki, I., 2000a. Optimal damper placement for critical excitation. Probabil. Eng. Mech. 15 (4), 317–325.

Takewaki, I., 2000b. A non-stationary random critical excitation method for MDOF linear structural models. J. Struct. Constr. Eng. 536, 71–77 (in Japanese).

CHAPTER FIVE

Critical Excitation for Elastic-Plastic Response

Contents

5.1. INTRODUCTION

As for critical excitation methods for nonlinear systems, the following papers may be relevant. Iyengar (1972) formulated critical excitation problems for autonomous nonlinear systems, e.g. Duffing oscillator. By applying the Schwarz inequality, he derived a response upper bound. He considered both deterministic and random inputs. Westermo (1985) examined critical excitations for nonlinear hysteretic and nonhysteretic systems by adopting the input energy as the objective function. He restricted the class of critical excitations to periodic ones. He presented several interesting points inherent in the critical excitation problems for nonlinear systems. Philippacopoulos (1980), Philippacopoulos and Wang (1984) utilized a deterministic equivalent linearization technique in critical excitation problems of nonlinear single-degree-of-freedom (SDOF) hysteretic systems. They derived several critical inelastic response spectra and compared them with inelastic response spectra for recorded motions.

The purpose of this chapter is to explain a new probabilistic critical excitation method for SDOF elastic-plastic structures. More specifically, this chapter is aimed at presenting a new measure for describing the degree of criticality of recorded ground motions. The power and the intensity of the excitations are fixed and the critical excitation is found under these restrictions. While transfer functions and unit impulse response functions can be defined in linear elastic structures, such analytical expressions cannot be utilized in elastic-plastic structures. This situation causes much difficulty in finding a critical excitation for elastic-plastic structures. To overcome such difficulty, a statistical equivalent linearization technique is used. Drenick (1977) proposed a concept to utilize an equivalent linearization technique in finding a critical excitation for nonlinear systems. However, he did not mention the applicability of the concept to actual problems and his concept is restricted to deterministic equivalent linearization as well as Philippacopoulos and Wang (1984). In view of the similarity to the theory for linear elastic systems (Takewaki 2000a, b), the shape of the critical power spectral density (PSD) function is restricted to a rectangular function attaining its upper bound in a certain frequency range in this chapter. The central frequency of the rectangular PSD function is regarded as a principal parameter and is varied in finding the critical PSD function. The critical excitations are obtained for two examples and their responses are compared with those to the corresponding recorded earthquake ground motions.

5.2. STATISTICAL EQUIVALENT LINEARIZATION FOR SDOF MODEL

Consider a single-story shear building model, as shown in Fig. 5.1(a), subjected to the horizontal base acceleration $\ddot{u}_g$ which is a stationary Gaussian

(a) **(b)**

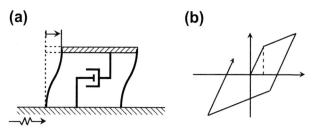

Figure 5.1 *(a) SDOF hysteretic shear building structure, (b) bilinear restoring-force characteristic.*

random process with zero mean. Let $S_g(\omega)$ denote the PSD function of the input $\ddot{u}_g$. The displacement of the mass relative to the base is denoted by $u(t)$. The restoring-force characteristic of the model is assumed to be the bilinear-type as shown in Fig. 5.1(b) (k: initial stiffness, α: ratio of the post-yield stiffness to the initial stiffness, u_y: yield displacement) and the damping of the model is assumed to be viscous. Let $m, c, f(u)$ denote the mass, the viscous damping coefficient and the restoring-force characteristic of the model, respectively. The equation of motion of the model may be written as

$$m\ddot{u}(t) + c\dot{u}(t) + f(u) = -m\ddot{u}_g(t) \tag{5.1}$$

Suppose that the response of the model can be approximated by the corresponding equivalent linear model which is governed by

$$m\ddot{u}(t) + c_{eq}\dot{u}(t) + k_{eq}u(t) = -m\ddot{u}_g(t) \tag{5.2}$$

where c_{eq} and k_{eq} denote the equivalent viscous damping coefficient and the equivalent stiffness. In this chapter, the statistical equivalent linearization method due to Caughey (1960), Kobori and Minai (1967) and Roberts and Spanos (1990) is employed. This method is pioneering in the statistical equivalent linearization field and various refined methods have been proposed (see, e.g., Roberts and Spanos 1990; Kobori et al. 1973; Spanos 1979, 1981; Grossmayer and Iwan 1981; Wen 1980, 1989). The principal objective of this chapter is to explain a critical excitation method for elastic-plastic structures and the detailed examination of the accuracy of the method will be made later. Many investigations have been made into the range of applicability of this statistical equivalent linearization method and only the model within such range of applicability will be treated in this chapter.

By minimizing the mean-square error between Eq. (5.1) and Eq. (5.2) with respect to the equivalent natural frequency ω_{eq} and the equivalent damping ratio h_{eq}, ω_{eq} and h_{eq} can be computed. The nondimensional equivalent natural frequency $\omega_{eq}^* = \omega_{eq}/\omega_1$ $(\omega_{eq}^2 = k_{eq}/m,\; \omega_1^2 = k/m)$ can be obtained from

$$1 - \omega_{eq}^{*2} = \frac{2(1-\alpha)}{\pi\lambda^2} \int_1^\infty F(z)\mathrm{d}z \tag{5.3}$$

where

$$F(z) = z^3\left\{ \pi - \cos^{-1}\left(1 - \frac{2}{z}\right) + \frac{2(z-2)}{z^2}\sqrt{z-1}\right\}e^{-z^2/\lambda} \tag{5.4}$$

In Eq. (5.3), the parameter λ can be defined by $\lambda = 2\sigma_u^2/u_y^2$ where σ_u^2 is the mean-square displacement of the model. The variable z in Eq. (5.3) represents the ratio of the amplitude of $u(t)$ to the yield displacement u_y. Let $h_0(= c/(2\omega_1 m))$ denote the initial damping ratio of the model. The equivalent damping ratio $h_{eq}(= c_{eq}/(2\omega_1 m))$ can be calculated from

$$h_{eq} = h_0 + \frac{1-\alpha}{\sqrt{\pi\lambda}\omega_{eq}^*}\left\{1 - \mathrm{erf}\left(1/\sqrt{\lambda}\right)\right\} \tag{5.5}$$

where erf( ) indicates the error function.

Once the nondimensional equivalent natural frequency and the equivalent damping ratio are obtained based on the assumption of σ_u, the mean-square displacement of the model can be evaluated in terms of the equivalent linear model by

$$\sigma_u^2 = \int_{-\infty}^\infty \left|H_{eq}(\omega)\right|^2 S_g(\omega)\mathrm{d}\omega \tag{5.6}$$

where

$$\left|H_{eq}(\omega)\right|^2 = \frac{1}{(\omega_{eq}^2 - \omega^2)^2 + (2h_{eq}\omega_1\omega)^2} \tag{5.7}$$

The procedure is repeated until the evaluated value σ_u in Eq. (5.6) coincides with the assumed value σ_u in Eqs. (5.3) and (5.5) (several cycles are sufficient for convergence).

Fig. 5.2 shows a sample of stationary random input accelerations and Fig. 5.3 presents the force-displacement relation in the elastic-plastic model and the equivalent linear model. Fig. 5.4 illustrates the force-displacement relation in the elastic-plastic model and the equivalent linear model in the case of trilinear restoring-force model (this is not used here).

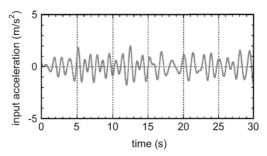

Figure 5.2 *Sample of stationary random input acceleration.*

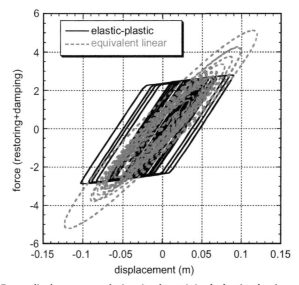

Figure 5.3 *Force-displacement relation in the original elastic-plastic model and the equivalent linear model.*

5.3. CRITICAL EXCITATION PROBLEM FOR SDOF MODEL

In contrast to linear elastic models, the properties of an elastic–plastic model change in time and for different excitations. This circumstance causes much difficulty in considering critical excitations for elastic–plastic models. The response of an elastic–plastic model can be approximated by that of the equivalent linear model, but the equivalent natural frequency and the equivalent damping ratio are still functions of excitations.

The problem of obtaining a critical excitation for stationary inputs may be stated as:

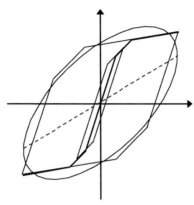

Figure 5.4 *Force-displacement relation in the original elastic-plastic model and the equivalent linear model in the case of trilinear restoring-force model.*

[Problem CESEP]

Given the floor mass m, the viscous damping coefficient c and the restoring-force characteristic f(u) of the model, find the critical PSD function $\tilde{S}_g(\omega)$ to maximize the mean-square displacement

$$\sigma_u^2 = \int_{-\infty}^{\infty} \left| H_{eq}(\omega) \right|^2 S_g(\omega) d\omega \tag{5.8}$$

subject to

$$\int_{-\infty}^{\infty} S_g(\omega) d\omega \leq \overline{S} \ (\overline{S}; \text{given power limit}) \tag{5.9}$$

$$\sup S_g(\omega) \leq \overline{s} \ (\overline{s}; \text{given PSD amplitude limit}) \tag{5.10}$$

Equation (5.9) limits the power of the excitation (Shinozuka 1970; Manohar and Sarkar 1995) and Eq. (5.10) is introduced to keep the present excitation model physically realistic (Takewaki 2000a, b). It is well known that a PSD function, a Fourier amplitude spectrum and an undamped velocity response spectrum of an earthquake have an approximate relationship (Hudson 1962). If the time duration of the earthquake is fixed, the PSD function corresponds to the Fourier amplitude spectrum and almost corresponds to the undamped velocity response spectrum. Therefore the present limitation on the peak of the PSD function indicates approximately the setting of a bound on the undamped velocity response spectrum.

Based on the knowledge that the critical PSD function for linear elastic models is a rectangular function attaining its upper bound $\overline{s}$ in a limited interval (Takewaki 2000a, b) (if $\bar{s}$ is infinite, the critical PSD function is reduced to the Dirac delta function), the critical PSD function for elastic-plastic models is restricted here to a rectangular function attaining its upper bound $\bar{s}$ in a limited interval. The noncriticality of the rectangular PSD function which does not attain its upper bound will be demonstrated later.

5.4. SOLUTION PROCEDURE

A procedure for finding the critical PSD function to Problem CESEP is explained. For a specified power $\overline{S}$ and intensity $\bar{s}$, the frequency band of the rectangular PSD function in the positive frequency range may be given by $\overline{S}/(2\bar{s})$. Statistical equivalent linearization can be applied in evaluating the standard deviation σ_u of displacement to each excitation with a specific central frequency ω_C of the rectangular PSD function (see Fig. 5.5(a)). The equivalent natural frequency, the equivalent damping ratio and the standard deviation σ_u of displacement are closely interrelated and are computed until the convergence criterion is satisfied. The standard deviation σ_u of displacement can be plotted with respect to ω_C/ω_1. The rectangular PSD function attaining the peak standard deviation of displacement can be regarded as the PSD function of the critical excitation (see Fig. 5.5(b)). In contrast to linear elastic models in which σ_u indicates its maximum at $\omega_C/\omega_1 \cong 1$ (Takewaki 2000a, b), σ_u attains its maximum in the range of $\omega_C/\omega_1 < 1$ in elastic–plastic models. This results from the fact that stiffness

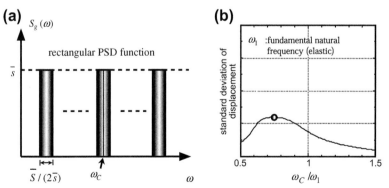

Figure 5.5 (a) Rectangular PSD functions with various central frequencies; (b) procedure for finding the critical PSD function maximizing the standard deviation of displacement for varied central frequencies.

reduction occurs in elastic–plastic models (equivalent stiffness) and the natural frequency of the model with the equivalent stiffness is resonant to the central frequency ω_C.

5.5. RELATION OF CRITICAL RESPONSE WITH INELASTIC RESPONSE TO RECORDED GROUND MOTIONS

In this section, the relation of the present critical response with the inelastic response to recorded ground motions is investigated. Two recorded earthquake ground motions, i.e. El Centro NS 1940 and Hyogoken–Nanbu 1995, Kobe University NS, are considered. Accelerations of these two ground motions are shown in Fig. 5.6. An SDOF elastic–plastic model with a damping ratio 0.02 is taken as the structural model. The initial elastic stiffness is varied to change the natural period of the model. The yield displacement is assumed to be constant ($u_y = 0.04m$). It is possible to consider a model with a varying yield displacement for a varied initial elastic stiffness, if desired. The ratio of the post-yield stiffness to the initial stiffness is assumed to be 0.5. Applicability of the present method to the model with a smaller ratio, i.e. 0.1 and 0.3, will be investigated in Sections 5.6 and 5.7.

Fig. 5.7 shows the PSD functions in a relaxed sense (Baratta et al. 1998) (approximate treatment for nonstationary motions) for these two ground motions. The duration has been employed as 40(s) in El Centro NS 1940 and 20(s) in Hyogoken–Nanbu 1995, Kobe University NS. It can be observed that a rather sharp peak exists around period $\cong$ 1.2(s) in Hyogoken–Nanbu 1995, Kobe University NS. The power $\overline{S} = 0.553(m^2/s^4)$ and the intensity $\overline{s} = 0.0661(m^2/s^3)$ have been computed for El Centro NS 1940 and the

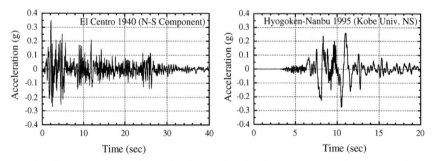

Figure 5.6 *Accelerations of El Centro NS 1940 and Kobe University NS 1995 (Hyogoken-Nanbu).*

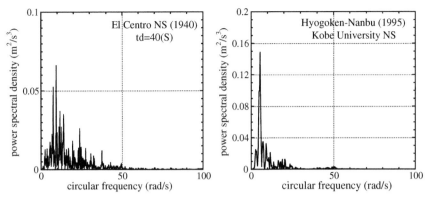

Figure 5.7 *Power spectral density functions of El Centro NS 1940 and Kobe University NS 1995 (Hyogoken-Nanbu).*

power $\overline{S} = 0.734(\mathrm{m^2/s^4})$ and the intensity $\overline{s} = 0.149(\mathrm{m^2/s^3})$ have been computed for Hyogoken-Nanbu 1995, Kobe University NS.

The solid and dotted line in Fig. 5.8 illustrates three times the standard deviation of displacement of the equivalent linear model to the critical excitation with the same power $\overline{S}$ and the intensity $\overline{s}$ plotted with respect to the undamped natural period of the model (initial elastic stiffness). The number "three" represents the assumed peak factor (Grossmayer and Iwan 1981; Der Kiureghian 1980) of the displacement. The maximum displacement of the elastic–plastic model (solid line) to the recorded ground

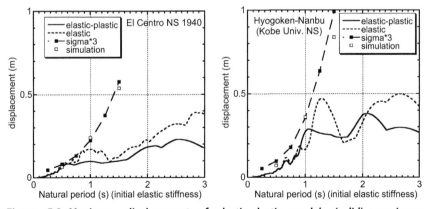

Figure 5.8 *Maximum displacement of elastic-plastic models (solid), maximum displacement of elastic models (dotted), three times the standard deviation of displacement of equivalent linear models (solid and dotted) and mean peak displacement of elastic-plastic models to 500 simulated motions ($\alpha = 0.5$).*

motions and the maximum displacement of the elastic model (dotted line) to the recorded ground motions are also shown in Fig. 5.8. It can be observed that, while the maximum elastic-plastic displacement is smaller than the maximum elastic response in almost all the range of natural period of interest in El Centro NS 1940, the maximum elastic-plastic displacement is larger than the maximum elastic displacement in some range of natural period (around 1.0 and 1.8(s)) in Hyogoken-Nanbu 1995, Kobe University NS. It is also understood from the response to Hyogoken-Nanbu 1995, Kobe University NS that, while the predominant period exists around 1.2–1.3(s) and the maximum elastic displacement exhibits a peak around such period range, the peak of the maximum elastic-plastic displacement shifts to a shorter natural period range (around 1.0(s)). This may result from the fact that the equivalent natural period of the elastic-plastic model with the initial natural period of around 1.0(s) coincides with 1.2–1.3(s) due to the stiffness reduction. This response amplification is smaller than that in elastic models. It can also be observed that the maximum elastic-plastic displacement is approaching the critical response $(3 \times \sigma_u)$ at $T_1 \cong 0.6$ in El Centro NS 1940.

It may be stated that the resonant characteristics of ground motions can be well represented by the present critical excitation method. However, care should be taken in that, while the response amplification can be observed clearly in elastic models, it can be observed only slightly in elastic-plastic models due to disappearance of clear natural periods of the elastic-plastic models.

Fig. 5.9(a) shows the standard deviation of displacement of the models (equivalent linear model) with various initial fundamental natural periods, 0.25(s), 0.5, 0.75, 1.0, 1.25, 1.5, for varied central frequencies of PSD functions (El Centro NS 1940) and Fig. 5.9(b) illustrates that for Hyogoken-Nanbu 1995 (Kobe University NS). It can be observed that, while, in the shorter natural period range, the plastic response level is small (stiffness reduction is then small) and the maximum value σ_u is attained around $\omega_C/\omega_1 \cong 1$, the plastic response level is large (stiffness reduction is then large) in the longer natural period range and the maximum value σ_u is attained at the range close to $\omega_C/\omega_1 = 0.5$.

Fig. 5.10 shows the plot of the nondimensional equivalent natural frequency with respect to the initial natural period of the model. The squared value of the ordinate in Fig. 5.10 represents the nondimensional equivalent stiffness k_{eq}/k. Fig. 5.11 illustrates the plot of the equivalent damping ratio with respect to the initial natural period of the model. It can be understood that the nondimensional equivalent natural frequency and the equivalent damping ratio are clearly dependent on excitations and the

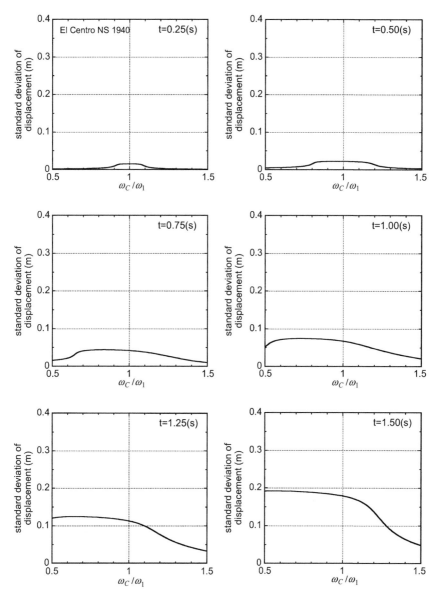

Figure 5.9(a) *Standard deviation of displacement of models (equivalent linear model) with various initial fundamental natural periods for varied central frequencies of PSD functions (El Centro NS 1940).*

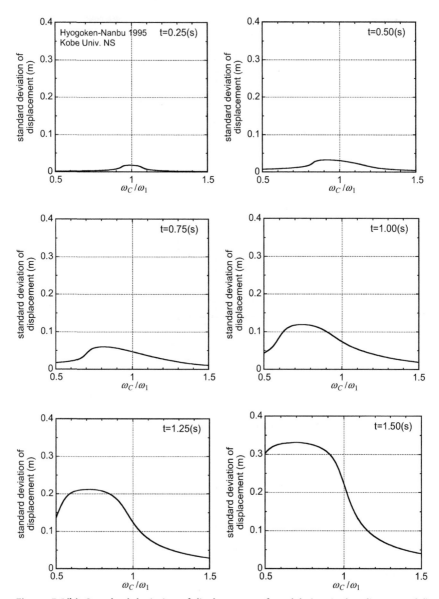

Figure 5.9(b) *Standard deviation of displacement of models (equivalent linear model) with various initial fundamental natural periods for varied central frequencies of PSD functions (Kobe University NS 1995 (Hyogoken-Nanbu)).*

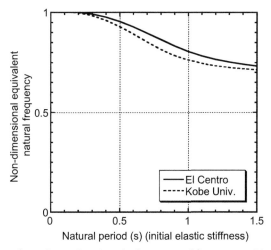

Figure 5.10 *Plot of nondimensional equivalent natural frequency with respect to initial natural period.*

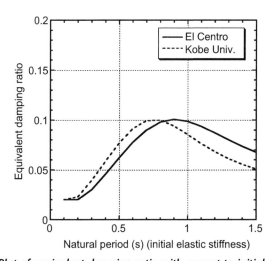

Figure 5.11 *Plot of equivalent damping ratio with respect to initial natural period.*

graph of the equivalent damping ratio with respect to the natural period of the model has its peak at a certain natural period dependent on excitations.

5.6. ACCURACY OF THE PROPOSED METHOD

In order to investigate the accuracy of the present critical excitation method, a numerical simulation analysis has been conducted. Three models with the initial natural period = 0.5, 1.0, 1.5(s) are considered. The critical PSD functions (rectangular PSD function) have been found for these three models using the method explained in Section 5.4. Five hundred simulated motions have been generated for each critical PSD function and elastic-plastic time history response analyses have been conducted to these 500 simulated motions. The simulated motions have been generated in terms of 100 cosine waves with uniform random numbers as the phase angles. The mean value of the maximum displacements is plotted in Fig. 5.8 by a hollow square mark. It can be observed that, while the peak factor 3 is appropriate for the model with $T_1 \cong 1.0(s)$, that is somewhat conservative for the models with $T_1 = 0.5, 1.5(s)$. However, the accuracy may be acceptable.

For the purpose of investigating another accuracy of the present equivalent linearization technique for finding the critical PSD function, Monte Carlo simulation for 500 simulated ground motions has been conducted. Figs. 5.12(a)–(c) show the comparison of the mean maximum displacement by means of time-history response analysis for nonlinear hysteretic models with three times the standard deviation of displacement of the equivalent linear models. Because the present method for finding the critical PSD function is based on the standard deviation of displacement of the equivalent linear models, the correspondence of the frequency ratios ω_C/ω_1 of these two quantity peaks is directly related to the accuracy of the present method. The simulation has been conducted for the excitation models with the same power $\bar{S}$ and intensity $\bar{s}$ of El Centro NS 1940 and Hyogoken-Nanbu 1995 (Kobe University NS). The post-yield stiffness ratios of $\alpha = 0.5$ (Fig. 5.12(a)), 0.3 (Fig. 5.12(b)) and 0.1 (Fig. 5.12(c)) have been considered. The damping ratio is set to 0.02 and the model fundamental natural period is selected as 1.0(s) or 0.75(s). It has been observed that as the post-yield stiffness ratio becomes smaller, the accuracy of the present equivalent linearization technique decreases slightly. A more rigorous equivalent linearization technique can be used if much computational resource is available.

5.7. CRITICALITY OF THE RECTANGULAR PSD FUNCTION AND APPLICABILITY IN WIDER PARAMETER RANGES

To investigate the criticality of the rectangular PSD function attaining its upper bound in a certain frequency range, some parametric analyses have been conducted for the model with $T_1 = 1.0(s)$. Six rectangular PSD

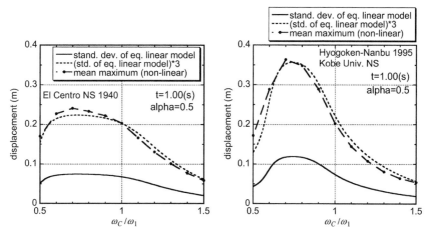

Figure 5.12(a) *Comparison of the mean maximum displacement by means of time-history response analysis for nonlinear hysteretic models to 500 simulated ground motions with three times the standard deviation of displacement of the equivalent linear models ($\alpha = 0.5$).*

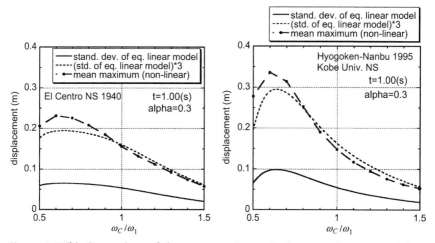

Figure 5.12(b) *Comparison of the mean maximum displacement by means of time-history response analysis for nonlinear hysteretic models to 500 simulated ground motions with three times the standard deviation of displacement of the equivalent linear models ($\alpha = 0.3$).*

functions with the common central frequency have been considered. These PSD functions have the same power $\bar{S}$ and have different maximum intensities $\max S_g(\omega) = \bar{s}, 0.9\bar{s}, 0.8\bar{s}, 0.7\bar{s}, 0.6\bar{s}, 0.5\bar{s}$. It has been found that the standard deviation σ_u of displacement decreases with the reduction of

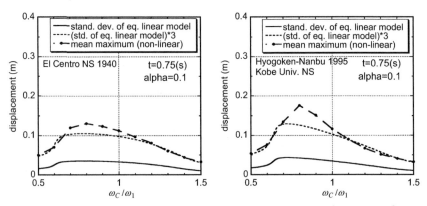

Figure 5.12(c) *Comparison of the mean maximum displacement by means of time-history response analysis for nonlinear hysteretic models to 500 simulated ground motions with three times the standard deviation of displacement of the equivalent linear models ($\alpha = 0.1$).*

the maximum intensity $\max S_g(\omega)$. This means that the rectangular PSD function attaining its upper bound $\bar{s}$ in a certain frequency range causes a larger standard deviation of displacement compared to the rectangular PSD function with the intensity smaller than the upper bound $\bar{s}$. It should be remarked that, while the transfer function in elastic models is independent of excitations, the equivalent natural frequency and the equivalent damping ratio are dependent on excitations. This property causes much difficulty in developing critical excitation methods for elastic-plastic models.

In order to demonstrate the applicability of the present critical excitation method to wider models, further simulation analyses have been conducted. The ratio of the post–yield stiffness to the initial stiffness is treated as the parameter to be discussed. Fig. 5.13(a) shows the maximum displacement of elastic–plastic models to 500 simulated motions (critical excitation) for 50, 84, 98% nonexceedance probability (solid and dotted) and the maximum displacement of equivalent linear models to 500 simulated motions (critical excitation) for 50, 84, 98% nonexceedance probability (solid marks) for $\alpha = 0.1$. The maximum displacement of elastic–plastic models (solid) and the maximum displacement of elastic models (dotted) are also plotted in the same figure. Fig. 5.13(b) illustrates the corresponding figure for $\alpha = 0.3$. The accuracy of the statistical equivalent linearization method has been checked for the models with $T_1 = 0.8, 1.0(s)$ to El Centro NS 1940 and for the models with $T_1 = 0.8, 0.9(s)$ to Hyogoken-Nanbu 1995

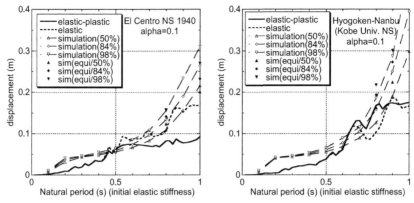

Figure 5.13(a) *Maximum displacement of E-P models (solid), maximum displacement of elastic models (dotted), maximum displacement of E-P models to 500 simulated motions (critical excitation) for 50, 84, 98% nonexceedance probability (solid and dotted) and maximum displacement of equivalent linear models to 500 simulated motions (critical excitation) for 50, 84, 98% nonexceedance probability (solid triangular and diamond marks) ($\alpha = 0.1$).*

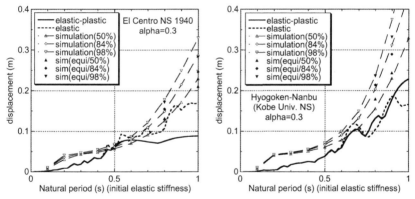

Figure 5.13(b) *Maximum displacement of E-P models (solid), maximum displacement of elastic models (dotted), maximum displacement of E-P models to 500 simulated motions (critical excitation) for 50, 84, 98% nonexceedance probability (solid and dotted) and maximum displacement of equivalent linear models to 500 simulated motions (critical excitation) for 50, 84, 98% nonexceedance probability (solid triangular and diamond marks) ($\alpha = 0.3$).*

(Kobe University NS). It may be stated that the accuracy of the present statistical equivalent linearization method may be acceptable in these parameter ranges (ductility factor = 3–7).

It is interesting to note that the critical response representation in terms of nonexceedance probabilities as shown in Figs. 5.13(a) and (b) can be an appropriate candidate for expressing the criticality of recorded ground motions. The distance between the critical response with a constant non-exceedance probability and the actual elastic-plastic response to a recorded motion can be regarded as a measure of criticality of the motion for the structural model with a selected natural period. It may be stated that El Centro NS 1940 has a high criticality for the structural model with the natural period of about 0.5(s) and Hyogoken-Nanbu 1995 (Kobe University NS) has a high criticality for the structural model with the natural period of about 0.7–0.8(s).

It should be remarked that the present critical excitation method is based on the stationary random vibration theory and recorded ground motions are nonstationary. Therefore the relation between the critical peak responses and the actual peak responses to recorded ground motions must be discussed carefully. The extension of the present theory to nonstationary random excitations may be necessary to take into account such factors and discuss the relation in more depth.

5.8. CRITICAL EXCITATION FOR MDOF ELASTIC-PLASTIC STRUCTURES

The purpose of the latter part in this chapter is to explain a probabilistic critical excitation method for MDOF elastic-plastic shear building structures on deformable ground. The power and intensity of the excitations are fixed and the critical excitation is found under these restrictions. While transfer functions and impulse response functions can be defined in linear elastic structures, such analytical expressions cannot be used in elastic-plastic structures. This situation causes much difficulty in finding a critical excitation for elastic-plastic structures. To overcome such difficulty, a statistical equivalent linearization technique is used. As stated before, Drenick (1977) proposed a concept to utilize an equivalent linearization technique in finding a critical excitation for nonlinear systems. However, he did not mention the applicability of the concept to actual problems and his concept is restricted to deterministic equivalent linearization. In this chapter, the shape of the critical PSD function is restricted to a rectangular function attaining its upper bound in a certain frequency range. The central frequency of the rectangular PSD function is regarded as a principal parameter and varied in finding the critical PSD function. The critical excitations are obtained for several examples and their responses are compared with those of the corresponding recorded earthquake ground motion. The definition of the critical excitation beneath a surface ground may be possible. However, the present definition at the ground surface level has the advantage of including uncertainties of wave propagation in the surface ground.

5.9. STATISTICAL EQUIVALENT LINEARIZATION FOR MDOF MODEL

Consider an f-story shear building model as shown in Fig. 5.14. This is supported by swaying and rocking springs and the corresponding dashpots. Let u_f and θ_F denote the horizontal displacement and angle of rotation of the ground floor. In addition, let $u_1, \cdots, u_f$ denote the horizontal displacements of the floors without a rigid-body mode component due to the ground floor motion. The set $\{u\} = \{u_1 \cdots u_f \; u_F \; \theta_F\}^T$ is treated as the displacement vector where $\{ \quad \}^T$ indicates the transpose of a vector. The parameters $\bar{k}_S$ and $\bar{k}_R$ denote the stiffnesses of the swaying and rocking springs and $\bar{c}_S$ and $\bar{c}_R$ denote the damping coefficients of the swaying and rocking dashpots. The relationship of the story shear force with the inter-story drift is assumed to follow a normal bilinear hysteretic rule as shown in Fig. 5.15. It is noted that the interstory drift $\delta_j = u_j - u_{j-1}$ does not include any rigid-body mode component. Let k_j denote the initial horizontal

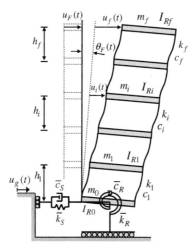

Figure 5.14 *MDOF elastic-plastic shear-building model supported by swaying and rocking springs and dashpots.*

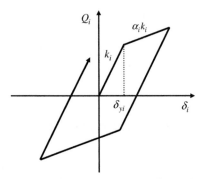

Figure 5.15 *Bilinear hysteretic restoring-force characteristic.*

stiffness in the j-th story. The ratio of the post-yield stiffness to the initial stiffness k_j in the j-th story is denoted by α_j and δ_{yj} denotes the yield interstory drift in the j-th story.

The parameters m_j, I_{Rj} and H_j denote the floor mass, the mass moment of inertia around its centroid and the height of the mass in the j-th floor from the ground. The story height in the j-th story is denoted by h_j. The system mass matrix may be written by

$$[M] = \begin{bmatrix} [M_B] & [M_{BS}] & [M_{BR}] \\ & E_1 & E_2 \\ sym. & & E_3 \end{bmatrix} \tag{5.11}$$

In Eq. (5.11), the component matrices and other parameters may be defined by

$$[M_B] = \mathrm{diag}(m_1 \cdots m_f),\ [M_{BS}] = \{m_1 \cdots m_f\}^T,$$
$$[M_{BR}] = \{m_1 H_1 \cdots m_f H_f\}^T \tag{5.12a–c}$$

$$E_1 = \sum_{i=0}^{f} m_i,\quad E_2 = \sum_{i=1}^{f} m_i H_i,\quad E_3 = \sum_{i=1}^{f} m_i H_i^2 + \sum_{i=0}^{f} I_{Ri},\quad H_i = \sum_{j=1}^{i} h_j$$

$$\tag{5.13a–d}$$

The system stiffness matrix in the elastic range may be expressed as

$$[K] = [K_B] + [K_S] + [K_R] \tag{5.14}$$

where the component matrices may be defined by

$$
[K_B] = \begin{bmatrix}
k_1 + k_2 & -k_2 & & & & & & & \\
& \ddots & \ddots & & & & & & 0 \\
& & \ddots & -k_i & & & & & \\
& & & k_i + k_{i+1} & \ddots & & & & \\
& & & & \ddots & -k_{f-1} & & & \\
& & & & & k_{f-1} + k_f & -k_f & & \\
& & & & & & k_f & 0 & \\
& & sym. & & & & & 0 & 0 \\
& & & & & & & & 0
\end{bmatrix},
$$

$$
[K_S] = \begin{bmatrix}
0 & 0 & 0 \\
& \overline{k}_S & 0 \\
sym. & & 0
\end{bmatrix}, \quad
[K_R] = \begin{bmatrix}
0 & 0 & 0 \\
& 0 & 0 \\
sym. & & \overline{k}_R
\end{bmatrix}
$$

$$(5.15a\text{--}c)$$

Assume that the damping matrix $[C]$ of the original model (before linearization) is expressed as the sum of the damping matrices of the three parts, i.e. the building dashpot, the swaying one and the rocking one. Let c_j denote the damping coefficient of the dashpot in the j-th story. It is also assumed that each structural portion has a stiffness–proportional damping matrix. The parameters h_B, h_S and h_R denote the so-called damping ratios of the shear building model, the swaying dashpot and the rocking dashpot, respectively. Because h_B, h_S and h_R are different, this system is a nonclassically damped system. Let ω_1 denote the undamped fundamental natural circular frequency in the elastic range of the original system. The system damping matrix $[C]$ may then be written as

$$[C] = [C_B] + [C_S] + [C_R] \tag{5.16}$$

$[C_B](= (2h_B/\omega_1)[K_B])$ is derived by replacing $\{k_j\}$ by $\{c_j\}$ in $[K_B]$ and

$$
[C_S] = \begin{bmatrix}
0 & 0 & 0 \\
& \overline{c}_S & 0 \\
sym. & & 0
\end{bmatrix}, \quad
[C_R] = \begin{bmatrix}
0 & 0 & 0 \\
& 0 & 0 \\
sym. & & \overline{c}_R
\end{bmatrix} \tag{5.17}
$$

This total system is subjected to the horizontal ground acceleration $\ddot{u}_g$ following a stationary Gaussian random process with zero mean. $S_g(\omega)$ is the PSD function of $\ddot{u}_g$. The equations of motion of the system may be described as

$$[M]\{\ddot{u}\} + [C]\{\dot{u}\} + \{F(u)\} = -[M]\{r\}\ddot{u}_g \qquad (5.18)$$

In Eq. (5.18), the components in $\{F(u)\}$ are expressed as superposition of the hysteretic restoring-force terms for the building and the elastic restoring-force terms for the supporting springs. The influence coefficient vector $\{r\}$ can be expressed by

$$\{r\} = \{0 \cdots 0 \ 1 \ 0\}^T \qquad (5.19)$$

The first f equations in Eq. (5.18) represent the equations of horizontal equilibrium of the building floors and the $(f+1)$-th equation and $(f+2)$-th equation represent the equation of horizontal equilibrium as a whole of the total system and that of rotational equilibrium as a whole of the total system around the ground floor, respectively.

To model the nonlinearity, the statistical equivalent linearization technique proposed by Kobori and Minai (1967) is used. That technique is based on the assumption of the slowly varying characteristics of the amplitude and phase in a harmonic vibration (Caughey 1960). Advanced statistical equivalent linearization techniques have been developed so far (Roberts and Spanos 1990; Kobori et al. 1973; Spanos 1979, 1981; Grossmayer and Iwan 1981; Wen 1980, 1989). Such techniques can be used in the present formulation if needed.

The ratio of the interstory drift in the i-th story to its yield drift is denoted by $\mu_i(t) = \delta_i(t)/\delta_{yi}$. Let z_i denote the amplitude of $\mu_i(t)$ and let σ_{μ_i} and $\sigma_{\dot{\mu}_i}$ denote the standard deviations of $\mu_i(t)$ and $\dot{\mu}_i(t)$. The parameters $\tilde{\omega}_i = \sigma_{\dot{\mu}_i}/\sigma_{\mu_i}$, $\overline{\omega}_i$ and $\omega_i^* = \tilde{\omega}_i/\overline{\omega}_i$ denote the mean circular frequency in the i-th story, the frequency for nondimensionalization and the nondimensional frequency of the interstory drift in the i-th story, respectively. In this case, the equivalent stiffness k_i^{eq} and equivalent damping coefficient c_i^{eq} may be described by

$$k_i^{eq} = k_i^* k_i, \quad c_i^{eq} = c_i^* (k_i/\overline{\omega}_i) \qquad (5.20a, b)$$

In Eq. (5.20), nondimensional parameters k_i^* and c_i^* are expressed by

$$k_i^* = \int_0^\infty k_{0i}(z_i)p(z_i, \sigma_{\mu_i})dz_i \qquad (5.21a)$$

$$c_i^* = \int_0^\infty c_{0i}(z_i, \omega_i^*) p(z_i, \sigma_{\mu_i}) dz_i \tag{5.21b}$$

The parameters $k_{0i}(z_i)$ and $c_{0i}(z_i, \omega_i^*)$ in Eqs. (5.21a, b) are derived as follows by minimizing the mean-squared errors of the restoring-force characteristic of the equivalent linear system from that of the original bilinear system (Kobori and Minai 1967).

$$k_{0i}(z_i) = \frac{1 - \alpha_i}{\pi} \cos^{-1}\left(1 - \frac{2}{z_i}\right)$$
$$+ \alpha_i - \frac{2}{\pi z_i^2}(1 - \alpha_i)(z_i - 2)\sqrt{z_i - 1} \quad (z_i \geq 1) \tag{5.22a}$$

$$k_{0i}(z_i) = 1.0 \quad (z_i \leq 1) \tag{5.22b}$$

$$c_{0i}(z_i, \omega_i^*) = \frac{4(1 - \alpha_i)(z_i^2 - 1)}{\pi \omega_i^* z_i^2} \quad (z_i \geq 1) \tag{5.23a}$$

$$c_{0i}(z_i, \omega_i^*) = 0 \quad (z_i \leq 1) \tag{5.23b}$$

The function $p(z_i, \sigma_{\mu_i})$ in Eqs. (5.21a, b) denotes the probability density function of the amplitude z_i and the following Rayleigh distribution is assumed here.

$$p(z_i, \sigma_{\mu_i}) = (z_i/\sigma_{\mu_i}^2)\exp\{-z_i^2/(2\sigma_{\mu_i}^2)\} \tag{5.24}$$

The squared values of the standard deviations σ_{μ_i} and $\sigma_{\dot{\mu}_i}$ are obtained by integrating the PSD function of the response. Those are expressed as the product of the squared value of the transfer function of the equivalent linear model and the PSD function $S_g(\omega)$ of the input. The objective function to be maximized in the present critical excitation problem is selected as

$$J = \sum_{i=1}^f \sigma_{\mu_i} \tag{5.25}$$

In the case where a common peak factor (Grossmayer and Iwan 1981; Der Kiureghian 1980) can be used in all the stories, the objective function, Eq. (5.25), is equivalent to the sum of the story ductility factors. The sum along the height of accumulated plastic deformations is an alternative candidate for the objective function. Such a problem may be challenging and desire to be developed.

5.10. CRITICAL EXCITATION PROBLEM FOR MDOF MODEL

Different from the case for linear elastic models, the tangent stiffness of an elastic-plastic model changes in time and for different excitations. This causes much difficulty in finding critical excitations for elastic-plastic models. The response of an elastic-plastic model can be simulated approximately by that of the equivalent linear model, but it should be remarked that the equivalent stiffnesses and damping coefficients are still functions of excitations.

The problem of obtaining a critical excitation for stationary vibration of elastic-plastic swaying-rocking (CESPSR) shear building models may be stated as:

[Problem CESEPSR]

Given the system mass matrix [M], the initial viscous damping matrix [C] and the restoring-force characteristics $\{F(u)\}$ of the model, find the critical PSD function $\tilde{S}_g(\omega)$ to maximize the objective function defined by Eq. (5.25), subject to

$$\int_{-\infty}^{\infty} S_g(\omega)d\omega \leq \overline{S} \ (\overline{S}; \text{given power limit}) \tag{5.26}$$

$$\sup S_g(\omega) \leq \overline{s} \ (\overline{s}; \text{given PSD amplitude limit}) \tag{5.27}$$

It is noted that Eq. (5.26) constrains the power of the excitation (Shinozuka 1970; Srinivasan et al. 1992; Manohar and Sarkar 1995) and Eq. (5.27) is introduced to keep the present input model physically realistic (Takewaki 2000a, b). It is also remarked that a PSD function, a Fourier amplitude spectrum and an undamped velocity response spectrum of an earthquake ground motion have an approximate relationship (Hudson 1962). If the time duration is fixed to be sufficiently long, the PSD function indicates the time average of the squared Fourier amplitude spectrum and has an approximate relationship with the undamped velocity response spectrum in the case where the maximum velocity occurs at the end of the duration. The present limitation on the peak of the PSD function implies approximately the bounding of the undamped velocity response spectrum.

Judging from the fact that the critical PSD function for linear elastic models is a rectangular function attaining its upper bound $\overline{s}$ in a finite interval (Takewaki 2000a, b), it seems natural to assume that the critical PSD function for elastic-plastic models found via the statistical equivalent linearization technique is also a rectangular function attaining its upper bound $\overline{s}$ in a finite interval.

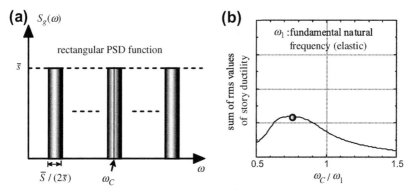

Figure 5.16 *(a) Rectangular PSD functions with various central frequencies; (b) procedure for finding the critical PSD function maximizing the sum of standard deviation values of story ductility for varied central frequencies.*

5.11. SOLUTION PROCEDURE

An algorithm is explained for finding the critical PSD function in the above problem. Given a specified power $\overline{S}$ and intensity $\overline{s}$, the frequency band of the rectangular PSD function in the positive frequency range may be obtained as $\overline{S}/(2\overline{s})$. The statistical equivalent linearization technique can be applied in the evaluation of the standard deviations σ_{μ_i} and $\sigma_{\dot{\mu}_i}$ to each excitation with a specific central frequency ω_C of the rectangular PSD function (see Fig. 5.16(a)). The equivalent stiffness, the equivalent damping coefficient and the response standard deviations σ_{μ_i}, $\sigma_{\dot{\mu}_i}$ are closely related and are updated until the convergence criterion is satisfied. The sum of response standard deviations σ_{μ_i} can be plotted with respect to ω_C/ω_1.

The rectangular PSD function causing the peak of the sum of σ_{μ_i} can be regarded as the critical PSD function (see Fig. 5.16(b)).

5.12. RELATION OF CRITICAL RESPONSE WITH INELASTIC RESPONSE TO RECORDED GROUND MOTIONS

The relation of the present critical response with the inelastic response to a recorded ground motion is investigated in this section. As a representative recorded ground motion, Hyogoken–Nanbu 1995 (Kobe University NS) is treated. The acceleration record is shown in Fig. 5.17(a) and its PSD function in a relaxed sense is plotted in Fig. 5.17(b) (Baratta et al. 1998)

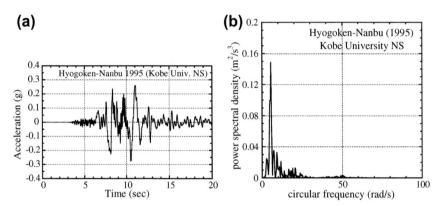

Figure 5.17 *(a) Acceleration of Kobe University NS 1995 (Hyogoken-Nanbu); (b) power spectral density function.*

(time–averaged approximate treatment for nonstationary motions). It can be seen that a sharp peak exists around the period of 1.2(s). The power $\bar{S} = 0.734(\text{m}^2/\text{s}^4)$ and the intensity $\bar{s} = 0.149(\text{m}^2/\text{s}^3)$ have been obtained from this motion. A 5-story shear building model has been investigated. The floor masses and mass moments of inertia are specified as $m_i = 30 \times 10^3(\text{kg})$, $I_{Ri} = 1.6 \times 10^5(\text{kg}\cdot\text{m}^2)$ $(i = 1, ..., 5)$. Those of the ground floor are $m_0 = 90 \times 10^3(\text{kg})$, $I_{R0} = 4.8 \times 10^5(\text{kg}\cdot\text{m}^2)$. The common story height is $h_i = 3.5(\text{m})$. The soil mass density and Poisson's ratio are $1.8 \times 10^3(\text{kg}/\text{m}^3)$ and 0.35, respectively. The shear wave velocities $V_S = 200(\text{m/s})$ and $V_S = 100(\text{m/s})$ have been assumed for the hard and soft soils, respectively. The swaying and rocking spring stiffnesses and the damping coefficients of the dashpots have been evaluated by the formula by Parmelee (1970) (see Appendix 5.1). The radius of the equivalent circular foundation is specified as 4(m).

The elastic story stiffnesses of the model have been determined so as to have the fundamental natural period of 1.0(s) *as an interaction model* and the uniform distribution of interstory drifts in the lowest eigenmode (Nakamura and Takewaki 1985; Takewaki 2000b) (see Appendix 5.2). This is based on the fact that the lowest-mode component is predominant in the response of linear elastic structures under critical input (Takewaki 2000a, b) and the present model has an almost uniform story–drift distribution. It should be noted that the elastic story stiffnesses of the models on the hard and soft soils have different distributions. The elastic story stiffness distributions are shown in Fig. 5.18. In this example, the common yield story drift has been chosen as 0.04(m) and the post-yield stiffness ratios to the initial stiffnesses have been

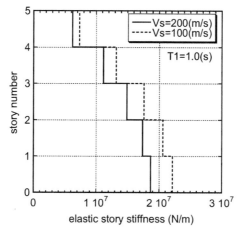

Figure 5.18 *Elastic story stiffness distributions of the models on hard soil (Vs = 200 m/s) and on soft soil (Vs = 100 m/s) which have the fundamental natural period of 1.0 s as an interaction model and the uniform distribution of interstory drifts in the lowest eigenmode.*

selected uniformly as 0.5, 0.3 and 0.1 (three cases). The initial viscous damping matrix has been specified as the initial stiffness-proportional one with a damping ratio 0.02.

Fig. 5.19(a) indicates the objective function with respect to central frequency of PSD functions, i.e. the sum of the standard deviations of story ductility of the models (equivalent linear model) on the hard soil

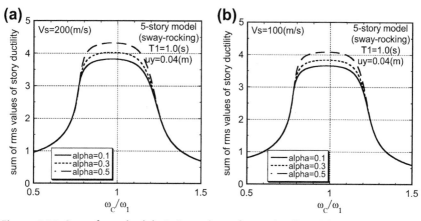

Figure 5.19 *Sum of standard deviation values of story ductility with respect to central frequency for post-yield stiffness ratios 0.1, 0.3 and 0.5 (PSD power $\bar{S}$ and intensity $\bar{s}$ of acceleration of Kobe University NS 1995 (Hyogoken-Nanbu) have been utilized): (a) Vs = 200 m/s; (b) Vs = 100 m/s.*

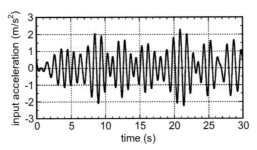

Figure 5.20 *Example of critical input acceleration.*

$(V_S = 200(\text{m/s}))$ with various post-yield stiffness ratios, $0.1, 0.3, 0.5$. It can be seen that the plastic response is small (stiffness reduction is then small) and the objective function exhibits a flat distribution around $\omega_C/\omega_1 \cong 1$. Furthermore, the maximum objective function J is observed around $\omega_C/\omega_1 \cong 1$ (exact value is 0.97). The corresponding figure is shown in Fig. 5.19(b) for the models on the soft soil ($V_S = 100(\text{m/s})$). A similar phenomenon can be observed in the models on the soft soil. Fig. 5.20 presents an example of the critical excitation for the model on the hard soil.

A set of 500 artificial accelerations has been generated in order to examine the relation of the critical response with the inelastic response to the recorded ground motion. The number 500 has been chosen judging from the convergence criterion. Artificial ground motions have been generated in terms of the sum of 100 cosine waves. The amplitudes of those cosine waves can be computed from the critical PSD function and the uniform random numbers between 0 and 2π have been used in the phase angles. Fig. 5.21(a) shows the sum of story ductilities computed for 50, 84, 98% nonexceedance probability of the model on the hard soil under the 500 simulated motions for post-yield stiffness ratios $0.1, 0.3, 0.5$ together with the sum of story ductilities to Hyogoken–Nanbu 1995 (Kobe University NS). The corresponding figure for the model on the soft soil is illustrated in Fig. 5.21(b). It can be seen that, in the model with the fundamental natural period of 1.0(s), the sum of story ductilities to Hyogoken–Nanbu 1995 (Kobe University NS) nearly corresponds to the sum of story ductilities computed for 50% nonexceedance probability. It should be remarked that the earthquake ground motion, Hyogoken–Nanbu 1995 (Kobe University NS), is close to the critical one around the period of 1.2(s) (Takewaki 2001). Fig. 5.22(a) illustrates the story ductilities computed for 50, 84, 98% nonexceedance probability of the model on the hard soil under the 500 simulated motions for post-yield stiffness ratios $0.5, 0.3, 0.1$. The corresponding figure for the model on the soft soil is shown

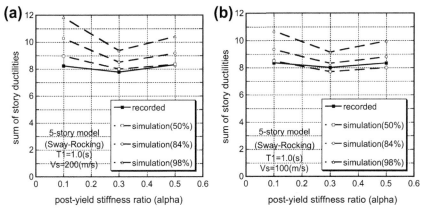

Figure 5.21 *Sum of story ductilities corresponding to 50, 84 and 98% nonexceedance probability under 500 simulated motions for post-yield stiffness ratios 0.1, 0.3, 0.5 and sum of story ductilities to Kobe University NS 1995 (Hyogoken-Nanbu) (a) Vs = 200 m/s; (b) Vs = 100 m/s.*

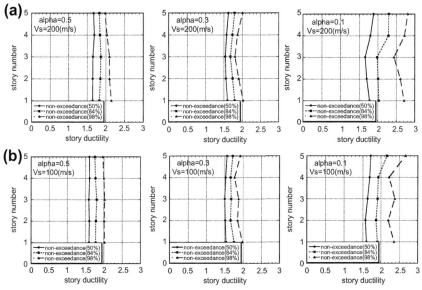

Figure 5.22 *Story ductilities corresponding to 50, 84 and 98% nonexceedance probability under 500 simulated motions for post-yield stiffness ratios 0.5, 0.3, 0.1: (a) Vs = 200 m/s; (b) Vs = 100 m/s.*

in Fig. 5.22(b). It can be seen that, while a fairly uniform distribution of story ductilities is observed in the models with larger post-yield stiffness ratios, a slight amplification is seen in upper stories in the models with smaller post-yield stiffness ratios. It is noted that stationary vibration is considered here and no special treatment has been taken on the modeling of the phase angles. This effect should be included in future rigorous and detailed investigations.

5.13. ACCURACY OF THE PROPOSED METHOD

An extensive numerical analysis has been conducted in order to investigate the accuracy of the explained critical excitation method. Fig. 5.23(a) shows the sum of story ductilities, with respect to central frequency, computed for 50, 84, 98% nonexceedance probability of the model on the hard soil under the 500 simulated motions for post-yield stiffness ratios 0.5, 0.3, 0.1. The sums of standard deviation values of story ductility in Fig. 5.19 multiplied by 2.5, i.e. $2.5\sum_{i=1}^{5}\sigma_{\mu_i}$, are also illustrated in Fig. 5.23(a) by solid circles. It can be observed that $2.5\sum_{i=1}^{5}\sigma_{\mu_i}$ exhibits a good correspondence to the value computed for 50% nonexceedance probability in case of $\alpha = 0.1$. In the present chapter, the sum of the RMS

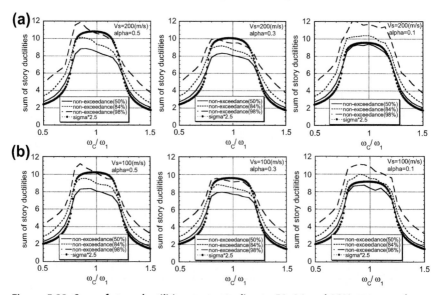

Figure 5.23 *Sum of story ductilities corresponding to 50, 84 and 98% nonexceedance probability under 500 simulated motions with respect to central frequency for post-yield stiffness ratios 0.5, 0.3, 0.1; (a) Vs = 200m/s; (b) Vs = 100m/s: Solid circles indicate 2.5×(sum of standard deviation values of story ductility shown in Fig. 5.19).*

values of story ductility is used for investigating the criticality. Therefore good correspondence of the sum of the RMS values of story ductility (times the peak factor) with the actual peak responses computed from the elastic-plastic response analysis is needed at least for the peak positions in order to validate the use of the sum of the RMS values of story ductility as the measure of the criticality of the ground motions. The corresponding figure for the model on the soft soil is shown in Fig. 5.23(b). It can be observed that the peak value of the sum of story ductilities is found in the range $\omega_C/\omega_1 < 1$. However, the sum of story ductilities is nearly constant around $\omega_C/\omega_1 = 1$ and the critical ratio ω_C/ω_1 can be determined from Fig. 5.19 in good approximation. The story shear force (restoring-force term) with respect to the story drift in the 1st, 3rd and 5th stories is shown in Fig. 5.24 under a simulated critical input acceleration ($Vs = 200(\text{m/s})$, post-yield stiffness ratio $= 0.3$). It should be noted that the drift of hysteretic centers is not observed clearly in the models with the post-yield stiffness ratios 0.3 and 0.5.

The statistics of elastic-plastic response of SDOF models under the simulated ground motions and other recorded motions have been investigated for validating the use of the statistical equivalent linearization technique in the critical excitation problem. It has been demonstrated that a similar tendency around the peak can be observed.

It is the fact that several hundred artificial ground motions are needed in investigating the validity of the peak factor introduced in Fig. 5.23. However, generation of these ground motions is not needed in finding the critical excitation. The present critical excitation can be obtained through the statistical equivalent linearization method. The validity of selection of the power $\overline{S}$ and intensity $\bar{s}$ should be discussed from the viewpoint of occurrence possibility of physical events. It should further be remarked that, while the present critical excitation method is based on the stationary

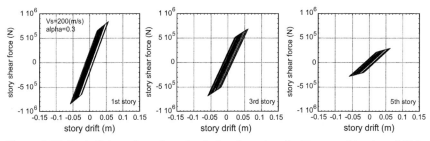

Figure 5.24 *Plot of story shear force (restoring-force term) with respect to story drift in the 1st, 3rd and 5th story under a simulated critical input acceleration (Vs = 200m/s, post-yield stiffness ratio = 0:3).*

random vibration theory, recorded ground motions are nonstationary. The relation between the critical peak responses and the actual peak responses to recorded ground motions must be discussed carefully from broader viewpoints. The extension of the present theory to nonstationary random excitations may be necessary to obtain a deeper and detailed relationship.

5.14. CONCLUSIONS

[SDOF Model]

The conclusions may be summarized as follows:

(1) A probabilistic critical excitation method for stationary random vibrations of SDOF *elastic-plastic models* can be developed by employing a statistical equivalent linearization method as a response simulator of the original elastic-plastic model. The power and the intensity of the excitations are constrained in the critical excitation problem. The shape of the critical PSD function is assumed to be a rectangular function attaining its upper bound in a certain frequency range in view of the similarity to the critical excitation problem for linear elastic models.

(2) The explained method regards the central frequency of the rectangular PSD function as a principal parameter for finding the critical PSD function. Since the equivalent natural frequency and the equivalent damping ratio are dependent on excitations, the technique using transfer functions proposed for elastic models cannot be utilized in the problem for elastic-plastic models.

(3) The simulation results by time–history response analysis for elastic–plastic models and the corresponding equivalent linear models revealed that the present critical excitation method is reliable in the models for which the validity of the statistical equivalent linearization method is guaranteed. The critical response representation in terms of nonexceedance probabilities can be an appropriate candidate for expressing the criticality of recorded ground motions.

(4) While the effect of the predominant frequency of a ground motion is reflected clearly in the comparison of the critical peak response with the actual peak response to the recorded ground motion in elastic models, such an effect is reflected only slightly in elastic-plastic models resulting from stiffness reduction and added damping ratio due to plastic responses.

[MDOF Model]

The conclusions may be summarized as follows:

(1) A probabilistic method of critical excitation for stationary random vibrations of MDOF *elastic-plastic* shear building models on deformable ground can be developed. A statistical equivalent linearization method for MDOF models is used as a response simulator of the original elastic-plastic hysteretic model. The excitation power and intensity are constrained in the critical excitation problem. The critical shape of the PSD function is assumed to possess a rectangular shape with its upper bound in a selected frequency range based on the similarity to the critical excitation problem for linear elastic models. The sum along the height of standard deviations of story ductilities can be an adequate candidate for the objective function to define the criticality.

(2) A solution algorithm for this critical excitation problem can be devised by regarding the central frequency of the rectangular PSD function as a principal parameter. Because the equivalent stiffnesses and damping coefficients depend strongly on properties of excitations, the procedure using transfer functions cannot be used in nature in the problem for elastic-plastic models.

(3) The results by elastic-plastic time-history response analysis disclosed that the introduced critical excitation method is effective and reliable in the models for which the validity of the statistical equivalent linearization is guaranteed. The representation of the critical response in terms of non-exceedance probabilities can be an appropriate candidate for expressing the degree of criticality of recorded ground motions.

APPENDIX 5.1 GROUND SPRING STIFFNESSES AND DASHPOT DAMPING COEFFICIENTS

Based on the results due to Parmelee (1970), the ground spring stiffnesses and dashpot damping coefficients are computed as

$$\bar{k}_S = \frac{6.77}{1.79 - \nu}\, Gr, \; \bar{k}_R = \frac{2.52}{1.00 - \nu}\, Gr^3,$$

$$\bar{c}_S = \frac{6.21}{2.54 - \nu}\rho V_S r^2, \; \bar{c}_R = \frac{0.136}{1.13 - \nu}\rho V_S r^4$$

$$(A5.1a-d)$$

The parameters G, ν, V_S, ρ and r are the soil shear modulus, Poisson's ratio, shear wave velocity, mass density of soil and radius of the equivalent circular footing plate, respectively.

APPENDIX 5.2 STORY STIFFNESSES OF SHEAR BUILDING WITH SPECIFIED LOWEST EIGENVALUE AND UNIFORM DISTRIBUTION OF LOWEST-MODE INTERSTORY DRIFTS

Let $\overline{\Omega}_1$ denote the specified lowest eigenvalue of the interaction model. It can also be expressed as $\overline{\Omega}_1 = (2\pi/T_1)^2$ in terms of the fundamental natural period T_1. The elastic story stiffnesses of the model with the specified lowest eigenvalue of the interaction model and a uniform distribution of lowest-mode interstory drifts can be obtained from the references (Nakamura and Takewaki 1985; Takewaki 2000b).

$$k_j = \overline{\Omega}_1 \sum_{i=j}^{f} m_i (U_F + \Theta_F H_i + i) \quad (j = 1, 2, \cdots, f) \tag{A5.2}$$

H_i is defined in Eq. (5.13d). U_F and Θ_F are defined by

$$U_F = \frac{D_2 D_5 - D_3 D_4}{D_1 D_4 - D_2^2}, \Theta_F = \frac{D_2 D_3 - D_1 D_5}{D_1 D_4 - D_2^2} \tag{A5.3a, b}$$

In Eqs. (A5.3a, b), $D_1, \cdots, D_5$ are expressed as follows.

$$D_1 = E_1 - \frac{\overline{k}_S}{\overline{\Omega}_1}, D_2 = E_2, D_3 = \sum_{i=1}^{f} i m_i,$$

$$\tag{A5.4a–e}$$

$$D_4 = E_3 - \frac{\overline{k}_R}{\overline{\Omega}_1}, D_5 = \sum_{i=1}^{f} i m_i H_i$$

E_1, E_2, E_3 in Eqs. (A5.4a–e) are defined in Eqs. (5.13a–c).

In the case where the coupling term $\overline{k}_{HR}$ of ground stiffness is not negligible, the present formulation can be extended directly by modifying the term D_2 in Eq. (A5.4b) to $D_2 = E_2 - (\overline{k}_{HR}/\overline{\Omega}_1)$. This modification depends on the exact governing equation of undamped lowest eigenvibration.

REFERENCES

Baratta, A., Elishakoff, I., Zuccaro, G., Shinozuka, M., 1998. A generalization of the Drenick–Shinozuka model for bounds on the seismic response of a Single-Degree-Of-Freedom system. Earthq. Engng. Struct. Dyn. 27 (5), 423–437.

Caughey, T.K., 1960. Random excitation of a system with bilinear hysteresis. J. Applied Mechanics 27 (4), 649–652.

Der Kiureghian, A., 1980. A response spectrum method for random vibrations. Report No. UCB/EERC 80-15, EERC. University of California, Berkeley, CA.

Drenick, R.F., 1977. The critical excitation of nonlinear systems. J. Applied Mech. 44 (2), 333–336.

Grossmayer, R.L., Iwan, W.D., 1981. A linearization scheme for hysteretic systems subjected to random excitation. Earthq. Engng. Struct. Dyn. 9 (2), 171–185.

Hudson, D.E., 1962. Some problems in the application of spectrum techniques to strong-motion earthquake analysis. Bull. Seism. Soc. Am. 52 (2), 417–430.

Iyengar, R.N., 1972. Worst inputs and a bound on the highest peak statistics of a class of non-linear systems. J. Sound and Vibration 25 (1), 29–37.

Kobori, T., Minai, R., 1967. Linearization technique for evaluating the elasto-plastic response of a structural system to non-stationary random excitations. Bull. Dis. Prev. Res. Inst., Kyoto University 10 (A), 235–260 (in Japanese).

Kobori, T., Minai, R., Suzuki, Y., 1973. Statistical linearization techniques of hysteretic structures to earthquake excitations. Bull. Dis. Prev. Res. Inst., Kyoto University 23 (3–4), 111–125.

Manohar, C.S., Sarkar, A., 1995. Critical earthquake input power spectral density function models for engineering structures. Earthq. Engng. Struct. Dyn. 24 (12), 1549–1566.

Nakamura, T., Takewaki, I., 1985. Optimum design of elastically supported shear buildings for constrained fundamental natural period. J. Structural Engineering, AIJ 31B, 93–102 (in Japanese).

Parmelee, R.A., 1970. The influence of foundation parameters on the seismic response of interation systems. Proc. of the Third Japan Earthquake Engineering Symposium.

Philippacopoulos, A.J., 1980. Critical Excitations for Linear and Nonlinear Structural Systems. Ph.D. Dissertation, Polytechnic Institute of New York.

Philippacopoulos, A., Wang, P., 1984. Seismic inputs for nonlinear structures. J. Eng. Mech. 110 (5), 828–836.

Roberts, J.B., Spanos, P.D., 1990. Random Vibration and Statistical Linearization. John Wiley & Sons, Chichester, UK.

Shinozuka, M., 1970. Maximum structural response to seismic excitations. J. Engng. Mech. Div. 96 (5), 729–738.

Spanos P-, T.D., 1979. Hysteretic structural vibrations under random load. J. Acoust. Soc. America 65 (2), 404–410.

Spanos, P.D., 1981. Stochastic linearization in structural dynamics. Applied Mechanics Reviews 34 (1), 1–8.

Srinivasan, M., Corotis, R., Ellingwood, B., 1992. Generation of critical stochastic earthquakes. Earthq. Engng. Struct. Dyn. 21 (4), 275–288.

Takewaki, I., 2000a. Optimal damper placement for critical excitation. Probabilistic Engineering Mechanics 15 (4), 317–325.

Takewaki, I., 2000b. Dynamic Structural Design: Inverse Problem Approach. WIT Press, Southampton, UK.

Takewaki, I., 2001. A new method for non-stationary random critical excitation. Earthq. Engng. Struct. Dyn. 30 (4), 519–535.

Wen, Y.K., 1980. Equivalent linearization for hysteretic systems under random excitation. J. Appl. Mech., ASME 47 (1), 150–154.

Wen, Y.K., 1989. Methods of random vibration for inelastic structures. Applied Mechanics Rev., 42 (2), 39–52.

Westermo, B.D., 1985. The critical excitation and response of simple dynamic systems. J. Sound and Vibration 100 (2), 233–242.

CHAPTER SIX

Critical Envelope Function for Nonstationary Random Earthquake Input

Contents

6.1. INTRODUCTION

The purpose of this chapter is to explain a new probabilistic critical excitation method for identifying the critical envelope function of ground motions. It is well known that the envelope shape of ground motions depends on various factors, for example arrival time and order of various kinds of waves, and the maximum structural responses of models with rather shorter natural periods are often induced by the intensive motions

© 2013 Elsevier Ltd.
All rights reserved.

existing mostly in the first half portion of ground motions. It is therefore of practical and scientific interest to investigate the most critical envelope shape in ground motions. Fig. 6.1 shows parts of time histories of four ground motions, El Centro NS 1940, Taft EW 1952, Hyogoken-Nanbu, Kobe University NS 1995 and Mexico Michoacan SCT1 EW 1985. It can be observed that a monotonically increasing function may be a candidate for the envelope function in the former half part of the total duration.

In this chapter, the nonstationary ground motion is assumed to be expressed as the product of a deterministic envelope function and another probabilistic function representing the frequency content. The former envelope function is determined in such a way that the corresponding mean-square drift of a single-degree-of-freedom (SDOF) model attains its maximum under the constraint on mean total energy. The critical excitation problem includes the double maximization procedure with respect to time and to the envelope function. The key to finding the critical envelope function is the order interchange in the double maximization procedure. An upper bound of the mean-square drift is also derived by the use of the Cauchy-Schwarz inequality. It is shown that the proposed technique is systematic and the upper bound can bound the exact response extremely efficiently within a reasonable accuracy.

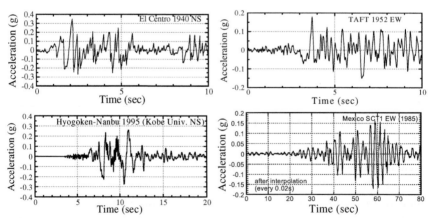

Figure 6.1 *Part of time histories of four ground motions, El Centro NS 1940, Taft EW 1952, Hyogoken-Nanbu, Kobe University NS 1995 and Mexico Michoacan SCT1 EW 1985.*

6.2. NONSTATIONARY RANDOM EARTHQUAKE GROUND MOTION MODEL

In this chapter, it is assumed that the input horizontal base acceleration can be described by the following uniformly modulated nonstationary random process.

$$\ddot{u}_g(t) = c(t)w(t) \tag{6.1}$$

where $c(t)$ is a deterministic envelope function to be obtained from the viewpoint of criticality and $w(t)$ is a stationary Gaussian process with zero mean. Let T denote the duration of $\ddot{u}_g(t)$. The power spectral density (PSD) function $S_w(\omega)$ of $w(t)$ is assumed to be given and its power (integration in the frequency domain) is determined by

$$\int_{-\infty}^{\infty} S_w(\omega)d\omega = \overline{S}_w \tag{6.2}$$

The auto-correlation function $R_w(t_1, t_2)$ of $w(t)$ can be defined by

$$R_w(t_1, t_2) = E[w(t_1)w(t_2)] = \int_{-\infty}^{\infty} S_w(\omega)e^{i\omega(t_1 - t_2)}d\omega \tag{6.3}$$

where $E[\cdot]$ indicates the ensemble mean. The time-dependent PSD function of $\ddot{u}_g(t)$ can then be expressed by

$$S_g(t;\omega) = c(t)^2 S_w(\omega) \tag{6.4}$$

The following constraint on the mean of the total energy (Drenick 1970; Shinozuka 1970) is taken into account.

$$C \equiv E\left[\int_0^T \ddot{u}_g(t)^2 dt\right] = \overline{C} \tag{6.5}$$

where $\overline{C}$ is a given value for the mean total energy. It can be shown (Housner and Jennings 1977) that this quantity C is related to the input energy to the SDOF model which has a certain relationship with the response spectrum. This relationship may be useful in setting C. Substitution of Eq. (6.1) into Eq. (6.5) leads to

$$\int_0^T c(t)^2 E[w(t)^2]dt = \overline{C} \qquad (6.6)$$

Substitution of Eq. (6.3) ($t_1 = t_2 = t$) into Eq. (6.6) provides

$$\int_0^T c(t)^2 \left(\int_{-\infty}^{\infty} S_w(\omega)d\omega \right) dt = \overline{C} \qquad (6.7)$$

Finally Eq. (6.7) is reduced to the following form with the use of Eq. (6.2).

$$\int_0^T c(t)^2 dt = \overline{C}/\overline{S}_w \qquad (6.8)$$

6.3. MEAN-SQUARE DRIFT

Consider an SDOF linear elastic model with ω_1 and h as the undamped natural circular frequency and the damping ratio, respectively. The damped natural circular frequency is then expressed by $\omega_{1d} = \omega_1\sqrt{1 - h^2}$.

Let $g(t) = H_e(t)(1/\omega_{1d})e^{-h\omega_1 t}\sin\omega_{1d}t$ ($H_e(t)$: Heaviside step function) denote the unit impulse response function multiplied by the mass. The mean-square drift of the SDOF model can then be expressed by (see Takewaki 2001a and Appendix 6.1)

$$\sigma_D(t)^2 = \int_{-\infty}^{\infty} \int_0^t c(\tau_1)g(t - \tau_1)e^{i\omega\tau_1}d\tau_1 \int_0^t c(\tau_2)g(t - \tau_2)e^{-i\omega\tau_2}d\tau_2 S_w(\omega)d\omega$$

$$= \int_{-\infty}^{\infty} \{A_C(t; \omega)^2 + A_S(t; \omega)^2\}S_w(\omega)d\omega$$

$$\equiv \int_{-\infty}^{\infty} H(t; \omega)^2 S_w(\omega)d\omega$$

$$(6.9)$$

The functions $A_C(t; \omega), A_S(t; \omega)$ are defined by

$$A_C(t; \omega) = \int_0^t g(t - \tau)\{-c(\tau) \cos \omega\tau\}d\tau \qquad (6.10a)$$

$$A_S(t; \omega) = \int_0^t g(t - \tau)\{-c(\tau) \sin \omega\tau\}d\tau \qquad (6.10b)$$

The functions $A_C(t; \omega)$ and $A_S(t; \omega)$ indicate the response of the SDOF model to the amplitude modulated cosine function and that to amplitude modulated sine function, respectively. It should be noted that the function $H(t; \omega)$ includes the effects both of the envelope function $c(t)$ and of zero initial conditions and its frequency content is time-dependent.

6.4. PROBLEM FOR FINDING CRITICAL ENVELOPE FUNCTION

The problem for finding the critical envelope function in nonstationary (CEFNS) random vibration may be described as:

[Problem CEFNS]
Given the natural frequency and damping ratio of an SDOF model and the PSD function $S_w(\omega)$ of $w(t)$ (its power is computed by Eq. (6.2)), find the non-negative critical envelope function $c(t)$ to maximize the specific mean-square drift $\sigma_D(t^*)^2$ (t^* is the time when the maximum mean-square drift to $\ddot{u}_g(t) = c(t)w(t)$ is attained) subject to the constraint on the mean of the total energy

$$C \equiv E\left[\int_0^T \ddot{u}_g(t)^2 dt\right] = \overline{C} \qquad (6.11)$$

6.5. DOUBLE MAXIMIZATION PROCEDURE

This problem includes the double maximization procedures, which may be described mathematically by

$$\max_{c(t)} \ \max_t \sigma_D(t)^2$$

The first maximization is performed with respect to time for a given envelope function $c(t)$ and the second maximization is done with respect to the envelope function $c(t)$. In the first maximization, the time t^* when the maximum mean–square drift is attained must be obtained for each envelope function. This original problem is complicated and it seems difficult to find a systematic solution procedure. To overcome this difficulty, a new sophisticated procedure based on the interchange of the order of the maximization procedures is proposed. The proposed procedure can be expressed by

$$\max_{t} \max_{c(t)} \sigma_D(t)^2$$

The first maximization process with respect to the envelope function for a given time (this problem is called a "subproblem" later) can be pursued by utilizing a nonlinear programming method. The second maximization procedure with respect to time can be implemented systematically by comparing the values at various times directly. It is important to note that the following property can be proved.

[Property A]

The envelope function $c(t)$ satisfying the constraint (6.11) and maximizing $\sigma_D(t_i)^2$ is the function with zero value in $t \geq t_i$.

The proof is shown in Appendix 6.2. This property will be utilized in the following sections where discretized values of the envelope function are obtained.

The proposed algorithm may be summarized as:

(i) Set a specific time $t = t_i$.

(ii) Find the critical envelope function at time $t = t_i$. When the envelope function is discretized as shown in the following section, the critical envelope function can be obtained by utilizing a nonlinear programming method.

(iii) Compute $\sigma_D(t_i)^2$ to the envelope function obtained in step (ii).

(iv) Repeat steps (i)–(iii) for various times and obtain $\sigma_D(t_m)^2 = \max \sigma_D(t_i)^2$.

(v) The envelope function as the solution to the subproblem for $t = t_m$ is determined as the critical envelope function.

It is important to note that the present algorithm based on the interchange of the order of the double maximization procedure is applicable to more sophisticated nonuniformly modulated nonstationary excitation models although the expression of Eq. (6.9) must be modified and a new critical excitation problem must be stated.

6.6. DISCRETIZATION OF ENVELOPE FUNCTION

For simple and essential presentation of the proposed solution procedure, a property on the PSD function $S_w(\omega)$ is used which has been discussed and found in Takewaki (2001a). Let $S_w(\omega)$ be a function with nonzero values at $\omega = \omega_{1d}$ only. This implies that the critical frequency content of excitations is a resonant one to the structural model. This treatment provides a good approximation for evaluation of $\sigma_D(t)^2$ of the model under the excitations with other PSD functions $S_w(\omega)$ with a fairly narrow band around $\omega = \omega_{1d}$ (Takewaki 2000). Because it has been shown in Takewaki (2001a) that the critical frequency content of the PSD function is a resonant one irrespective of envelope functions, the present treatment of the PSD function appears to be reasonable and this fact is expected to play an important role in finding the critical excitation for simultaneous variation of frequency contents and envelope functions (see Section 6.9). If we do not use this property on the PSD function $S_w(\omega)$, a tremendous amount of computational tasks may be necessary to conduct the simultaneous sensitivity analysis for frequency contents and envelope functions.

With the use of Eq. (6.2), the mean-square drift expressed by Eq. (6.9) is then reduced to

$$\sigma_D(t)^2 = H(t; \omega_{1d})^2 \overline{S}_w \qquad (6.12)$$

For discretization of $c(t)$, let c_i denote the value of $c(t)$ at $t = t_i$ ($c_0 = 0$). The following linear approximation of $c(t)$ is introduced.

$$c(t) = c_{j-1} + \frac{c_j - c_{j-1}}{\Delta t}\{t - (j-1)\Delta t\} \quad ((j-1)\Delta t \le t \le j\Delta t) \qquad (6.13)$$

where Δt is the time interval for discretization of $c(t)$. Let N denote the number of time intervals until time t ($t = N\Delta t$). Eq. (6.12) can then be expressed by

$$\sigma_D(t)^2 = \left(\left\{ \sum_{j=1}^{N} \int_{(j-1)\Delta t}^{j\Delta t} c(\tau)g(t-\tau)\cos\omega_{1d}\tau d\tau \right\}^2 \right.$$
$$\left. + \left\{ \sum_{j=1}^{N} \int_{(j-1)\Delta t}^{j\Delta t} c(\tau)g(t-\tau)\sin\omega_{1d}\tau d\tau \right\}^2 \right) \overline{S}_w \qquad (6.14)$$

Substitution of Eq. (6.13) into Eq. (6.14) leads to

$$\sigma_D(t)^2 = \overline{S}_w \left\{ \left(\sum_{j=1}^{N} B_{Cj}(t)c_j \right)^2 + \left(\sum_{j=1}^{N} B_{Sj}(t)c_j \right)^2 \right\} \qquad (6.15)$$

where $B_{Cj}(t)$ and $B_{Sj}(t)$ are the coefficients on c_j and can be expressed in terms of given parameters. The expressions of $B_{Cj}(t)$ and $B_{Sj}(t)$ are shown in Appendix 6.3.

6.7. UPPER BOUND OF MEAN-SQUARE DRIFT

Introduction of the discretization of $c(t)$, i.e. Eq. (6.13), into the constraint (6.8) provides

$$\frac{2}{3} \sum_{j=1}^{N_T} c_j^2 + \frac{1}{3} \sum_{j=1}^{N_T} c_{j-1}c_j = \frac{\overline{C}}{\overline{S}_w \Delta t} \qquad (6.16)$$

where $N_T = T/\Delta t$ is the number of time intervals in the whole duration. Another interpolation of $c(t)$ may be possible from the viewpoint of simplification. Constant interpolation of $c(t)$ in every interval may provide the following expression of the constraint (6.8).

$$\sum_{j=1}^{N_T} c_j^2 = \frac{\overline{C}}{\overline{S}_w \Delta t} \qquad (6.17)$$

For simplicity of expression, the above constant interpolation of $c(t)$ is used only for the constraint (6.8). This is due to the fact that, while the response (6.9) is sensitive to the shape of $c(t)$, the constraint (6.8) is not so sensitive to the shape of $c(t)$ so long as the time interval is small enough.

As shown in Section 6.5 (Property A), N can be used in place of N_T in the constraint (6.17) when the maximization of $\sigma_D(t_i)^2$ at time t_i is considered. In other words, $c_j = 0$ can be assumed for $j > i$ when the maximization of $\sigma_D(t_i)^2$ at time t_i is considered. Application of the Cauchy–Schwarz inequality to Eq. (6.15) and substitution of Eq. (6.17) into the resulting equation provide

$$\sigma_D(t)^2 = \overline{S}_w \left\{ \left(\sum_{j=1}^{N} B_{Cj}(t) c_j \right)^2 + \left(\sum_{j=1}^{N} B_{Sj}(t) c_j \right)^2 \right\}$$

$$\leq \overline{S}_w \left\{ \left(\sum_{j=1}^{N} B_{Cj}(t)^2 \right) \left(\sum_{j=1}^{N} c_j^2 \right) \right.$$

$$\left. + \left(\sum_{j=1}^{N} B_{Sj}(t)^2 \right) \left(\sum_{j=1}^{N} c_j^2 \right) \right\} \qquad (6.18)$$

$$= \overline{S}_w \left(\sum_{j=1}^{N} c_j^2 \right) \left\{ \sum_{j=1}^{N} (B_{Cj}(t)^2 + B_{Sj}(t)^2) \right\}$$

$$= \frac{\overline{C}}{\Delta t} \left\{ \sum_{j=1}^{N} (B_{Cj}(t)^2 + B_{Sj}(t)^2) \right\}$$

Every term in the last line of inequality (6.18) does not include any term dependent on $\{c_i\}$ and inequality (6.18) implies that the upper bound of $\sigma_D(t)^2$ in the subproblem stated in Section 6.5 can be given by the value in the last line of inequality (6.18). It should be noted that $B_{Cj}(t)$ and $B_{Sj}(t)$ include terms of Δt (see Appendix 6.3). It can therefore be shown that the leading order of the last line of inequality (6.18) for Δt is zero-th.

When the linear interpolation of $c(t)$ is used also for the constraint (6.8), the following bound can be derived with the use of the non-negative property of $\{c_i\}$.

$$\sigma_D(t)^2 \leq \frac{3\overline{C}}{2\Delta t} \left\{ \sum_{j=1}^{N} (B_{Cj}(t)^2 + B_{Sj}(t)^2) \right\} \qquad (6.19)$$

It can be seen that the upper bound (6.19) is larger than the upper bound (6.18). It is therefore preferable to use the upper bound (6.18) in order to bound the exact value strictly.

It is interesting to note that an upper bound of $\sigma_D(t)^2$ under the constraint (6.8) can be derived for continuous functions $c(t)$. That expression is shown in Appendix 6.4.

6.8. NUMERICAL EXAMPLES

Consider an SDOF linear elastic model that has the undamped natural period 0.5(s) and the damping ratio 0.02. The time interval for discretization of $c(t)$ is 0.05(s). The specified mean total energy, the power of $w(t)$ and the excitation duration are given by $\overline{C} = 2.28(m^2/s^3)$, $\overline{S}_w = 1.51(m^2/s^3)$ and 40(s), respectively.

The maximization problem of Eq. (6.15) under the constraints on Eq. (6.17) $(N_T \to N)$ and on the non-negative property of $\{c_i\}$ constitutes a nonlinear programming problem where both the objective function and the constraints are nonlinear. This nonlinear programming problem is solved here by the Powell's method (Vaderplaats 1984).

For the purpose of comparing the response to the critical envelope function with those to other envelope functions, the following envelope functions are considered.

$$\text{constant}: c(t) = a$$

$$\text{linear}: c(t) = at$$

$$\text{exponential}: c(t) = a\exp(h\omega_1 t)$$

$$\text{realistic}: c(t) = a\{\exp(-0.13t) - \exp(-0.45t)\} \qquad (6.20\text{a--d})$$

In Eqs. (6.20a–d), the parameter a is to be determined from the constraint (6.17) $(N_T \to N)$. Fig. 6.2 shows the plots of these envelope functions in which the parameter a has been determined for the total time duration 40(s). The terminology "realistic" is used here because the

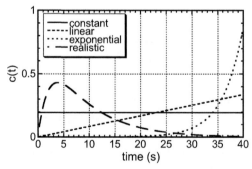

Figure 6.2 *Four examples of envelope functions satisfying the constraint (6.17) (constant, linear, exponential-function and realistic distributions).*

expression (6.20d) has been proposed by Shinozuka and Sato (1965) to simulate actual ground motions realistically using this envelope function.

The lower four curves in Fig. 6.3 show $\sigma_D(t)$, the square root of Eq. (6.15), for four envelope functions in (6.20a–d). It should be noted that Fig. 6.3 does not represent the root–mean–square drift to a given envelope function. The parameter a has been determined for each parameter N so as to satisfy the constraint (6.17) $(N_T \rightarrow N)$ and $\sigma_D(t)$ $(t = N\Delta t)$ for these different envelope functions is plotted. The root–mean–square drift to the critical envelope function obtained by the above-mentioned nonlinear programming method for each parameter N is also shown in Fig. 6.3. The subproblem stated in Section 6.5 has been solved for each parameter N. Furthermore, the upper bound given by inequality (6.18) is also plotted in Fig. 6.3. It can be observed from Fig. 6.3 that the value $\sigma_D(t)$ to the critical envelope function, the solution of the subproblem, obtained for each parameter N appears to converge to a constant value at a certain time. Strictly speaking, the exact maximum occurs at the end of the duration and the exact solution to the original problem CEFNS can be obtained for the subproblem in case of $N = N_T$. It can also be understood from Fig. 6.3 that the root–mean–square drift corresponding to the solution of each subproblem exists between the value for other noncritical envelope functions and the upper bound. This demonstrates the validity of the nonlinear

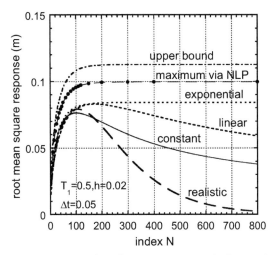

Figure 6.3 *Root-mean-square drift to four excitations with the amplitude-modulated envelope functions (6.20a–d) which have been obtained for each parameter N, the root-mean-square drift (6.15) to the critical excitation as the solution to the subproblem in Section 6.5 and the upper bound of the root-mean-square drift by (6.18). (Nonlinear programming, (NLP.).*

programming method employed in this chapter. It can be seen from Fig. 6.3 that the upper bound of the root–mean–square drift can bound the exact value within about 10% error. It is noteworthy that, while it takes a few hours in ordinary workstations to obtain the solution to the nonlinear programming problem for a fairly large number N, the computation of the upper bound requires extremely less time (a second or less).

Fig. 6.4(a) shows the envelope function as the solution to the subproblem $N = 200$ and Fig. 6.4(b) shows that to the subproblem $N = N_T = 800$. As stated before, Fig. 6.4(b) indicates the exact solution to the original problem CEFNS. It is interesting to note that, while an increasing exponential function is the critical one in the deterministic problem treated by Drenick (1970), the superimposed envelope function of

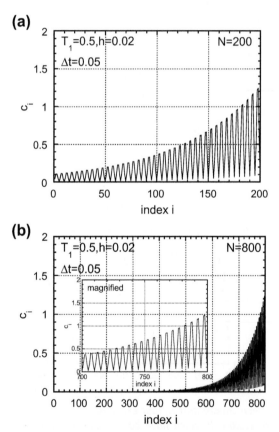

Figure 6.4 *(a) Envelope function as solution to subproblem for $N = 200$ (approximate solution to original problem); (b) envelope function as solution to subproblem for $N = N_T = 800$ (exact solution to original problem).*

the original envelope function, to be found in the present problem, of the critical excitation can be a function similar to an increasing exponential function in the probabilistic problem. This result may be related to the condition $\omega = \omega_{1d}$ used in Section 6.6 and further examination of the envelope function will be necessary.

Fig. 6.5(a) illustrates an approximate time-history sample of the critical acceleration for $N = N_T = 800$. In the computation, a narrow band-limited white noise with the band $\Delta\omega = 0.2(rad/s)$ is used for $S_w(\omega)$ and $w(t)$ has been expressed as the sum of 100 cosine waves. The amplitudes of the cosine waves have been computed from the PSD function $S_w(\omega)$ and the phase angles have been assumed to be uniformly random between 0 and 2π. Fig. 6.5(b) shows the magnified plot of the time-history sample between 35(s) and 40(s).

It should be remarked that, when the duration of ground motions becomes shorter compared to the natural period of the SDOF model and the damping becomes rather small, the discussion on impulsive loading must be made carefully. A similar discussion was made by Drenick (1970) for deterministic critical excitation problems. This problem is open to future discussion.

In order to examine the accuracy of the numerical procedure, the time interval has been shortened from 0.05(s) to 0.025(s). Fig. 6.6 shows the

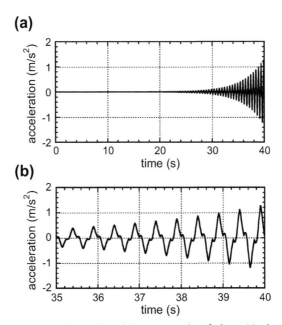

Figure 6.5 *(a) An approximate time-history sample of the critical acceleration for* $N = N_T = 800$; *(b) magnified plot of the time-history sample between 35(s) and 40(s).*

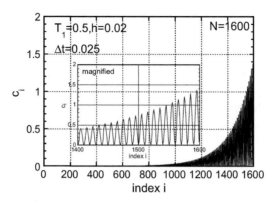

Figure 6.6 *Envelope function as the solution to the subproblem for N = N_T = 1600 (exact solution to the original problem).*

envelope function as the solution to the subproblem $N = 1600$. Due to the shortening of the time interval, the number of discretization has been increased. It can be observed that a smoother envelope function has been obtained by the shortening of the time interval. However, the required computer time is extremely longer.

Furthermore, to investigate the effect of the natural period of the SDOF structural model and its damping ratio on the upper bound of response, two new models have been considered. Those two models have the same natural period of 2.0(s) and different damping ratios of 0.02 and 0.10. The lower solid curve in Fig. 6.7 shows $\sigma_D(t)$, the square root of Eq. (6.15), for the realistic envelope function in (6.20d). It should be noted again that Fig. 6.7 does not represent the square root of the mean-square drift to a given envelope function. The parameter a has been determined for each parameter N so as to satisfy the constraint (6.17) $(N_T \to N)$, and $\sigma_D(t)$ $(t = N\Delta t)$ for these different envelope functions is plotted. The square root of the mean-square drift to the critical envelope function obtained by the above-mentioned nonlinear programming method for each parameter N is also plotted in Fig. 6.7. The subproblem stated in Section 6.5 has been solved for each parameter N. The upper bound given by the inequality (6.18) (square root of (6.18)) is also plotted in Fig. 6.7. It can be observed from Fig. 6.7 that the value $\sigma_D(t)$ to the critical envelope function, the solution of the subproblem, obtained for each parameter N appears to converge to a constant value at a certain time. Strictly speaking, the exact maximum occurs at the end of the duration and the exact solution to the original problem CEFNS can be obtained for the subproblem in case of $N = N_T$. It can also be understood from Fig. 6.7 that the square root of the mean-square drift corresponding to

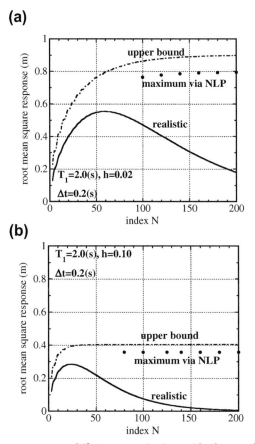

Figure 6.7 *Root-mean-square drift to an excitation with the amplitude-modulated envelope function (6.20d) which has been obtained for each parameter N, the root-mean-square drift (6.15) to the critical excitation as the solution to the subproblem in Section 6.5 and the upper bound of the root-mean-square drift by (6.18): (a) damping ratio = 0.02, (b) damping ratio = 0.10. NLP*

the solution of each subproblem exists between the value for other noncritical envelope functions and the upper bound. This demonstrates the validity of the nonlinear programming method employed in this chapter.

6.9. CRITICAL EXCITATION FOR VARIABLE ENVELOPE FUNCTIONS AND VARIABLE FREQUENCY CONTENTS

The critical frequency content has been found in the references (Takewaki 2000, 2001a) as a resonant one to structural models for any fixed envelope function. In contrast to the previous study, the critical envelope function has been revealed in this chapter for the fixed frequency content, i.e. the resonant critical one. It may therefore be concluded that the combination of the critical frequency content and the critical envelope function is the critical set for a more general problem with variable frequency contents and variable envelope functions.

6.10. CONCLUSIONS

The conclusions may be summarized as follows:

(1) A new probabilistic critical excitation method can be developed for nonstationary responses of an SDOF structural system that is subjected to a nonstationary ground acceleration described as the product of an envelope function and another function representing the frequency content of a stationary random process. The non–negative critical envelope function maximizing the displacement response is determined under the constraint on the mean total energy of the input motion.

(2) It has been found that the order interchange of the double maximization procedure with respect to time and to the envelope function can be an excellent and powerful solution algorithm for finding the critical envelope function. A nonlinear programming procedure can be utilized in one of the maximization procedures (with respect to the envelope function).

(3) An upper bound of the mean–square drift can be derived by using the Cauchy-Schwarz inequality and the constraint on the input motion (mean total energy). The upper bound can bound the mean–square drift efficiently and effectively within a reasonable accuracy. The required time for this procedure based on the upper bound is shorter than one second in this case and is superior to the nonlinear programming procedure, stated above, requiring several hours.

The practicality and reality of the explained critical excitation method should be discussed carefully (see, e.g., Srinivasan et al. 1991; Pirasteh et al. 1988). The recorded ground motions, e.g. Mexico (1985), Hyogoken-Nanbu (1995) and Chi-Chi (1999), imply that unexpected ground motions could occur especially in the near fields. The critical excitation can differ in each structure. This is because each structure has a different natural period. In defining a critical excitation for a group of structures, the excitation with a wide-band PSD function may be a candidate for the critical excitation. Furthermore, for an inelastic structure with a time-varying principal vibration frequency, the excitation with a rather wide-band PSD function may also be critical (Drenick 1977; Takewaki 2001b). Extension of the present method to inelastic structures may be possible with the help of the equivalent linearization method (Takewaki 2001b).

It is expected that the critical excitation method can provide useful information for the design of important structures of which functional and structural damage must be absolutely avoided during severe earthquakes. In such a situation, two-dimensional (2-D) and three-dimensional (3-D) inputs may be required. Extension of the present method to such 2-D and 3-D input models is a challenging problem.

APPENDIX 6.1 MEAN-SQUARE DRIFT OF SDOF SYSTEM IN STATIONARY STATE

The mean-square drift $\sigma_D{}^2$ of the SDOF system in the stationary state may be expressed as

$$\sigma_D{}^2 = \int_{-\infty}^{\infty} |H_D(\omega)|^2 S_g(\omega) d\omega \tag{A6.1}$$

where $H_D(\omega)$ is the transfer function of the drift to the input acceleration and $S_g(\omega)$ is the time-averaged PSD function of the input acceleration.

APPENDIX 6.2 PROOF OF PROPERTY A

Consider an excitation with an envelope function shown in Fig. A1(a). It is assumed that the PSD function $S_w(\omega)$ of $w(t)$ is given and does not change. Consider the mean-square drift at time t. It is proven here that the excitation with nonzero value of $c(\tau)$ in $\tau > t$ cannot be the excitation satisfying the constraint (6.8) and maximizing the response $\sigma_D(t)^2$ at time t. Call the excitation with nonzero value of $c(\tau)$ in $\tau > t$ "excitation A." On the other hand, call the excitation with zero value of $c(\tau)$ in $\tau > t$ "excitation B." It is noted that excitation B is constructed in such a way that the area of $c(\tau)^2$ in $\tau > t$ in excitation A is moved to the region $\tau < t$ and $c(\tau)$ in $\tau < t$ is amplified as $c^*(\tau)^2 = \alpha c(\tau) \, (\alpha > 1)$ (see Fig. A1(b)). From Eq. (6.9), the mean-square drift $\sigma_D^*(t)^2$ to excitation B has the following relationship with the mean-square drift $\sigma_D(t)^2$ to excitation A.

$$\sigma_D^*(t)^2 = \alpha^2 \sigma_D(t)^2 > \sigma_D(t)^2 \tag{A6.2}$$

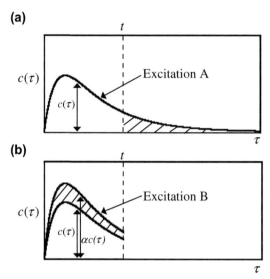

Figure A1 *(a) Excitation with nonzero value of $c(\tau)$ in $\tau > t$ (Excitation A); (b) excitation with zero value of $c(\tau)$ in $\tau > t$ (Excitation B) in which the area of $c(\tau)^2$ in $\tau > t$ in excitation A is moved to the region $\tau < t$ and $c(\tau)$ in $\tau < t$ is amplified as $c^*(\tau) = \alpha c(\tau)$ $(\alpha > 1)$.*

Eq. (A6.2) implies that excitation A with nonzero value of $c(\tau)$ in $\tau > t$ cannot be the excitation satisfying the constraint (6.8) and maximizing the response $\sigma_D(t)^2$ at time t.

APPENDIX 6.3 EXPRESSIONS OF $B_{Cj}(t)$ AND $B_{Sj}(t)$

The expressions of $B_{Cj}(t)$ and $B_{Sj}(t)$ in Eq. (6.15) may be given as follows:

$$B_{Cj}(t) = P_j(t) - P_{j+1}(t) - (j-1)Q_j(t) + (j+1)Q_{j+1}(t)$$

$$(j = 1, 2, \cdots, N-1)$$

$$B_{CN}(t) = P_N(t) - (N-1)Q_N(t)$$

$$B_{Sj}(t) = p_j(t) - p_{j+1}(t) - (j-1)q_j(t) + (j+1)q_{j+1}(t)$$

$$(j = 1, 2, \cdots, N-1)$$

$$B_{SN}(t) = p_N(t) - (N-1)q_N(t) \tag{A6.3a-d}$$

where

$$P_j(t) = \frac{e^{-h\omega_1 t}}{2\omega_{1d}\Delta t}(F_{S1} \sin \omega_{1d}t + F_{C1} \cos \omega_{1d}t)$$

$$Q_j(t) = \frac{e^{-h\omega_1 t}}{2\omega_{1d}}(F_{S2} \sin \omega_{1d}t + F_{C2} \cos \omega_{1d}t)$$

$$p_j(t) = \frac{e^{-h\omega_1 t}}{2\omega_{1d}\Delta t}(F_{S3} \sin \omega_{1d}t + F_{C3} \cos \omega_{1d}t)$$

$$q_j(t) = \frac{e^{-h\omega_1 t}}{2\omega_{1d}}(F_{S4} \sin \omega_{1d}t + F_{C4} \cos \omega_{1d}t) \qquad (A6.4a-d)$$

$F_{S1}, \ldots, F_{C4}$ in Eqs. (A6.4a–d) may be expressed by

$$F_{S1} = \frac{1}{r}\left[2\omega_{1d}\left\{j\Delta t D_{Sj} - (j-1)\Delta t D_{Sj-1}\right\} + (1 + jh\omega_1\Delta t)D_{Cj}\right.$$

$$- \{1 + (j-1)h\omega_1\Delta t\}D_{Cj-1}$$

$$\left. - \frac{2h\omega_1}{r}\left\{h\omega_1(D_{Cj} - D_{Cj-1}) + 2\omega_{1d}(D_{Sj} - D_{Sj-1})\right\}\right]$$

$$+ \frac{1}{(h\omega_1)^2}\left[(jh\omega_1\Delta t - 1)e^{jh\omega_1\Delta t} - \{(j-1)h\omega_1\Delta t - 1\}e^{(j-1)h\omega_1\Delta t}\right]$$

$$F_{C1} = \frac{1}{r}\left[2\omega_{1d}\left\{j\Delta t D_{Cj} - (j-1)\Delta t D_{Cj-1}\right\} - (1 + jh\omega_1\Delta t)D_{Sj}\right.$$

$$+ \{1 + (j-1)h\omega_1\Delta t\}D_{Sj-1} + \frac{2h\omega_1}{r}\left\{h\omega_1(D_{Sj} - D_{Sj-1})\right.$$

$$\left.\left. - 2\omega_{1d}(D_{Cj} - D_{Cj-1})\right\}\right]$$

$$F_{S2} = \frac{1}{r}\left\{2\omega_{1d}(D_{Sj} - D_{Sj-1}) + h\omega_1(D_{Cj} - D_{Cj-1})\right\}$$

$$+ \frac{1}{h\omega_1}\left\{e^{jh\omega_1\Delta t} - e^{(j-1)h\omega_1\Delta t}\right\}$$

$$F_{C2} = \frac{1}{r}\{2\omega_{1d}(D_{Cj} - D_{Cj-1}) - h\omega_1(D_{Sj} - D_{Sj-1})\}$$

$$F_{S3} = \frac{1}{r}\left[-2\omega_{1d}\Big\{ j\Delta t D_{Cj} - (j-1)\Delta t D_{Cj-1} \Big\} + (1 + jh\omega_1\Delta t)D_{Sj} \right.$$

$$- \{1 + (j-1)h\omega_1\Delta t\}D_{Sj-1}$$

$$\left. - \frac{2h\omega_1}{r}\Big\{ h\omega_1(D_{Sj} - D_{Sj-1}) - 2\omega_{1d}(D_{Cj} - D_{Cj-1}) \Big\} \right]$$

$$F_{C3} = \frac{1}{r}\left[2\omega_{1d}\Big\{ j\Delta t D_{Sj} - (j-1)\Delta t D_{Sj-1} \Big\} + (1 + jh\omega_1\Delta t)D_{Cj} \right.$$

$$- \{1 + (j-1)h\omega_1\Delta t\}D_{Cj-1} - \frac{2h\omega_1}{r}\Big\{ h\omega_1(D_{Cj} - D_{Cj-1})$$

$$\left. + 2\omega_{1d}(D_{Sj} - D_{Sj-1}) \Big\} \right]$$

$$- \frac{1}{(h\omega_1)^2}\left[\Big(jh\omega_1\Delta t - 1\Big)e^{jh\omega_1\Delta t} - \{(j-1)h\omega_1\Delta t - 1\}e^{(j-1)h\omega_1\Delta t} \right]$$

$$F_{S4} = \frac{1}{r}\Big\{ -2\omega_{1d}(D_{Cj} - D_{Cj-1}) + h\omega_1(D_{Sj} - D_{Sj-1}) \Big\}$$

$$F_{C4} = \frac{1}{r}\Big\{ 2\omega_{1d}(D_{Sj} - D_{Sj-1}) + h\omega_1(D_{Cj} - D_{Cj-1}) \Big\}$$

(A6.5a–h)

$$- \frac{1}{h\omega_1}\Big\{ e^{jh\omega_1\Delta t} - e^{(j-1)h\omega_1\Delta t} \Big\}$$

D_{Sj}, D_{Cj} and r are given by

$$D_{Sj} = e^{jh\omega_1\Delta t}\sin(2j\omega_{1d}\Delta t), \quad D_{Cj} = e^{jh\omega_1\Delta t}\cos(2j\omega_{1d}\Delta t)$$

$$r = (h\omega_1)^2 + (2\omega_{1d})^2$$

(A6.6a–c)

APPENDIX 6.4 UPPER BOUND FOR CONTINUOUS FUNCTIONS C(t)

Application of the Cauchy-Schwarz inequality to Eq. (6.9) leads to

$$\sigma_D(t)^2 = \int\limits_{-\infty}^{\infty} \{A_C(t;\omega)^2 + A_S(t;\omega)^2\} S_w(\omega) d\omega$$

$$= \int\limits_{-\infty}^{\infty} \left[\left(\int\limits_0^t g(t-\tau)\{c(\tau) \cos \omega\tau\} d\tau \right)^2 \right.$$

$$+ \left. \left(\int\limits_0^t g(t-\tau)\{c(\tau) \sin \omega\tau\} d\tau \right)^2 \right] S_w(\omega) d\omega \qquad \text{(A6.7)}$$

$$\leq \int\limits_{-\infty}^{\infty} \left[\left(\int\limits_0^t g(t-\tau)^2 d\tau \int\limits_0^t \{c(\tau) \cos \omega\tau\}^2 d\tau \right) \right.$$

$$+ \left. \left(\int\limits_0^t g(t-\tau)^2 d\tau \int\limits_0^t \{c(\tau) \sin \omega\tau\}^2 d\tau \right) \right] S_w(\omega) d\omega$$

Substitution of the identity $\cos^2 \omega\tau + \sin^2 \omega\tau = 1$ into inequality (A6.7) provides

$$\sigma_D(t)^2 \leq \int\limits_{-\infty}^{\infty} \left[\left(\int\limits_0^t g(t-\tau)^2 d\tau \int\limits_0^t \{c(\tau) \cos \omega\tau\}^2 d\tau \right) \right.$$

$$+ \left. \left(\int\limits_0^t g(t-\tau)^2 d\tau \int\limits_0^t \{c(\tau) \sin \omega\tau\}^2 d\tau \right) \right] S_w(\omega) d\omega \qquad \text{(A6.8)}$$

$$= \left(\int\limits_0^t g(t-\tau)^2 d\tau \right) \left(\int\limits_0^t c(\tau)^2 d\tau \right) \left(\int\limits_{-\infty}^{\infty} S_w(\omega) d\omega \right)$$

Application of the constraint (6.8) and the relation (6.2) to inequality (A6.8) yields the following upper bound for continuous functions $c(t)$.

$$\sigma_D(t)^2 \leq \left(\int\limits_0^t g(t-\tau)^2 d\tau \right) \overline{C} \qquad \text{(A6.9)}$$

The expression $\int_0^t g(t-\tau)^2 d\tau$ in (A6.9) can be expressed by

$$
\int_0^t g(t-\tau)^2 d\tau = \frac{1}{2\omega_{1d}^2}\left[\frac{1}{2h\omega_1}\left(1-e^{-2h\omega_1 t}\right)\right.
$$

$$
\left.-\frac{h}{2\omega_1}\left\{1+e^{-2h\omega_1 t}\left(\frac{\sqrt{1-h^2}}{h}\sin 2\omega_{1d}t - \cos 2\omega_{1d}t\right)\right\}\right]
$$

(A6.10)

REFERENCES

Drenick, R.F., 1970. Model-free design of aseismic structures. J. Engng. Mech. Div. 96 (4), 483–493.

Drenick, R.F., 1977. The critical excitation of nonlinear systems. J. Applied Mech. 44 (2), 333–336.

Housner, G.W., Jennings, P.C., 1977. The capacity of extreme earthquake motions to damage structures. In: Hall, W.J. (Ed.), Structural and Geotechnical Mechanics, pp. 102–116. Prentice-Hall Englewood Cliff, NJ.

Pirasteh, A.A., Cherry, J.L., Balling, R.J., 1988. The use of optimization to construct critical accelerograms for given structures and sites. Earthq. Engng. Struct. Dyn. 16 (4), 597–613.

Shinozuka, M., 1970. Maximum structural response to seismic excitations. J. Engng. Mech. Div. 96 (5), 729–738.

Shinozuka, M., Sato, Y., 1965. Simulation of non-stationary random process. J. Eng. Mech. Div. 91 (EM4), 11–1422.

Srinivasan, M., Ellingwood, B., Corotis, R., 1991. Critical base excitations of structural systems. J. Eng. Mech. 117 (6), 1403–1422.

Takewaki, I., 2000. Optimal damper placement for critical excitation. Probabilistic Engineering Mechanics 15 (4), 317–325.

Takewaki, I., 2001a. A new method for non-stationary random critical excitation. Earthq. Engng. Struct. Dyn. 30 (4), 519–535.

Takewaki, I., 2001b. Probabilistic critical excitation for MDOF elastic–plastic structures on compliant ground. Earthq. Engng. Struct. Dyn. 30 (9), 1345–1360.

Vaderplaats, G.N., 1984. Numerical Optimization Techniques for Engineering Design with Applications. McGraw-Hill Book Company, New York.

CHAPTER SEVEN

Robust Stiffness Design for Structure-Dependent Critical Excitation

Contents

7.1. INTRODUCTION

The purpose of this chapter is to explain a new robust stiffness design method for building structures taking into account the dependence of critical excitations on the design of building structures. A global stiffness parameter is maximized (a global flexibility parameter is minimized actually) under the constraints on excitations (power and intensity) and on total structural cost. This design problem includes a min–max procedure (maximization of structural flexibility with respect to excitation and minimization of structural flexibility with respect to story stiffnesses). It is shown that the technique for a fixed design explained in Chapter 2 (Takewaki 2001) can be used effectively in the maximization procedure and this leads to extreme simplification in an accurate manner of the present complicated, highly nonlinear problem. The elastic-plastic responses of the designed building structures to a broader class of excitations and to code-specified design earthquakes are clarified through a statistical equivalent linearization technique.

7.2. PROBLEM FOR FIXED DESIGN

A probabilistic critical excitation method has been developed by Takewaki (2001) for building structures with fixed mechanical properties and member sizes. For the formulation in the following sections, a summary of the method is described.

Consider an n-story shear building model subjected to a horizontal acceleration $\ddot{u}_g(t)$. This input is characterized by a stationary Gaussian process with zero mean. Let $S_g(\omega)$ denote the power spectral density (PSD) function of the ground acceleration $\ddot{u}_g(t)$. The system mass, damping and stiffness matrices of the building model and the influence coefficient vector by the base input are denoted by $\mathbf{M}, \mathbf{C}, \mathbf{K}, \mathbf{r}$, respectively. The equations of motion of the model in the frequency domain can be expressed as

$$(-\omega^2 \mathbf{M} + i\omega \mathbf{C} + \mathbf{K})\mathbf{U}(\omega) = -\mathbf{M}\mathbf{r}\ddot{U}_g(\omega) \tag{7.1}$$

In Eq. (7.1), i denotes the imaginary unit. $\mathbf{U}(\omega)$ and $\ddot{U}_g(\omega)$ are the Fourier transforms of the horizontal floor displacement vector $\mathbf{u}(t)$ relative to the ground and $\ddot{u}_g(t)$, respectively. Eq. (7.1) can also be expressed as

$$\mathbf{A}\mathbf{U}(\omega) = \mathbf{B}\ddot{U}_g(\omega) \tag{7.2}$$

where $\mathbf{A} = -\omega^2 \mathbf{M} + i\omega \mathbf{C} + \mathbf{K}$ and $\mathbf{B} = -\mathbf{M}\mathbf{r}$.

Let $\delta_i(t)$ denote the interstory drift in the i-th story. The Fourier transform $\mathbf{D}(\omega)$ of $\{\delta_i(t)\}$ can be related to $\mathbf{U}(\omega)$ by

$$\mathbf{D}(\omega) = \mathbf{T}\mathbf{U}(\omega) \tag{7.3}$$

$\mathbf{T}$ consists of 1, -1 and 0. Substitution of Eq. (7.2) into Eq. (7.3) provides

$$\mathbf{D}(\omega) = \mathbf{T}\mathbf{A}^{-1}\mathbf{B}\ddot{U}_g(\omega) \tag{7.4}$$

Rewrite Eq. (7.4) as

$$\mathbf{D}(\omega) = \mathbf{H}_D(\omega)\ddot{U}_g(\omega) \tag{7.5}$$

$\mathbf{H}_D(\omega) = \{H_{D_i}(\omega)\} = \mathbf{TA}^{-1}\mathbf{B}$ is the deformation transfer function. The mean-square interstory drift in the i-th story may be described as

$$\sigma_{D_i}^{2} = \int_{-\infty}^{\infty} |H_{D_i}(\omega)|^2 S_g(\omega)\mathrm{d}\omega = \int_{-\infty}^{\infty} H_{D_i}(\omega)H_{D_i}^{*}(\omega)S_g(\omega)\mathrm{d}\omega \quad (7.6)$$

where $()^*$ denotes a complex conjugate.

The sum of the mean-square interstory drifts can be regarded as "an inverse of the global stiffness" (global flexibility) of the shear building model and is described as

$$f = \sum_{i=1}^{n} \sigma_{D_i}^{2} = \int_{-\infty}^{\infty} F(\omega)S_g(\omega)\mathrm{d}\omega \quad (7.7)$$

The function $F(\omega)$ (called F-function) in Eq. (7.7) is defined by

$$F(\omega) = \sum_{i=1}^{n} |H_{D_i}(\omega)|^2 \quad (7.8)$$

The critical excitation problem for a fixed design may be described as:

[Problem for Fixed Design]

Find the critical PSD function $\tilde{S}_g(\omega)$ to maximize f defined by Eq. (7.7)

$$\text{subject to the constraint on input power} \int_{-\infty}^{\infty} S_g(\omega)d\omega \leq \overline{S} \quad (7.9)$$

$$\text{and to the constraint on intensity} \sup S_g(\omega) \leq \bar{s} \quad (7.10)$$

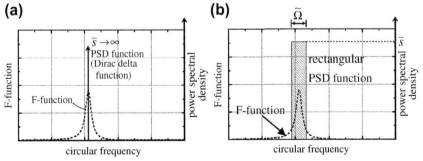

Figure 7.1 *Power spectral density function of critical excitation: (a) critical excitation for infinite power spectral density amplitude $\bar{s}$ and (b) critical excitation for finite power spectral density amplitude $\bar{s}$.*

The PSD function derived as the solution to this problem can be found to be a Dirac delta function for the infinite PSD amplitude or a rectangular shape for the finite PSD amplitude (Takewaki 2001). This fact is described in Fig. 7.1.

7.3. PROBLEM FOR STRUCTURE-DEPENDENT CRITICAL EXCITATION

Consider a critical excitation problem and a structural design problem for maximum stiffness simultaneously. The constraints on design earthquakes are the input power, i.e. integration of the PSD function in the frequency domain, and the intensity, i.e. the maximum value of the PSD function. Let k_i and $\mathbf{k} = \{k_i\}$ denote the story stiffness in the i-th story and the set of those story stiffnesses. The constraint for structures is on the sum of story stiffnesses. The sum of story stiffnesses may be regarded approximately as a measure of the total quantity or weight of structural members. The problem considered in this chapter may be stated as follows:

[Problem of Stiffness Design for Critical Excitation]

Find the optimal set $\tilde{\mathbf{k}}$ of stiffnesses and the corresponding critical PSD function $\tilde{S}_g(\omega)$

such that the procedure $\min_{\mathbf{k}} \max_{S_g(\omega)} f(\mathbf{k}; S_g(\omega))$ is achieved $\qquad$ (7.11)

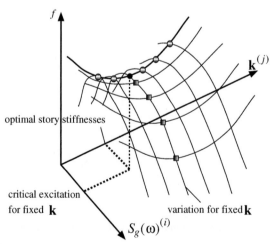

Figure 7.2 *Schematic diagram of performance upgrading via concept of system-dependent critical excitation.*

$$\text{subject to the constraint on input power} \quad \int_{-\infty}^{\infty} S_g(\omega)\,d\omega \leq \overline{S} \quad (7.12)$$

$$\text{to the constraint on input intensity sup. } S_g(\omega) \leq \overline{s} \quad (7.13)$$

$$\text{to the constraint on total structural quantity (cost)} \sum_{i=1}^{n} k_i = \overline{K} \quad (7.14)$$

$$\text{and to the positivity of story stiffnesses } k_i > 0 \; (i = 1, \cdots, n) \quad (7.15)$$

Fig. 7.2 shows a schematic diagram of the present problem. The solution of the min-max problem corresponds to a saddle point.

7.4. SOLUTION PROCEDURE

The procedure for solving the problem in the previous section is explained in this section. Assume that the system damping matrix $\mathbf{C}$ is not a function of story stiffnesses $\mathbf{k}$. The critical excitation characterized by its PSD function $S_g(\omega)$ for a fixed set $\mathbf{k}$ of stiffnesses can be found by using the method described above (Takewaki 2001). This fact can be used effectively in the maximization in the procedure (7.11).

$\tilde{\Omega}$ denotes the frequency bandwidth in the frequency range of the critical rectangular PSD function $S_g(\omega)$, i.e. $\overline{S}/(2\overline{s}) = \tilde{\Omega}$. Assume that the upper and lower frequency limits of the rectangular PSD function can be expressed by

$$\omega_U = \omega_1 + \frac{1}{2}\tilde{\Omega} \quad (7.16a)$$

$$\omega_L = \omega_1 - \frac{1}{2}\tilde{\Omega} \quad (7.16b)$$

In Eq. (7.16), ω_1 is the undamped fundamental natural circular frequency of the shear building model. It should be remembered that ω_1 is a function of $\mathbf{k}$ and the critical rectangular PSD function $S_g(\omega)$ is a function of $\mathbf{k}$. The objective function (7.7) for the critical rectangular PSD function can then be approximated by

$$\hat{f} = \int_{-\infty}^{\infty} F(\omega;\mathbf{k})S_g(\omega)\,d\omega$$

$$\cong 2\overline{s} \int_{\omega_L}^{\omega_U} F(\omega;\mathbf{k})\,d\omega \quad (7.17)$$

$$= 2\overline{s}\{\Phi(\omega_U;\mathbf{k}) - \Phi(\omega_L;\mathbf{k})\}$$

The function $\Phi(\overline{\omega}; \mathbf{k})$ denotes the following function.

$$\Phi(\overline{\omega}; \mathbf{k}) = \int_0^{\overline{\omega}} F(\omega; \mathbf{k}) d\omega \qquad (7.18)$$

In Eq. (7.17), a nearly symmetric property of the function $F(\omega; \mathbf{k})$ has been utilized. This insightful approximate manipulation enables an analytical treatment of the present complicated strongly nonlinear problem.

The minimization of the objective function f with respect to $\mathbf{k}$ in the procedure (7.11) will be considered next. The following Lagrangian is defined in terms of a Lagrange multiplier λ.

$$\begin{aligned}
L &= \hat{f} + \lambda \left(\sum_{i=1}^n k_i - \overline{K} \right) \\
&= 2\overline{s}\{\Phi(\omega_U; \mathbf{k}) - \Phi(\omega_L; \mathbf{k})\} + \lambda \left(\sum_{i=1}^n k_i - \overline{K} \right)
\end{aligned} \qquad (7.19)$$

The expression (7.17) has been used in Eq. (7.19). The stationarity condition of the Lagrangian defined in Eq. (7.19) with respect to story stiffnesses $\mathbf{k}$ may be written by

$$\frac{\partial L}{\partial k_i} = \hat{f}_{,i} + \lambda$$

$$= 2\overline{s} \left[\frac{\partial \omega_1}{\partial k_i} \{F(\omega_U; \mathbf{k}) - F(\omega_L; \mathbf{k})\} + \int_{\omega_L}^{\omega_U} \frac{\partial_2 F(\omega; \mathbf{k})}{\partial k_i} d\omega \right] + \lambda = 0$$

$$(7.20)$$

where $\hat{f}_{,i} \equiv \partial \hat{f}/\partial k_i$. Eq. (7.20) represents the optimality condition to be satisfied in optimization. For exact differentiation, the following expressions are used in this chapter.

$$\frac{\partial_1 F}{\partial x} \equiv \frac{\partial F(x, y)}{\partial x}, \frac{\partial_2 F}{\partial y} \equiv \frac{\partial F(x, y)}{\partial y} \qquad (7.21)$$

It is the case that x is an implicit function of y. However, Eq. (7.21) implies that $\partial_2 F/\partial y$ indicates the partial differentiation only for the terms including y explicitly.

The first-order sensitivity of the undamped fundamental natural circular frequency ω_1 in Eq. (7.20) with respect to story stiffnesses $\mathbf{k}$ can be obtained from $\omega_1 = \phi^T \mathbf{K}_{,i} \phi/(2\omega_1)$ where ϕ is the lowest eigenmode. The sensitivity $\partial_2 F/\partial k_i$ in Eq. (7.20) can be computed by

$$\frac{\partial_2 F}{\partial k_i} = \sum_l H_{D_l}(\omega), {}_i H_{D_l}{}^*(\omega) + \sum_l H_{D_l}(\omega) H_{D_l}{}^*(\omega)_{,i} \qquad (7.22)$$

$H_{D_i}(\omega) = \mathbf{T}_i \mathbf{A}^{-1} \mathbf{B}$ is the deformation transfer function and the well-known expression $\mathbf{A}^{-1}{}_{,i} = -\mathbf{A}^{-1} \mathbf{A}_{,i} \mathbf{A}^{-1}$ can be used for computing design sensitivities of transfer functions. The vector $\mathbf{T}_i$ denotes the i-th row of the matrix $\mathbf{T}$. It should be remarked that operations of complex conjugate and partial differentiation are exchangeable.

Next, define the following quantity for a simpler expression.

$$\alpha_j = \hat{f}_{,j+1}/\hat{f}_{,1} \qquad (7.23)$$

It is noted that the optimality condition described by Eq. (7.20) is equivalent to

$$\alpha_j = 1 \ (j = 1, \cdots, n-1) \qquad (7.24)$$

The linear increment of α_j due to the story stiffness change $\Delta \mathbf{k}$ may be expressed by

$$\Delta \alpha_j = \left(\frac{1}{\hat{f}_{,1}} \frac{\partial \hat{f}_{,j+1}}{\partial \mathbf{k}} - \frac{\hat{f}_{,j+1}}{\hat{f}_{,1}{}^2} \frac{\partial \hat{f}_{,1}}{\partial \mathbf{k}} \right) \Delta \mathbf{k} = \frac{1}{\hat{f}_{,1}} \left(\frac{\partial \hat{f}_{,j+1}}{\partial \mathbf{k}} - \frac{\partial \hat{f}_{,1}}{\partial \mathbf{k}} \alpha_j \right) \Delta \mathbf{k}$$
$$(7.25)$$

The constraint expressed by Eq. (7.14) may be reduced to

$$\sum_{j=1}^{n} \Delta k_j = 0 \qquad (7.26)$$

In Eq. (7.26), Δk_j denotes a small increment of the j-th story stiffness and $\Delta \mathbf{k} = \{\Delta k_j\}$. The combination of Eq. (7.25) with Eq. (7.26) yields the following simultaneous linear equation for $\Delta \mathbf{k} = \{\Delta k_j\}$.

$$
\begin{bmatrix}
\dfrac{1}{\hat{f}_{,1}}\left(\dfrac{\partial \hat{f}_{,2}}{\partial \mathbf{k}} - \dfrac{\partial \hat{f}_{,1}}{\partial \mathbf{k}}\alpha_1\right) \\
\vdots \\
\dfrac{1}{\hat{f}_{,1}}\left(\dfrac{\partial \hat{f}_{,n}}{\partial \mathbf{k}} - \dfrac{\partial \hat{f}_{,1}}{\partial \mathbf{k}}\alpha_{n-1}\right) \\
1\cdots 1
\end{bmatrix}
\Delta \mathbf{k} =
\left\{
\begin{array}{c}
\Delta \alpha_1 \\
\vdots \\
\Delta \alpha_{n-1} \\
0
\end{array}
\right\}
\tag{7.27}
$$

Assume that the increment of α_j is defined by $\{\Delta\alpha_j\} = \{(\alpha_{Fj} - \alpha_{0j})/N\}$ where $\{\alpha_{Fj}\} = \{1\cdots1\}^T$ (see Eq. (7.24)) and the quantities for the initial design are used for the specification of $\{\alpha_{0j}\}$. Sequential application of Eq. (7.27) to each updated design $\mathbf{k}(\mathbf{k}_{old} + \Delta\mathbf{k} \to \mathbf{k}_{new})$ provides the final design satisfying the optimality conditions (7.24).

The second-order design sensitivity of the objective function in the left-hand side of Eq. (7.27) can be computed by

$$
\begin{aligned}
\frac{\partial^2 \hat{f}}{\partial k_i \partial k_j} = 2\bar{s}\Bigg[&\frac{\partial^2 \omega_1}{\partial k_i \partial k_j}\Big\{F(\omega_U;\mathbf{k}) - F(\omega_L;\mathbf{k})\Big\} \\
&+ \frac{\partial \omega_1}{\partial k_i}\left\{\frac{\partial_1 F(\omega_U;\mathbf{k})}{\partial \omega_1}\frac{\partial \omega_1}{\partial k_j} + \frac{\partial_2 F(\omega_U;\mathbf{k})}{\partial k_j}\right. \\
&\qquad\left. - \frac{\partial_1 F(\omega_L;\mathbf{k})}{\partial \omega_1}\frac{\partial \omega_1}{\partial k_j} - \frac{\partial_2 F(\omega_L;\mathbf{k})}{\partial k_j}\right\} \\
&+ \frac{\partial \omega_1}{\partial k_j}\left\{\frac{\partial_2 F(\omega_U;\mathbf{k})}{\partial k_i} - \frac{\partial_2 F(\omega_L;\mathbf{k})}{\partial k_i}\right\} + \int_{\omega_L}^{\omega_U}\frac{\partial_2^2 F(\omega;\mathbf{k})}{\partial k_i \partial k_j}\,d\omega\Bigg]
\end{aligned}
\tag{7.28}
$$

It is noted that the second-order sensitivity of the undamped fundamental natural circular frequency in Eq. (7.28) with respect to story stiffnesses can be obtained from

$$
\begin{aligned}
\omega_{1,ij} &= \frac{1}{2\omega_1}\left(\boldsymbol{\phi}_{,j}^T \mathbf{K}_{,i}\boldsymbol{\phi} + \boldsymbol{\phi}^T \mathbf{K}_{,i}\boldsymbol{\phi}_{,j}\right) - \frac{\omega_{1,i}\omega_{1,j}}{\omega_1} \\
&= \frac{1}{\omega_1}\boldsymbol{\phi}_{,j}^T \mathbf{K}_{,i}\boldsymbol{\phi} - \frac{\omega_{1,i}\omega_{1,j}}{\omega_1}
\end{aligned}
\tag{7.29}
$$

In addition, the following relation is used.

$$\partial \mathbf{A}(\omega_U(\omega_1))/\partial \omega_1 = -2\omega_U \mathbf{M} + i\mathbf{C} \tag{7.30}$$

The second-order design sensitivity of the function F in Eq. (7.28) can be computed by

$$\frac{\partial_2^2 F(\omega; \mathbf{k})}{\partial k_i \partial k_j} = \sum_l H_{D_l}(\omega)_{,ij} H_{D_l}{}^*(\omega) + \sum_l H_{D_l}(\omega)_{,i} H_{D_l}{}^*(\omega)_{,j}$$
$$+ \sum_l H_{D_l}(\omega)_{,j} H_{D_l}{}^*(\omega)_{,i} + \sum_l H_{D_l}(\omega) H_{D_l}{}^*(\omega)_{,ij} \tag{7.31}$$

The sensitivity expression $\mathbf{A}^{-1}{}_{,ij} = \mathbf{A}^{-1}\mathbf{A}_{,j}\mathbf{A}^{-1}\mathbf{A}_{,i}\mathbf{A}^{-1} + \mathbf{A}^{-1}\mathbf{A}_{,i}\mathbf{A}_{,j}\mathbf{A}^{-1}$ is used in the computation of the second-order design sensitivity of the transfer function. Eq. (7.31) has to be substituted into Eq. (7.28) and numerical integration must be performed finally.

7.5. NUMERICAL DESIGN EXAMPLES

Design examples are shown for 6-story and 12-story models in order to demonstrate the validity of the method described in the previous section. For comparison, a design with a straight-line lowest-mode, i.e. uniform lowest-mode interstory drift components, is taken into account. This design is called the "frequency constraint (FC) design" because this design is an optimal design under a fundamental natural frequency constraint (Nakamura and Yamane 1986). On the other hand, the design obtained by the method described in the previous section is called the "maximum performance (MP) design." This design seeks the maximization of global stiffness performance, in other words, the minimization of global flexibility.

The power of the PSD function is specified as $\overline{S} = 0.553 (\mathrm{m}^2/\mathrm{s}^4)$ and the amplitude of the PSD function has been given by $\bar{s} = 0.0661 (\mathrm{m}^2/\mathrm{s}^3)$. These parameters are equivalent to those for the ground motion of El Centro NS (1940) (duration = 40(s)). In nature, earthquake ground motions are nonstationary random processes and the present treatment of time-averaging is approximate. However, reasonable application may be possible so long as the range of applicability is clearly indicated (Lai 1982). See, (for example, Lai 1982, Nigam 1983 and Vanmarcke 1983) for the time-averaged approximate PSD functions.

Assume that the floor masses have the same value in all the stories and are given by $m_i = 32.0 \times 10^3 (\mathrm{kg})$. The damping ratio is given by 0.05 (stiffness proportional). In the design method described in the previous section, the FC design has been used as the initial design. The total sum of story stiffnesses is specified as $\overline{K} = 0.319 \times 10^9 (\text{N/m})$ for 6-story models and $\overline{K} = 0.570 \times 10^9 (\text{N/m})$ for 12-story models. Note that these values were computed from the FC design with the fundamental natural period $T_1 = 0.600(\text{s})$ for 6-story models and with $T_1 = 1.20(\text{s})$ for 12-story models. The total step number for incremental computation is $N = 100$ and the numerical integration with respect to frequency has been implemented with 100 frequency intervals.

Fig. 7.3 illustrates the story stiffness distributions of the FC design and the MP design of 6-, 12- and 18-story models. It can be understood that the story

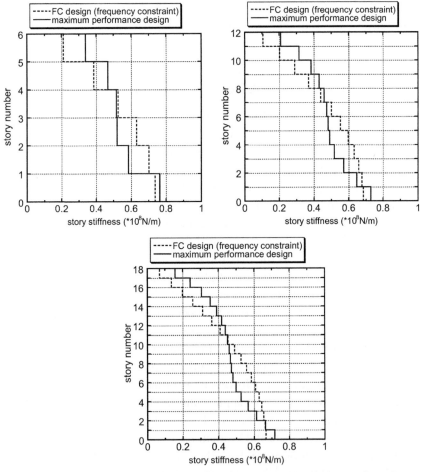

Figure 7.3 *Story stiffness distributions (frequency constraint design and maximum performance design).*

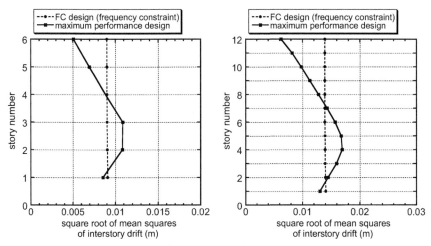

Figure 7.4 *RMS distributions of interstory drifts (frequency constraint design and maximum performance design).*

stiffnesses in the lower stories of the FC design were moved to the upper stories in the MP design. This is because the reduction in the interstory drifts in the upper stories is cost-effective for the upgrading of global stiffness performance. It was found that the fundamental natural period of the optimal design (MP design) for the 6-story model is $T_1 = 0.612(\text{s})$ and that for the 12-story model is $T_1 = 1.22(\text{s})$. Then the objective function is $f = 0.4901 \times 10^{-3}(\text{m}^2)$ for the initial design (FC design) and is $f = 0.4615 \times 10^{-3}(\text{m}^2)$ for the optimal

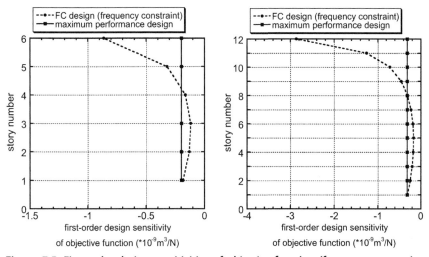

Figure 7.5 *First-order design sensitivities of objective function (frequency constraint design and maximum performance design).*

design (MP design) in 6-story models. In addition, the objective function found for 12-story models is $f = 0.2327 \times 10^{-2} (\text{m}^2)$ for the initial design (FC design) and is $f = 0.2145 \times 10^{-2} (\text{m}^2)$ for the optimal design (MP design).

Fig. 7.4 shows the square roots of the mean–square interstory drifts (RMS values) for the FC design and the MP design of 6- and 12-story models subjected to the critical excitation. It can be observed that the RMS values of the FC design are nearly uniform. Note that this distribution is obtained for the critical excitation and more detailed investigation is necessary for robust design. Fig. 7.5 illustrates the first-order design sensitivities of the objective functions for both FC design and MP design. It can be seen that the optimality conditions (7.20) are satisfied accurately in the MP design.

7.6. RESPONSE TO A BROADER CLASS OF EXCITATIONS

In view of structural robustness, it may be meaningful to investigate the ultimate and limit state in the plastic range of the building structure under a broader class of disturbance. The normal trilinear hysteretic restoring-force characteristic as shown in Fig. 7.6 is used for the story restoring-force characteristic. The maximum elastic–plastic responses under random excitations have been evaluated by the statistical equivalent linearization technique (see Appendix). It is well known that nonstationary characteristics of excitations and responses are very important in the analysis of the energy input into structures and the damage of structures. However, it is also well understood that, as far as only the maximum displacement is concerned, the stationary treatment leads to sufficient accuracy. For this reason, stationary treatment is

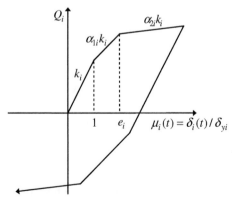

Figure 7.6 Normal trilinear hysteretic restoring-force characteristics in i-th story.

employed here. To evaluate the mean peak story ductility factors, the so-called peak factors have been multiplied on the RMS interstory drifts. The power and the amplitude of the PSD function have been specified as $\overline{S} = 0.553 \times 2.0 (\mathrm{m}^2/\mathrm{s}^4)$ and $\overline{s} = 0.0661 \times 2.0 (\mathrm{m}^2/\mathrm{s}^3)$. The excitation duration is specified by 40(s). The yield interstory drifts have been specified by $\delta_{y_i} = 0.01 (\mathrm{m})$ in all stories for 6-story models and by $\delta_{y_i} = 0.02 (\mathrm{m})$ in all stories for 12-story models. The ratio of the second yield-point interstory drift to the yield interstory drift, the ratio of the second-branch stiffness to the initial elastic stiffness and the ratio of the third-branch stiffness to the initial elastic stiffness have been specified as $e_i = 2.0$, $\alpha_{1i} = 0.5, \alpha_{2i} = 0.05$, respectively in all stories. The fundamental natural circular frequency of the model in the elastic range is denoted by ω_1.

In order to generate various input motions, the central frequency ω_C of the rectangular PSD function has been changed so as to be $\omega_C/\omega_1 = 0.5, 0.6, \cdots, 1.3$. The rectangular PSD function has been chosen even for the elastic-plastic models in view of the criticality for elastic responses. Since the equivalent stiffness and damping coefficients are functions of excitations, this restriction seems to be too restrictive. However, more general and arbitrary treatment of excitations is difficult due to the nonlinear nature of structural systems and this issue should be discussed in more depth. The frequency bandwidth is given by $\tilde{\Omega} = 0.553/(0.0661 \times 2.0) (\mathrm{rad/s})$. Note that the rectangular PSD function is resonant to the natural period of the equivalent linear model in the range of $\omega_C/\omega_1 \leq 1.0$. Fig. 7.7 presents the mean peak

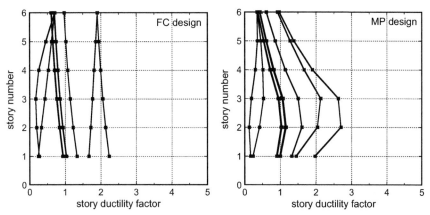

Figure 7.7 *Mean peak story ductility factors of 6-story models subjected to various excitations with different frequency ranges (frequency constraint design and maximum performance design).*

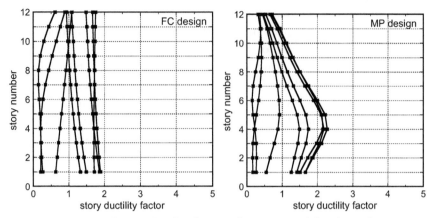

Figure 7.8 *Mean peak story ductility factors of 12-story models subjected to various excitations with different frequency ranges (frequency constraint design and maximum performance design).*

story ductility factors in 6-story models subjected to various excitations with the rectangular PSD function of different central frequencies $\omega_C/\omega_1 = 0.5, 0.6, \cdots, 1.3$ (FC design and MP design). The maximum responses have been obtained for the excitation with the parameter $\omega_C/\omega_1 = 1.0$ in 6-story models. It can be seen from Fig. 7.7 that, while the MP design exhibits larger mean peak story ductility factors in the lower stories, the FC design exhibits a fairly uniform distribution of the mean peak story ductility factors.

However, a slightly whipping (amplified) phenomenon in the upper stories can be seen in the FC design. A similar phenomenon can be seen in Fig. 7.8 for 12-story models. The maximum story ductility factors have been observed for the excitation with the parameter $\omega_C/\omega_1 = 0.8$ in 12-story models.

7.7. RESPONSE TO CODE-SPECIFIED DESIGN EARTHQUAKES

For investigating the response characteristics of FC and MP designs to code-specified design earthquakes, elastic–plastic responses under code-specified design earthquakes have been computed by use of the above-mentioned statistical equivalent linearization technique (see Appendix). The earthquake ground motions specified in the Japanese seismic resistant

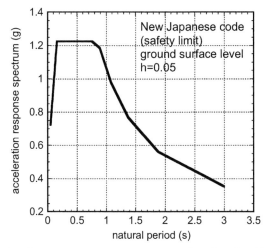

Figure 7.9 *Acceleration response spectrum at ground surface level by new Japanese seismic-resistant design code corresponding to safety limit (critical damping ratio = 0.05).*

design code revised in 1998 have been employed. The second-type soil condition has been assumed. The acceleration response spectrum for damping ratio = 0.05 at the ground surface corresponding to the safety limit has been computed as shown in Fig. 7.9. The power and intensity are given by $\int_{-\infty}^{\infty} S_g(\omega)d\omega = 2.84(\text{m}^2/\text{s}^4)$ and $\max S_g(\omega) = 0.0798(\text{m}^2/\text{s}^3)$. The portions in $T < 0.05(\text{s})$ and $T > 3.0(\text{s})$ have been neglected in the computation of the power. The duration of the excitation has been specified to be 40(s). Fig. 7.10 illustrates the mean peak story ductility factors under the design earthquakes. It can be seen that a slight whipping (amplified) phenomenon is observed in the FC design of 6-story models.

In order to investigate the accuracy of the statistical equivalent linearization, an elastic-plastic time-history response analysis has been conducted for 12-story models. The acceleration response spectrum shown in Fig. 7.9 has been transformed to the PSD function and a set of one hundred artificial ground motions has been generated. The line with marks in Fig. 7.10 shows the mean peak story ductility factors in the MP design. It can be observed that the present statistical equivalent linearization technique can predict the mean peak elastic-plastic responses within a reasonable accuracy.

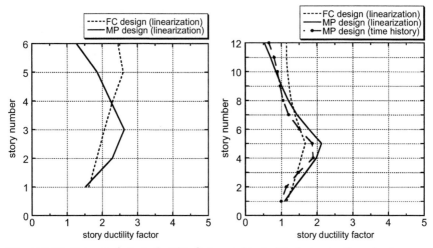

Figure 7.10 *Mean peak story ductility factors subjected to design earthquakes by new Japanese seismic code (safety limit) (frequency constraint design and maximum performance design).*

7.8. CONCLUSIONS

The conclusions may be stated as follows:

(1) A robust method of stiffness design can be developed. In this method, both a critical excitation problem depending on structural parameters and a stiffness design problem for better structural performances are taken into account simultaneously. A global and generalized stiffness parameter has been chosen as the objective function to be maximized. In another viewpoint, a global flexibility parameter has been minimized. The optimality conditions to this robust stiffness design problem have been derived.

(2) Effective approximation has been introduced in the computation of the first- and second-order design sensitivities of the objective function defined only for the critical excitation. An efficient numerical procedure has been devised to solve this robust stiffness design problem. It has been demonstrated that the procedure of finding the critical excitation for a fixed design can be extremely simplified by taking advantage of the procedure by the present author (Takewaki 2001).

(3) Design examples demonstrated that the present method can certainly upgrade the global stiffness performance in a robust way. Stochastic equivalent linearization analysis for the so-designed structures under various candidate design earthquakes has also clarified that the building structure designed by the present method exhibits a favorable elastic-plastic envelope response. A design with a straight-line lowest mode also exhibits a favorable elastic-plastic response to a broader class of disturbances except a slight whipping (amplified) phenomenon in the upper stories.

APPENDIX EVALUATION OF MAXIMUM ELASTIC-PLASTIC RESPONSES

The statistical equivalent linearization technique by Kobori and Minai (1967) for MDOF structures has been modified here for normal trilinear hysteretic restoring-force characteristics. Following the approach by Caughey (1960), the mean-square minimization technique has also been introduced in contrast to the simple mean procedure employed in the paper by Kobori and Minai (1967). The technique employs the assumption of the slowly varying characteristics of the amplitude and phase in a sinusoidal vibration. More advanced statistical equivalent linearization techniques have been developed so far and such techniques can be used if desired.

The parameters $\delta_{yi}, \Delta_{yi}, e_i = \Delta_{yi}/\delta_{yi}, \alpha_{1i}, \alpha_{2i}$ denote the yield interstory drift, the second yield-point interstory drift, its ratio to the yield interstory drift, the ratio of the second-branch stiffness to the initial elastic stiffness and the ratio of the third-branch stiffness to the initial elastic stiffness in the i-th story, respectively.

Consider the ratio (ductility factor) $\mu_i(t) = \delta_i(t)/\delta_{yi}$ of the interstory drift to its yield drift. The amplitude of $\mu_i(t)$ is denoted by z_i. The standard deviations of $\mu_i(t)$ and its time derivative $\dot{\mu}_i(t)$ are denoted by σ_{μ_i} and $\sigma_{\dot{\mu}_i}$, respectively. The parameters $\tilde{\omega}_i = \sigma_{\dot{\mu}_i}/\sigma_{\mu_i}, \overline{\omega}_i$ and $\omega_i = \tilde{\omega}_i/\overline{\omega}_i$ denote the mean circular frequency in the i-th story, the frequency for non-dimensionalization and the nondimensional frequency in the i-th story, respectively. Then the equivalent stiffness k_i^{eq} and equivalent damping coefficient c_i^{eq} are expressed by

$$k_i^{eq} = k_i^* k_i, \quad c_i^{eq} = c_i^* (k_i/\overline{\omega}_i) \qquad \text{(A7.1a, b)}$$

In Eq. (A7.1), nondimensional stiffness and damping coefficient k_i^* and c_i^* are described as

$$k_i^* = \frac{\int\limits_0^\infty z_i^2 k_{0i}(z_i) p(z_i, \sigma_{\mu_i}) dz_i}{\int\limits_0^\infty z_i^2 p(z_i, \sigma_{\mu_i}) dz_i} \qquad \text{(A7.2a)}$$

$$c_i^* = \frac{\int_0^\infty z_i^2 c_{0i}(z_i, \omega_i^*) p(z_i, \sigma_{\mu_i}) dz_i}{\int_0^\infty z_i^2 p(z_i, \sigma_{\mu_i}) dz_i} \qquad \text{(A7.2b)}$$

The quantities $k_{0i}(z_i)$ and $c_{0i}(z_i, \omega_i)$ defined by Eqs. (A7.2a, b) can be derived by minimizing the mean-squared errors of the restoring-force characteristic of the equivalent system from that of the original trilinear hysteretic system.

$$k_{0i}(z_i) = 1.0 \qquad (z_i \leq 1) \qquad \text{(A7.3a)}$$

$$k_{0i}(z_i) = \frac{1 - \alpha_{1i}}{\pi} \cos^{-1}\left(1 - \frac{2}{z_i}\right) + \alpha_{1i} - \frac{2}{\pi z_i^2}(1 - \alpha_{1i})(z_i - 2)\sqrt{z_i - 1}$$

$$(1 \leq z_i \leq e_i)$$

$$\text{(A7.3b)}$$

$$k_{0i}(z_i) = \frac{2}{\pi z_i^2}\left[\frac{z_i^2}{2}\left\{(1 - \alpha_{1i})\cos^{-1}\left(1 - \frac{2}{z_i}\right)\right.\right.$$

$$+ (\alpha_{1i} - \alpha_{2i})\cos^{-1}\left(1 - \frac{2e_i}{z_i}\right) + \alpha_{2i}\pi\Big\}$$

$$\left. - (1 - \alpha_{1i})(z_i - 2)\sqrt{z_i - 1} - (\alpha_{1i} - \alpha_{2i})(z_i - 2e_i)\sqrt{e_i(z_i - e_i)}\right]$$

$$(e_i \leq z_i) \qquad \text{(A7.3c)}$$

$$c_{0i}(z_i, \omega_i^*) = 0 \qquad (z_i \leq 1) \qquad \text{(A7.4a)}$$

$$c_{0i}(z_i, \omega_i^*) = \frac{4(1 - \alpha_{1i})(z_i - 1)}{\pi \omega_i^* z_i^2} \qquad (1 \leq z_i \leq e_i) \qquad \text{(A7.4b)}$$

$$c_{0i}(z_i, \omega_i^*) = \frac{4}{\pi \omega_i^* z_i^2}\{(1 - \alpha_{1i})(z_i - 1) + e_i(\alpha_{1i} - \alpha_{2i})(z_i - e_i)\} \qquad (e_i \leq z_i)$$

$$\text{(A7.4c)}$$

The quantity $p(z_i, \sigma_{\mu_i})$ in Eqs. (A7.2a, b) is the probability density function of the amplitude z_i and the following Rayleigh distribution is adopted.

$$p(z_i, \sigma_{\mu_i}) = (z_i / \sigma_{\mu i}^2)\exp\{-z_i^2 / (2\sigma_{\mu i}^2)\} \qquad \text{(A7.5)}$$

The standard deviations σ_{μ_i} and $\sigma_{\dot{\mu}_i}$ are derived by integrating the PSD function of the response. This response is expressed as the product of the squared value of the transfer function of the equivalent linear model and the PSD function $S_g(\omega)$ of the input excitation.

For computing the mean peak story ductility factors, the peak factors proposed by Der Kiureghian (1981) are introduced. These peak factors are determined from the first three moments of response PSD functions and the excitation duration.

REFERENCES

Caughey, T.K., 1960. Random excitation of a system with bilinear hysteresis. J. Applied Mechanics 27 (4), 649–652.

Der Kiureghian, A., 1981. A response spectrum method for random vibration analysis of MDF systems. Earthquake Engng. & Struct. Dyn. 9 (5), 419–435.

Kobori, T., Minai, R., 1967. Linearization technique for evaluating the elasto-plastic response of a structural system to non-stationary random excitations. Bull. Dis. Prev. Res. Inst., Kyoto University. 10 (A), 235–260 (in Japanese).

Lai, S.P., 1982. Statistical characterization of strong ground motions using power spectral density function. Bulletin of the Seismological Soc. of America 72 (1), 259–274.

Nakamura, T., Yamane, T., 1986. Optimum design and earthquake-response constrained design of elastic shear buildings. Earthquake Engng. & Struct. Dyn. 14 (5), 797–815.

Nigam, N.C., 1983. Introduction to Random Vibrations. The MIT Press, Cambridge. pp. 68–69.

Takewaki, I., 2001. A new method for non-stationary random critical excitation. Earthquake Engrg. and Struct. Dyn. 30 (4), 519–535.

Vanmarcke, E., 1983. Random Fields: Analysis and Synthesis. The MIT Press, Cambridge. pp. 339–340.

CHAPTER EIGHT

Critical Excitation for Earthquake Energy Input in SDOF System

Contents

8.1. INTRODUCTION

The purpose of this chapter is to explain a new general critical excitation method for a damped linear elastic single-degree-of-freedom (SDOF)
system as shown in Fig. 8.1. The input energy to the SDOF system during an
earthquake is introduced as a new measure of criticality. It is shown that the
formulation of the earthquake input energy in the frequency domain is
essential for solving the critical excitation problem and deriving a bound on
the earthquake input energy for a class of ground motions. The criticality is
expressed in terms of degree of concentration of input motion components
on the maximum portion of the characteristic function defining the earthquake input energy. It is remarkable that no mathematical programming
technique is required in the solution procedure. The constancy of earthquake input energy (Housner 1956, 1959) for various natural periods and
damping ratios is discussed from a new point of view based on an original
sophisticated mathematical treatment. It is shown that the constancy of
earthquake input energy is directly related to the uniformity of "the Fourier
amplitude spectrum" of ground motion acceleration, not the uniformity of
the velocity response spectrum. The bounds under acceleration and velocity

Critical Excitation Methods in Earthquake Engineering
© 2013 Elsevier Ltd.
All rights reserved.
175

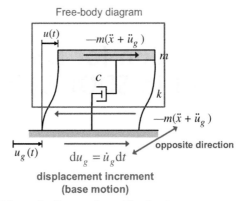

Figure 8.1 *Schematic diagram for evaluating input energy to SDOF model.*

constraints (time integral of the squared base acceleration and time integral of the squared base velocity) are clarified through numerical examinations for recorded ground motions to be meaningful in the short and intermediate/long natural period ranges, respectively.

Fig. 8.1 shows a schematic diagram for evaluating the input energy to an SDOF model. It can be observed that the restoring force equilibrating with the inertial force has an opposite direction to the base motion increment. In the last century, the power of ground motions was frequently discussed in building seismic resistant design. The time integral (acceleration power) of the squared ground acceleration and the time integral (velocity power) of the squared ground velocity may be two representative measures of power for damage potential. Fig. 8.2 illustrates the total input energy of Kobe University NS,

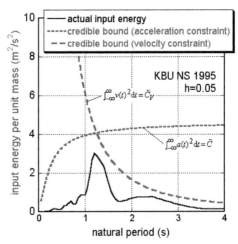

Figure 8.2 *Total input energy of Kobe University NS, Hyogoken-Nanbu 1995 with respect to natural period of structures and its upper bounds.*

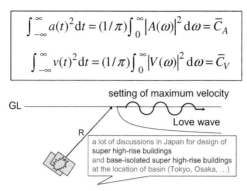

$$\int_{-\infty}^{\infty} a(t)^2 \, dt = (1/\pi) \int_0^{\infty} |A(\omega)|^2 \, d\omega = \overline{C}_A$$

$$\int_{-\infty}^{\infty} v(t)^2 \, dt = (1/\pi) \int_0^{\infty} |V(\omega)|^2 \, d\omega = \overline{C}_V$$

setting of maximum velocity

GL

Love wave

R

a lot of discussions in Japan for design of super high-rise buildings and base-isolated super high-rise buildings at the location of basin (Tokyo, Osaka, …)

Figure 8.3 *Acceleration power and velocity power for defining earthquake power.*

Hyogoken–Nanbu 1995 with respect to natural period of structures. In Fig. 8.2, the upper bound of the input energy for specified acceleration power and the upper bound of the input energy for specified velocity power are also shown (see Fig. 8.3 for the acceleration and velocity powers for defining the earthquake power). The derivation of these upper bounds will be made in this chapter.

8.2. EARTHQUAKE INPUT ENERGY TO SDOF SYSTEM IN THE FREQUENCY DOMAIN

A lot of work has been conducted on the topics of earthquake input energy. For example, (see Tanabashi 1935; Housner 1956, 1959; Berg and Thomaides 1960; Goel and Berg 1968; Housner and Jennings 1975; Kato and Akiyama 1975; Takizawa 1977; Mahin and Lin 1983; Zahrah and Hall 1984; Akiyama 1985; Ohi et al. 1985; Uang and Bertero 1990; Leger and Dussault 1992; Kuwamura et al. 1994; Fajfar and Vidic 1994; Ogawa et al. 2000; Riddell and Garcia 2001; Ordaz et al. 2003). Different from most of the previous works, the earthquake input energy is formulated here in the frequency domain (Page 1952; Lyon 1975; Takizawa 1977; Ohi et al. 1985; Ordaz et al. 2003) to facilitate the formulation of critical excitation methods and the derivation of bounds of the earthquake input energy.

Consider a damped, linear elastic SDOF system of mass m, stiffness k and damping coefficient c as shown in Fig. 8.1. Let $\Omega = \sqrt{k/m}$, $h = c/(2\Omega m)$ and x denote the undamped natural circular frequency, the damping ratio and the displacement of the mass relative to the ground. An over-dot indicates the time derivative. The input energy to the SDOF system by a horizontal ground acceleration $\ddot{u}_g(t) = a(t)$ from $t = 0$ to $t = t_0$ (end of input) can be defined by the work of the ground on the SDOF structural system and is expressed by

$$E_I = \int_0^{t_0} m(\ddot{u}_g + \ddot{x})\dot{u}_g dt \tag{8.1}$$

The term $m(\ddot{u}_g + \ddot{x})$ in Eq. (8.1) indicates the inertial force (although $-m(\ddot{u}_g + \ddot{x})$ is the exact inertial force) and is equal to the sum of the restoring force kx and the damping force $c\dot{x}$ in the model. Integration by parts of Eq. (8.1) yields

$$\begin{aligned} E_I &= \int_0^{t_0} m(\ddot{x} + \ddot{u}_g)\dot{u}_g dt = \int_0^{t_0} m\ddot{x}\dot{u}_g dt + [(1/2)m\dot{u}_g^2]_0^{t_0} \\ &= [m\dot{x}\dot{u}_g]_0^{t_0} - \int_0^{t_0} m\dot{x}\ddot{u}_g dt + [(1/2)m\dot{u}_g^2]_0^{t_0} \end{aligned} \tag{8.2}$$

Assume that $\dot{x} = 0$ at $t = 0$ and $\dot{u}_g = 0$ at $t = 0$ and $t = t_0$. Then the input energy can be simplified to the following form.

$$E_I = -\int_0^{t_0} m\ddot{u}_g \dot{x} dt \tag{8.3}$$

It is known (Page 1952; Lyon 1975; Takizawa 1977; Ohi et al. 1985; Ordaz et al. 2003) that the input energy per unit mass to the SDOF system can also be expressed in the frequency domain by use of Fourier transformation.

$$\begin{aligned} E_I/m &= -\int_{-\infty}^{\infty} \dot{x}a dt = -\int_{-\infty}^{\infty} \left[(1/2\pi)\int_{-\infty}^{\infty} \dot{X}e^{i\omega t}d\omega\right]a dt \\ &= -(1/2\pi)\int_{-\infty}^{\infty} A(-\omega)\{H_V(\omega; \Omega, h)A(\omega)\}d\omega \\ &= \int_0^{\infty} |A(\omega)|^2\{-\mathrm{Re}[H_V(\omega; \Omega, h)]/\pi\}d\omega \\ &\equiv \int_0^{\infty} |A(\omega)|^2 F(\omega)d\omega \end{aligned} \tag{8.4}$$

In Eq. (8.4), the function $H_V(\omega; \Omega, h)$ is the transfer function defined by $\dot{X}(\omega) = H_V(\omega; \Omega, h)A(\omega)$ and $F(\omega) = -\mathrm{Re}[H_V(\omega; \Omega, h)]/\pi$. $\dot{X}(\omega)$ and $A(\omega)$ denote the Fourier transforms of $\dot{x}$ and $\ddot{u}_g(t) = a(t)$. The symbol i indicates the imaginary unit. The velocity transfer function $H_V(\omega; \Omega, h)$ can be expressed by

$$H_V(\omega; \Omega, h) = -i\omega/(\Omega^2 - \omega^2 + 2ih\Omega\omega) \qquad (8.5)$$

Eq. (8.4) implies that the earthquake input energy to damped, linear elastic SDOF systems does not depend on the phase of input motions and this is well known (Page 1952; Lyon 1975; Takizawa 1977; Ohi et al. 1985; Kuwamura et al. 1994; Ordaz et al. 2003). It can also be observed from Eq. (8.4) that the function $F(\omega)$ plays a central role in the evaluation of the earthquake input energy and may have some influence on the investigation of constancy of the earthquake input energy to structures with various model parameters. The property of the function $F(\omega)$ in Eq. (8.4) will therefore be clarified in the following section in detail.

8.3. PROPERTY OF ENERGY TRANSFER FUNCTION AND CONSTANCY OF EARTHQUAKE INPUT ENERGY

The functions $F(\omega)$ for various model natural periods $T = 0.5, 1.0, 2.0$s and damping ratios $h = 0.05, 0.20$ are plotted in Fig. 8.4. It should be remarked that the area of $F(\omega)$ can be proved to be constant regardless of Ω and h. This property for any damping ratio has already been pointed out by

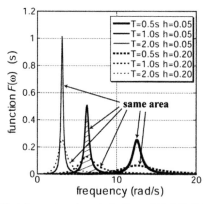

Figure 8.4 *Function F(ω) for natural periods T = 0.5, 1.0, 2.0s and damping ratios h = 0.05, 0.20.*

Ordaz et al. (2003). However, its proof has never been presented and the proof is shown here.

The function $F(\omega)$, called the energy transfer function, can be expressed by

$$F(\omega) = \frac{2h\Omega\omega^2}{\pi\{(\Omega^2 - \omega^2)^2 + (2h\Omega\omega)^2\}} \tag{8.6}$$

It can be understood that $F(\omega)$ is a positive function. With the use of complex variables, four singular points of $F(\omega)$ can be obtained as $z_1 = (hi + \sqrt{1-h^2})\Omega$, $z_2 = (hi - \sqrt{1-h^2})\Omega$, $z_3 = (-hi + \sqrt{1-h^2})\Omega$, $z_4 = (-hi - \sqrt{1-h^2})\Omega$. Consider an integration path of a semi-circle in the complex plane as shown in Fig. 8.5. The singular points inside the integration path are indicated by z_1 and z_2. The residues for the two singular points z_1 and z_2 can be computed as

$$\text{Res}[z = z_1] = \frac{2hz_1^2\Omega}{\pi(z_1 - z_2)(z_1 - z_3)(z_1 - z_4)} = \frac{z_1}{4\pi\sqrt{1-h^2}\Omega i} \tag{8.7a}$$

$$\text{Res}[z = z_2] = \frac{2hz_2^2\Omega}{\pi(z_2 - z_1)(z_2 - z_3)(z_2 - z_4)} = \frac{-z_2}{4\pi\sqrt{1-h^2}\Omega i} \tag{8.7b}$$

The integration path in Cauchy integral consists of one on the real axis and the other on the semi-circle. The integral along the path on the semi-circle will vanish as the radius becomes infinite. On the other hand, the

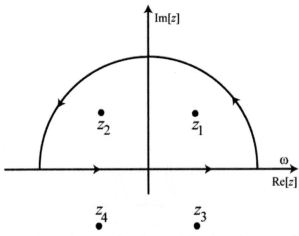

Figure 8.5 *Integration path in complex plane and singular points z_1, z_2 of function $F(\omega)$ inside the path and those z_3, z_4 outside the path.*

integral on the real axis with infinite lower and upper limits will be reduced to the residue theorem. The residue theorem yields the following relation.

$$\int_{-\infty}^{\infty} F(\omega)d\omega = 2\pi i \times (\text{Res}[z = z_1] + \text{Res}[z = z_2]) \qquad (8.8)$$

Substitute Eqs. (8.7a, b) into Eq. (8.8) and the property of the energy transfer function $F(\omega)$ as an even function results in the following relation.

$$2\int_{0}^{\infty} F(\omega)d\omega = 1 \qquad (8.9)$$

Eq. (8.9) means that the area of $F(\omega)$ is constant regardless of Ω and h.

It can be derived from Eqs. (8.4) and (8.9) that, if the Fourier amplitude spectrum of an input ground acceleration is uniform with respect to frequency, the earthquake input energy to a damped, linear elastic SDOF system per unit mass is exactly constant irrespective of model natural frequency and damping ratio. Let $S_V(h = 0)$ denote the velocity response spectrum of the input for null damping ratio. If $A(\omega)$ is exactly constant with respect to frequency and an assumption $A(\Omega) \cong S_V(h = 0)$ holds (Hudson 1962), Eqs. (8.4) and (8.9) provide

$$E_I \cong \frac{1}{2}m\{S_V(h = 0)\}^2 \qquad (8.10)$$

Eq. (8.10) is somewhat similar to the maximum total energy proposed by Housner (1956, 1959). It is noted that Housner (1959) discussed the maximum total energy defined by $E_H = \max_t \{-\int_0^t m\ddot{u}_g \dot{x}dt\}$ instead of E_I defined by Eq. (8.3) and introduced some assumptions, e.g. the slowly-varying property of the total energy. A general inequality $S_V(h \neq 0) \leq S_V(h = 0)$ for any damping ratio and a more exact relation $A(\Omega) \leq S_V(h = 0)$ can also be demonstrated for most cases. If $S_V(h \neq 0) \cong A(\Omega) =$ constant holds for a specific damping ratio denoted by h_a, Eq. (8.10) may be replaced by $E_I \cong (1/2)m\{S_V(h_a)\}^2$. While Housner discussed the constancy of earthquake input energy (maximum total energy) only with respect to natural period by paying special attention to the uniformity of a velocity response spectrum with respect to natural period (Housner 1956, 1959), another viewpoint based on exact mathematical treatment has been introduced in this chapter. It should be noted that the constancy of input energy defined by Eq. (8.3) is *directly* related to the constancy of "the Fourier amplitude spectrum" of a ground motion acceleration, not the constancy of the velocity response spectrum. This problem will be investigated numerically for recorded ground motions afterwards.

8.4. CRITICAL EXCITATION PROBLEM FOR EARTHQUAKE INPUT ENERGY WITH ACCELERATION CONSTRAINT

It is explained in this section that a critical excitation method for the earthquake input energy can provide upper bounds on the input energy. Westermo (1985) has tackled a similar problem for the maximum input energy to an SDOF system subjected to external forces. His solution is restrictive because it includes the velocity response containing the solution itself implicitly. A more general solution procedure will be explained here.

The capacity of ground motions is often discussed in terms of the time integral of a squared ground acceleration $a(t)^2$ (Arias 1970; Housner and Jennings 1975; Riddell and Garcia 2001). This quantity is well known as the Arias intensity measure with a different coefficient. The constraint on this quantity can be expressed by

$$\int_{-\infty}^{\infty} a(t)^2 dt = (1/\pi) \int_{0}^{\infty} |A(\omega)|^2 d\omega = \overline{C}_A \qquad (8.11)$$

$\overline{C}_A$ is the specified value of the time integral of a squared acceleration. Another index called a "power" (average rate of energy supplied) has been defined by dividing that quantity by its significant duration (Housner 1975). It is also clear that the maximum value of the Fourier amplitude spectrum of the input is finite. The infinite Fourier amplitude spectrum may represent a perfect harmonic function or that multiplied by an exponential function (Drenick 1970). This is unrealistic as an actual ground motion. The constraint on this property may be expressed by

$$|A(\omega)| \leq \overline{A} \, (\overline{A}\text{: Specified Value}) \qquad (8.12)$$

The critical excitation problem may then be stated as follows:

[Critical Excitation Problem for Acceleration]
Find $|A(\omega)|$ that maximizes the earthquake input energy per unit mass expressed by Eq. (8.4) subject to $\overline{A}$ Specified the constraints (8.11) and (8.12) on acceleration.

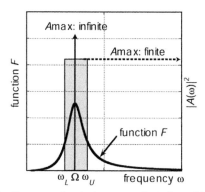

Figure. 8.6 *Schematic diagram of solution procedure for critical excitation problem with acceleration constraints.*

It is clear from the present author's work (Takewaki 2001a, b, 2002a, b) on power spectral density functions that, if $\overline{A}$ is infinite, the critical function of $|A(\omega)|^2$ turns out to be the Dirac delta function which has a nonzero value only at the point maximizing $F(\omega)$. On the other hand, if $\overline{A}$ is finite, the critical one of $|A(\omega)|^2$ yields a rectangular function attaining $\overline{A}^2$. The bandwidth of the frequency can be obtained as $\Delta\omega = \pi\overline{C}_A/\overline{A}^2$. The location of the rectangular function, i.e. the lower and upper limits, can be computed by maximizing $\overline{A}^2 \int_{\omega_L}^{\omega_U} F(\omega)d\omega$. It should be reminded that $\omega_U - \omega_L = \Delta\omega$. It can be shown that a good and simple approximation is given by $(\omega_U + \omega_L)/2 = \Omega$. The essential feature of the solution procedure explained in this section is shown in Fig. 8.6. It is interesting to note that Westermo's periodic solution (Westermo 1985) may correspond to the case of infinite value of $\overline{A}$.

The absolute bound may be computed for the infinite value of $\overline{A}$. This absolute bound can be evaluated as $\overline{C}_A/(2h\Omega)$ by introducing an assumption that $F(\omega)$ attains its maximum at $\omega = \Omega$ and substituting Eq. (8.11) into Eq. (8.4).

8.5. CRITICAL EXCITATION PROBLEM FOR EARTHQUAKE INPUT ENERGY WITH VELOCITY CONSTRAINT

It has often been discussed that the maximum ground acceleration controls the behavior of structures with short natural periods and the maximum ground velocity controls the behavior of structures with intermediate or rather long natural periods (see, for example, Tanabashi 1956). Consider the following constraint on the ground motion velocity $\ddot{u}_g(t) = v(t)$.

$$\int_{-\infty}^{\infty} v(t)^2 dt = (1/\pi) \int_{0}^{\infty} |V(\omega)|^2 d\omega = \overline{C}_V \ (\overline{C}_V\text{: Specified Value})$$

(8.13)

$V(\omega)$ is the Fourier transform of the ground velocity. With the relation $A(\omega) = i\omega V(\omega)$ kept in mind, Eq. (8.4) can be reduced to

$$E_I/m = \int_{0}^{\infty} |V(\omega)|^2 \omega^2 F(\omega) d\omega$$

(8.14)

The maximum value of $|V(\omega)|$ is certainly finite in a realistic situation. The constraint on the upper limit of $V(\omega)$ may be expressed by

$$|V(\omega)| \le \overline{V} \ (\overline{V}\text{: Upper limit of } V(\omega))$$

(8.15)

The critical excitation problem for velocity constraints may be described as follows:

[Critical Excitation Problem for Velocity]
Find $|V(\omega)|$ that maximizes the earthquake input energy per unit mass expressed by Eq. (8.14) subject to the constraints (8.13) and (8.15) on velocity.

It is apparent that almost the same solution procedure as for acceleration constraints can be used by replacing $A(\omega)$ and $F(\omega)$ by $V(\omega)$ and $\omega^2 F(\omega)$, respectively. The functions $\omega^2 F(\omega)$ for three natural periods $T = 0.5, 1.0, 2.0$s and two damping ratios $h = 0.05, 0.20$ are plotted in Fig. 8.7. It can be

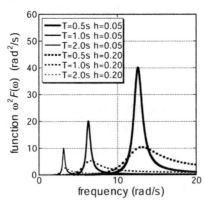

Figure 8.7 *Function $\omega^2 F(\omega)$ for natural periods $T = 0.5, 1.0, 2.0$s and damping ratios $h = 0.05, 0.20$.*

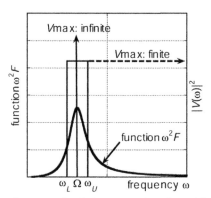

Figure 8.8 *Schematic diagram of solution procedure for critical excitation problem with velocity constraints.*

seen that $\omega^2 F(\omega)$ becomes larger in peak and wider with increase of the model natural frequency. In case of a finite value of $\overline{V}$, the bandwidth of the critical rectangular function $|V(\omega)|^2$ can be derived from $\Delta\omega = \pi\overline{C}_V/\overline{V}^2$. The upper and lower frequency limits of the rectangular function can be specified by maximizing $\overline{V}^2 \int_{\omega_L}^{\omega_U} \omega^2 F(\omega)\mathrm{d}\omega$ where $\omega_U - \omega_L = \Delta\omega$. A good and simple approximation can be given by $(\omega_U + \omega_L)/2 = \Omega$. The essential feature of the solution procedure is shown in Fig. 8.8.

The absolute bound may be computed for the infinite value of $\overline{V}$. This absolute bound can be evaluated as $\Omega\overline{C}_V/(2h)$ by using an assumption that $\omega^2 F(\omega)$ attains its maximum at $\omega = \Omega$ and substituting Eq. (8.13) into Eq. (8.14).

8.6. ACTUAL EARTHQUAKE INPUT ENERGY AND ITS BOUND FOR RECORDED GROUND MOTIONS

In order to examine the relation of the proposed upper bounds of the earthquake input energy with actual ones, the results of numerical calculation are shown for some recorded ground motions. The ground motions were taken from the PEER motions (Abrahamson et al. 1998). Four types of ground motions are considered, i.e. (1) one at rock site in near-fault earthquake (near-fault rock motion), (2) one at soil site in near-fault earthquake (near-fault soil motion), (3) one of long-duration at rock site (long-duration rock motion) and (4) one of long-duration at soil site (long-duration soil motion). The profile of the selected motions is shown in Table 8.1. The Fourier amplitude spectra of these ground motion accelerations are plotted in Figs. 8.9(a)–(d). $A_{\max} = \max|A(\omega)|$ and $V_{\max} = \max|V(\omega)|$ have

Table 8.1 Ground motions selected from PEER motions (Abrahamson et al. 1998)

earthquake	site and component	magnitude in Mw	max acc in G	max vel in m/s	max dis in m
(Near fault motion/rock records)					
Loma Prieta 1989	Los Gatos NS	6.9	0.570	0.988	0.379
Hyogoken–Nanbu 1995	JMA Kobe NS	6.9	0.833	0.920	0.206
(Near fault motion/soil records)					
Cape Mendocino 1992	Petrolia NS	7.0	0.589	0.461	0.265
	Petrolia EW		0.662	0.909	0.268
Northridge 1994	Rinaldi NS	6.7	0.480	0.795	0.505
	Rinaldi EW		0.841	1.726	0.487
	Sylmar NS		0.842	1.288	0.306
	Sylmar EW		0.604	0.778	0.203
Imperial Valley 1979	Meloland NS	6.5	0.317	0.711	1.242
	Meloland EW		0.297	0.943	3.124
(Long duration motion/rock records)					
Michoacan 1985	Caleta de Campos NS	8.1	0.141	0.255	1.464
Miyagiken–oki 1978	Ofunato NS	7.4	0.211	0.131	0.163
(Long duration motion/soil records)					
Chile 1985	Vina del Mar NS	8.0	0.362	0.337	2.400
	Vina del Mar EW		0.214	0.267	1.212
Olympia 1949	Seattle Army Base NS	6.5	0.0678	0.0785	0.192
	Seattle Army Base EW		0.0673	0.0777	0.0278

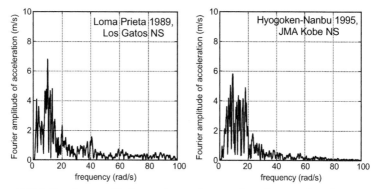

Figure 8.9(a) *Fourier amplitude spectrum of ground motion acceleration: near-fault rock motion.*

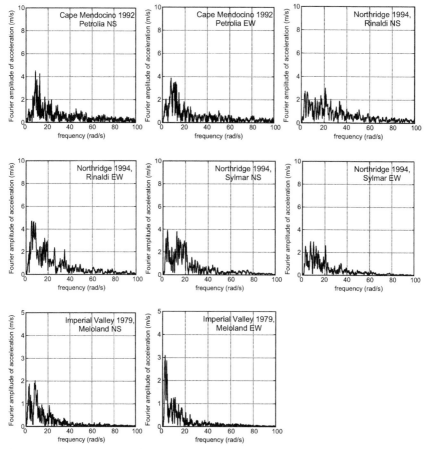

Figure 8.9(b) *Fourier amplitude spectrum of ground motion acceleration: near-fault soil motion.*

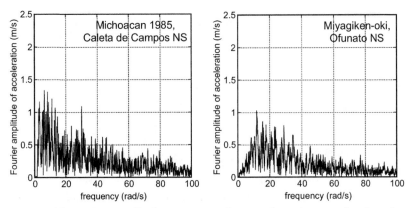

Figure 8.9(c) *Fourier amplitude spectrum of ground motion acceleration: long-duration rock motion.*

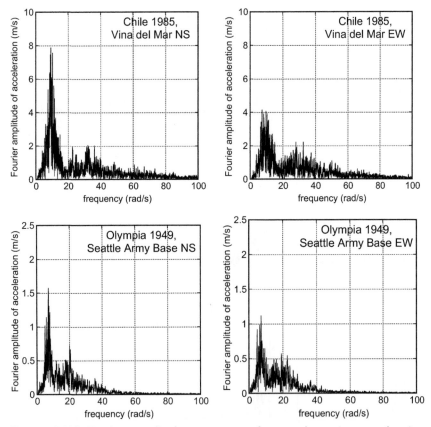

Figure 8.9(d) *Fourier amplitude spectrum of ground motion acceleration: long-duration soil motion.*

been used as $\overline{A}$ and $\overline{V}$, respectively. Due to this treatment of $\overline{A}$ and $\overline{V}$, the bounds, shown in the previous sections, for acceleration and velocity constraints are called "credible bounds" in the following. The selection of $\overline{A}$ and $\overline{V}$ may be arguable. However, it may be clear at least that, if $\overline{A}$ is chosen between $A_{\max} = \max|A(\omega)|$ and infinity, the corresponding bound of the earthquake input energy exists between the credible bound and the absolute bound $\overline{C}_A/(2h\Omega)$. A similar fact can be seen for velocity constraints. If $\overline{V}$ is chosen between $V_{\max} = \max|V(\omega)|$ and infinity, the corresponding bound of the earthquake input energy exists between the credible bound and the absolute bound $\Omega\overline{C}_V/(2h)$. The quantities $A_{\max}, \overline{C}_A, \Delta\omega$ corresponding to the critical excitation problem for acceleration constraints are shown in Table 8.2 and those $V_{\max}, \overline{C}_V, \Delta\omega$ corresponding to the critical excitation problem for velocity constraints are shown in Table 8.3.

Fig. 8.10(a) indicates the actual earthquake input energy to SDOF models with various natural periods and the corresponding credible bounds for near-fault rock motions. The damping ratio is given by 0.05. It can be seen that, because the Fourier amplitude spectrum of the ground motion acceleration is not uniform in the frequency range in almost all the ground motions, the constancy of the earthquake input energy is not seen. As far as the bound of input energy is concerned, it should be noted that the monotonic increase of the credible bound for acceleration constraints in the shorter natural period range depends mainly on the characteristic of the function $F(\omega)$ as a monotonically increasing function with respect to natural period (Fig. 8.4 is arranged with respect to natural frequency). This fact provides mathematical explanation of actual phenomena for most ground motions. It can also be seen that the actual input energy in the shorter natural period range is bounded appropriately by the bound for acceleration constraints and that in the intermediate and longer natural period range is bounded properly by the bound for velocity constraints. These characteristics correspond to the well-known fact (Tanabashi 1956) that the maximum ground acceleration governs the behavior of structures with shorter natural periods and the maximum ground velocity controls the behavior of structures with intermediate or longer natural periods. From another viewpoint, it may be stated from Fig. 8.10(a) that, while the behavior of structures with shorter natural periods is controlled by an hypothesis of "constant energy," that of structures with intermediate or longer natural periods is governed by a hypothesis of "constant maximum displacement." In the previous studies on earthquake input energy, this property of "constant maximum displacement" in the longer natural period

Table 8.2 Maximum Fourier amplitude spectrum of ground motion acceleration, time integral of squared ground motion acceleration and frequency bandwidth of critical rectangular Fourier amplitude spectrum of ground motion acceleration

earthquake	site and component	Amax in m/s	C_A in m^2/s^3	$\Delta\omega$ in rad/s
(Near fault motion/rock records)				
Loma Prieta 1989	Los Gatos NS	6.80	49.5	3.36
Hyogoken–Nanbu 1995	JMA Kobe NS	5.81	52.3	4.87
(Near fault motion/soil records)				
Cape Mendocino 1992	Petrolia NS	4.49	21.5	3.35
	Petrolia EW	3.85	23.9	5.07
Northridge 1994	Rinaldi NS	2.98	25.0	8.84
	Rinaldi EW	4.70	46.3	6.58
	Sylmar NS	3.92	31.3	6.40
	Sylmar EW	2.95	16.3	5.88
Imperial Valley 1979	Meloland NS	2.01	5.43	4.22
	Meloland EW	3.09	6.93	2.28
(Long duration motion/rock records)				
Michoacan 1985	Caleta de Campos NS	1.33	3.97	7.05
Miyagiken-oki 1978	Ofunato NS	1.03	2.35	6.96
(Long duration motion/soil records)				
Chile 1985	Vina del Mar NS	7.87	34.3	1.74
	Vina del Mar EW	4.14	18.7	3.43
Olympia 1949	Seattle Army Base NS	1.57	1.28	1.63
	Seattle Army Base EW	1.12	0.877	2.20

Table 8.3 Maximum Fourier amplitude spectrum of ground motion velocity, time integral of squared ground motion velocity and frequency bandwidth of critical rectangular Fourier amplitude spectrum of ground motion velocity

earthquake	site and component	Vmax in m	C_V in m²/s	$\Delta\omega$ rad/s
(Near fault motion/ rock records)				
Loma Prieta 1989	Los Gatos NS	1.81	1.49	1.43
Hyogoken-Nanbu 1995	JMA Kobe NS	0.746	0.854	4.82
(Near fault motion/soil records)				
Cape Mendocino 1992	Petrolia NS	0.531	0.253	2.82
	Petrolia EW	0.697	0.509	3.29
Northridge 1994	Rinaldi NS	1.01	0.62	1.90
	Rinaldi EW	1.02	1.13	3.42
	Sylmar NS	1.22	0.858	1.81
	Sylmar EW	0.968	0.45	1.51
Imperial Valley 1979	Meloland NS	0.738	0.356	2.05
	Meloland EW	1.44	1.06	1.61
(Long duration motion/ rock records)				
Michoacan 1985	Caleta de Campos NS	0.408	0.0759	1.44
Miyagiken-oki 1978	Ofunato NS	0.087	0.0119	4.89
(Long duration motion/ soil records)				
Chile 1985	Vina del Mar NS	0.865	0.455	1.91
	Vina del Mar EW	0.563	0.199	1.97
Olympia 1949	Seattle Army Base NS	0.224	0.0232	1.45
	Seattle Army Base EW	0.189	0.0154	1.35

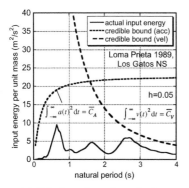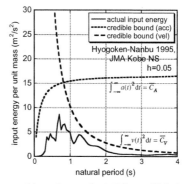

Figure 8.10(a) *Actual earthquake input energy (damping ratio 0.05), credible bound for acceleration constraints and credible bound for velocity constraints: near-fault rock motion.*

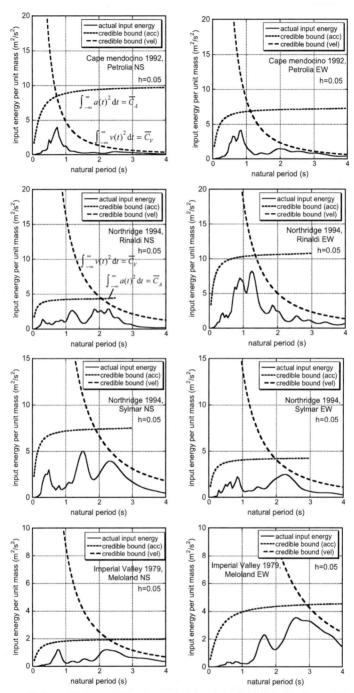

Figure 8.10(b) *Actual earthquake input energy (damping ratio 0.05), credible bound for acceleration constraints and credible bound for velocity constraints: near-fault soil motion.*

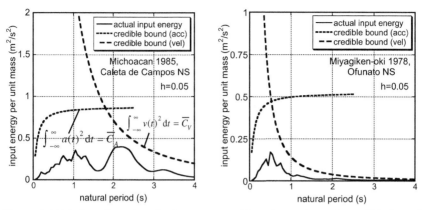

Figure 8.10(c) *Actual earthquake input energy (damping ratio 0.05), credible bound for acceleration constraints and credible bound for velocity constraints: long-duration rock motion.*

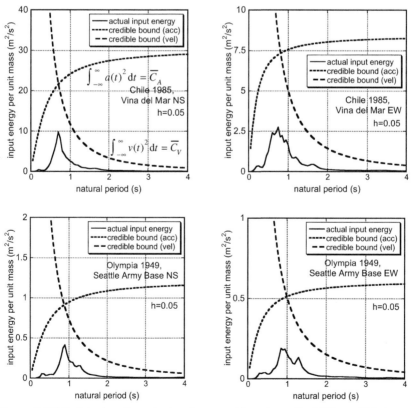

Figure 8.10(d) *Actual earthquake input energy (damping ratio 0.05), credible bound for acceleration constraints and credible bound for velocity constraints: long-duration soil motion.*

range has never been considered explicitly. This should be discussed in detail in the case of structures subjected to long-period ground motions.

Fig. 8.10(b) illustrates the actual earthquake input energy for various natural periods and the corresponding credible bounds for near-fault soil motions. As in near-fault rock motions, the actual input energy is bounded by the two kinds of bound (acceleration-constrained bound and velocity-constrained bound). It can also be seen that most of the bounds in the intermediate natural period range are nearly constant. This property can be explained by Eqs. (8.4), (8.9), (8.11) and the critical shape of $|A(\omega)|^2$ as rectangular. It is noted that the frequency limits ω_L and ω_U are varied so as to coincide with the peak of $F(\omega)$ for various natural periods.

Fig. 8.10(c) shows those for long-duration rock motions and Fig. 8.10(d) presents those for long-duration soil motions. It can be observed that the actual input energy is bounded properly and exactly by the two kinds of bound pointed out earlier. It may be concluded that the two kinds of bound proposed in this chapter provide a physically meaningful upper limit on the earthquake input energy for various types of recorded ground motions.

8.7. CONCLUSIONS

The conclusions may be stated as follows:

(1) The energy transfer function $F(\omega)$ characterizing the earthquake input energy in the frequency domain to a damped linear elastic SDOF system has been proved by the Residue theorem to have an equi-area property irrespective of natural period and damping ratio. This property guarantees that, if the Fourier amplitude spectrum of a ground acceleration is uniform, the constancy of the input energy defined by Eq. (8.3) holds strictly. Otherwise, its constancy is not guaranteed. The constancy of the earthquake input energy defined by Eq. (8.3) is *directly* related to the constancy of "the Fourier amplitude spectrum" of the ground acceleration, not the constancy of the velocity response spectrum.

(2) A critical excitation method has been formulated that has the earthquake input energy as a new measure of criticality and has acceleration and/or velocity constraints (the time integral of a squared ground acceleration and the time integral of a squared ground velocity). No mathematical programming technique is necessary in this method and structural engineers can find the solution, i.e. upper bounds of earthquake input energy, without difficulty.

(3) The solution to this critical excitation problem provides a useful bound of the earthquake input energy for a class of ground motions satisfying intensity constraints. The solution with acceleration constraints can properly bound the earthquake input energy in a shorter natural period range and that with velocity constraints and can properly limit the input energy in an intermediate or longer natural period range.

The manipulation of the earthquake input energy in the time domain is suitable for the evaluation of the time history of the input energy, especially for nonlinear systems. Dual use of the frequency-domain and time-domain techniques may be preferable in the sophisticated seismic analysis of structures for more robust design.

REFERENCES

Abrahamson, N., Ashford, S., Elgamal, A., Kramer, S., Seible, F., Somerville, P., 1998. Proc. of 1st PEER Workshop on Characterization of Special Source Effects, San Diego. Pacific Earthquake Engineering Research Center, Richmond.

Akiyama, H., 1985. Earthquake Resistant Limit-State Design for Buildings. University of Tokyo Press, Tokyo, Japan.

Arias, A., 1970. A measure of earthquake intensity. In: Hansen, R.J. (Ed.), Seismic Design for Nuclear Power Plants. The MIT Press, Cambridge, MA, pp. 438–469.

Berg, G.V., Thomaides, T.T., 1960. Energy consumption by structures in strong-motion earthquakes. Proc. of 2nd World Conf. on Earthquake Engineering, Tokyo and Kyoto. Gakujutsu Bunken Fukyu-kai, Tokyo 681–696.

Drenick, R.F., 1970. Model-free design of aseismic structures. Journal of Engineering Mechanics Division 96 (4), 483–493.

Fajfar, P., Vidic, T., 1994. Consistent inelastic design spectra: hysteretic and input energy. Earthquake Engineering and Structural Dynamics 23 (5), 523–537.

Goel, S.C., Berg, G.V., 1968. Inelastic earthquake response of tall steel frames. Journal of Structural Division 94, 1907–1934.

Housner, G.W., 1956. Limit design of structures to resist earthquakes. Proc. of the First World Conference on Earthquake Engineering. University of California, Berkeley, CA,. Berkeley, 5:1–11.

Housner, G.W., 1959. Behavior of structures during earthquakes. Journal of the Engineering Mechanics Division 85 (4), 109–129.

Housner, G.W., 1975. Measures of severity of earthquake ground shaking. Proc. of the US National Conf. on Earthquake Engineering, Ann Arbor, Michigan, 25–33.

Housner, G.W., Jennings, P.C., 1975. The capacity of extreme earthquake motions to damage structures. In: Hall, W.J. (Ed.), Structural and Geotechnical Mechanics: A volume honoring N.M. Newmark, pp. 102–116. Prentice-Hall Englewood Cliff, NJ.

Hudson, D.E., 1962. Some problems in the application of spectrum techniques to strong-motion earthquake analysis. Bulletin of Seismological Society of America 52 (2), 417–430.

Kato, B., Akiyama, H., 1975. Energy input and damages in structures subjected to severe earthquakes. Journal of Structural and Construction Eng., Archi. Inst. of Japan 235, 9–18 (in Japanese).

Kuwamura, H., Kirino, Y., Akiyama, H., 1994. Prediction of earthquake energy input from smoothed Fourier amplitude spectrum. Earthquake Engineering and Structural Dynamics 23 (10), 1125–1137.

Leger, P., Dussault, S., 1992. Seismic-energy dissipation in MDOF structures. Journal of Structural Engineering 118 (5), 1251–1269.

Lyon, R.H., 1975. Statistical Energy Analysis of Dynamical Systems. The MIT Press, Cambridge, MA.

Mahin, S.A., Lin, J., 1983. Construction of inelastic response spectrum for single-degree-of-freedom system. Report No. UCB/EERC-83/17, Earthquake Engineering Research Center. University of California, Berkeley, CA.

Ogawa, K., Inoue, K., Nakashima, M., 2000. A study on earthquake input energy causing damages in structures. Journal of Structural and Construction Eng. (Transactions of AIJ) 530, 177–184 (in Japanese).

Ohi, K., Takanashi, K., Tanaka, H., 1985. A simple method to estimate the statistical parameters of energy input to structures during earthquakes. Journal of Structural and Construction Eng., Archi. Inst. of Japan 347, 47–55 (in Japanese).

Ordaz, M., Huerta, B., Reinoso, E., 2003. Exact computation of input-energy spectra from Fourier amplitude spectra. Earthquake Engineering and Structural Dynamics 32 (4), 597–605.

Page, C.H., 1952. Instantaneous power spectra. Journal of Applied Physics 23 (1), 103–106.

Riddell, R., Garcia, J.E., 2001. Hysteretic energy spectrum and damage control. Earthquake Engineering and Structural Dynamics 30 (12), 1791–1816.

Takewaki, I., 2001a. A new method for nonstationary random critical excitation. Earthquake Engineering and Structural Dynamics 30 (4), 519–535.

Takewaki, I., 2001b. Probabilistic critical excitation for MDOF elastic–plastic structures on compliant ground. Earthquake Engineering and Structural Dynamics 30 (9), 1345–1360.

Takewaki, I., 2002a. Seismic critical excitation method for robust design: A review. J. Struct. Eng. 128 (5), 665–672.

Takewaki, I., 2002b. Robust building stiffness design for variable critical excitations. J. Struct. Eng. 128 (12), 1565–1574.

Takizawa, H., 1977. Energy-response spectrum of earthquake ground motions. Proc. of the 14th Natural disaster science symposium, Hokkaido (in Japanese), 359–362.

Tanabashi, R., 1935. Personal view on destructiveness of earthquake ground motions and building seismic resistance. Journal of Architecture and Building Science, Archi. Inst. of Japan 48 (599) (in Japanese).

Tanabashi, R., 1956. Studies on the nonlinear vibrations of structures subjected to destructive earthquakes. Proc. of the First World Conference on Earthquake Engineering. University of California, Berkeley, CA. Berkeley, 6:1–16.

Uang, C.M., Bertero, V.V., 1990. Evaluation of seismic energy in structures. Earthquake Engineering and Structural Dynamics 19 (1), 77–90.

Westermo, B.D., 1985. The critical excitation and response of simple dynamic systems. Journal of Sound and Vibration 100 (2), 233–242.

Zahrah, T., Hall, W., 1984. Earthquake energy absorption in SDOF structures. J. Struct. Eng. 110 (8), 1757–1772.

CHAPTER NINE

Critical Excitation for Earthquake Energy Input in MDOF System

Contents

9.1. INTRODUCTION

The purpose of this chapter is to explain a new complex modal analysis-based method in the frequency domain for computation of earthquake input energy to highly damped linear elastic passive control structures. It can be shown that the formulation of the earthquake input energy in the frequency domain is essential for deriving a bound on the earthquake input energy for a class of ground motions. This is because, in contrast to the formulation in the time domain, which requires time–history response analysis, the formulation in the frequency domain only requires the computation of the Fourier amplitude spectrum of the input motion acceleration and the use of time-invariant velocity and acceleration transfer functions. Importance of over-damped modes in the energy computation of specific nonproportionally damped models is demonstrated by comparing the energy transfer functions and the displacement transfer functions.

Through numerical examinations for four recorded ground motions, it is demonstrated that the modal analysis-based method in the frequency domain is very efficient in the computation of earthquake input energy. Furthermore, it is shown that the formulation of earthquake input energy in the frequency domain is essential for understanding the robustness of passively controlled structures to disturbances with various frequency contents.

9.2. EARTHQUAKE INPUT ENERGY TO PROPORTIONALLY DAMPED MULTI-DEGREE-OF-FREEDOM (MDOF) SYSTEM (FREQUENCY-DOMAIN MODAL ANALYSIS)

Consider first a proportionally damped, linear elastic MDOF shear building model, as shown in Fig. 9.1, of mass matrix $[M]$ subjected to a horizontal ground acceleration $\ddot{u}_g(t) = a(t)$. The parameters Ω_i, h_i and $\{x\}$ denote the i-th undamped natural circular frequency, the i-th damping ratio and a set of the horizontal floor displacements relative to the ground, respectively. The over-dot indicates the time derivative. The input energy to such a system by the ground motion from $t = 0$ to $t = t_0$ (end of input) can be defined by the work of the ground on that system (Uang and Bertero 1990; see Fig. 9.1) and is expressed by

$$E_I = \int_0^{t_0} \{1\}^T [M] \left(\{1\} \ddot{u}_g + \{\ddot{x}\} \right) \dot{u}_g dt \qquad (9.1)$$

where $\{1\} = \{1 \cdots 1\}^T$. The term $\{1\}^T [M](\{1\} \ddot{u}_g + \{\ddot{x}\})$ in the integrand indicates the sum of the horizontal inertial forces acting on the system with

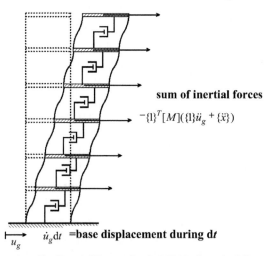

sum of inertial forces

$-\{1\}^T [M](\{1\} \ddot{u}_g + \{\ddot{x}\})$

$\dot{u}_g dt$ =**base displacement during dt**

u_g

Figure 9.1 *Proportionally damped linear elastic MDOF shear building model subjected to unidirectional horizontal ground acceleration.*

minus sign. Double minus signs result in the positive sign. Integration by parts of Eq. (9.1) yields

$$E_I = \left[(1/2)\{1\}^T[M]\{1\}\dot{u}_g^2\right]_0^{t_0} + \left[\{\dot{x}\}^T[M]\{1\}\dot{u}_g\right]_0^{t_0} - \int_0^{t_0} \{\dot{x}\}^T[M]$$

$$\times \{1\}\ddot{u}_g dt$$

(9.2)

Assume that $\{\dot{x}\} = \{0\}$ at $t = 0$ and $\dot{u}_g = 0$ at $t = 0$ and $t = t_0$. Then the input energy E_I can be reduced finally to the following form

$$E_I = -\int_0^{t_0} \{\dot{x}\}^T[M]\{1\}\ddot{u}_g dt \qquad (9.3)$$

It is known that the input energy can also be expressed in the frequency domain. Let $\{\dot{X}\}$ denote the Fourier transform of $\{\dot{x}\}$. Substitution of the Fourier inverse transformation of the relative nodal velocities $\{\dot{x}\}$ into Eq. (9.3) provides

$$E_I = -\int_{-\infty}^{\infty} \left[\frac{1}{2\pi} \int_{-\infty}^{\infty} \{\dot{X}\}^T e^{i\omega t} d\omega\right][M]\{1\}\ddot{u}_g dt$$

$$= -\frac{1}{2\pi} \int_{-\infty}^{\infty} \{\dot{X}\}^T[M]\{1\} \left[\int_{-\infty}^{\infty} \ddot{u}_g e^{i\omega t} dt\right] d\omega \qquad (9.4)$$

$$= -\frac{1}{2\pi} \int_{-\infty}^{\infty} \{\dot{X}\}^T[M]\{1\}A(-\omega) d\omega$$

In Eq. (9.4), $A(\omega)$ is the Fourier transform of $\ddot{u}_g(t) = a(t)$ and the symbol i denotes the imaginary unit. Introduce the following coordinate transformation

$$\{x\} = [\Phi]\{q\} \qquad (9.5)$$

$[\Phi]$ is the modal matrix and $\{q\}$ is defined by

$$\{q\} = \{\cdots q_i \cdots\}^T,$$

$$q_i = \int_0^t \{-\ddot{u}_g(\tau)\} g_i(t - \tau) d\tau,$$

$$g_i(t) = (1/\Omega_{Di}) \exp(-h_i \Omega_i t) \sin \Omega_{Di} t,$$

$$\Omega_{Di} = \Omega_i \sqrt{1 - h_i^2} \qquad (9.6a\text{--}d)$$

In Eqs. (9.6a–d), the parameters q_i, $g_i(t)$ and Ω_{Di} denote the normal coordinate, the unit impulse response function for under-damped vibration and the i-th damped natural circular frequency, respectively. In the case where over-damped vibration modes appear, the corresponding unit impulse response function has to be modified. This case will be explained later for nonproportionally damped models. The input energy can be derived as follows by substituting the coordinate transformation $\{\dot{x}\} = [\Phi]\{\dot{q}\}$, its Fourier transform $\{\dot{X}\} = [\Phi]\{\dot{Q}\}$ and the velocity transfer relation $\{\dot{Q}\} = \{H_V\} A$ into Eq. (9.4).

$$E_I = -\frac{1}{2\pi} \int_{-\infty}^{\infty} \{H_V\}^T [\Phi]^T [M] \{1\} |A(\omega)|^2 d\omega$$

$$= -\frac{1}{\pi} \int_0^{\infty} \{H_V\}^T [\Phi]^T [M] \{1\} |A(\omega)|^2 d\omega \qquad (9.7)$$

$$= \int_0^{\infty} F_{MP}(\omega) |A(\omega)|^2 d\omega$$

$\{H_V\}$ and $F_{MP}(\omega)$ in Eq. (9.7) are the velocity transfer function vector with respect to base acceleration and the energy transfer function, respectively, and are expressed by

$$\{H_V\} = \{\cdots H_{Vi} \cdots\}^T,$$

$$H_{Vi}(\omega; \Omega_i, h_i) = -i\omega/(\Omega_i^2 - \omega^2 + 2ih_i \Omega_i \omega),$$

$$F_{MP}(\omega) = -\frac{1}{\pi} \text{Re}\left[\{H_V\}^T [\Phi]^T [M] \{1\}\right] \qquad (9.8a\text{--}c)$$

The symbol $\text{Re}[\cdot]$ indicates the real part of a complex number.

Eq. (9.7) implies that the earthquake input energy to damped, linear elastic MDOF systems does not depend on the phase of input motions.

This fact is well known (Page 1952; Lyon 1975; Takizawa 1977; Ohi et al. 1985; Kuwamura et al. 1994; Ordaz et al. 2003). A similar result has been shown for SDOF models. It can be understood from Eq. (9.7) that the energy transfer function $F_{MP}(\omega)$ plays a central role in the evaluation of input energy. Eq. (9.7) implies also that, if the envelope of the Fourier amplitude spectrum of input ground motions is available (Shinozuka 1970; Takewaki 2001a, b, 2002a, b), the upper bound of input energy can be derived directly from Eq. (9.7) by substituting that envelope function into $|A(\omega)|$. This manipulation is completely valid because the function $F_{MP}(\omega)$ is a non-negative function of frequency which can be proved by using Eqs. (9.8b, c). Note that while an envelope of the Fourier amplitude spectrum of input ground motions has been considered in Shinozuka (1970), the critical shape of the Fourier amplitude spectrum or the power spectrum density function has been treated and obtained in Takewaki (2001a, b, 2002a, b).

9.3. EARTHQUAKE INPUT ENERGY TO NONPROPORTIONALLY DAMPED MDOF SYSTEM (FREQUENCY-DOMAIN MODAL ANALYSIS)

Consider next a nonproportionally damped, linear elastic shear building model. The state-space formulation in the frequency domain (Veletsos and Ventura 1986) is effective and essential for deriving the earthquake input energy to nonproportionally damped systems via modal analysis.

Let $[M], [C], [K]$ denote the system mass, damping and stiffness matrices, respectively, of the nonproportionally damped system. The eigenvalue problem in terms of the state space of the nonproportionally damped system may be stated as

$$([B] + s[A])\{z\} = \{0\} \tag{9.9}$$

where $[A], [B]$ and $\{z\}$ are defined by

$$[A] = \begin{bmatrix} [0] & [M] \\ [M] & [C] \end{bmatrix}, \quad [B] = \begin{bmatrix} -[M] & [0] \\ [0] & [K] \end{bmatrix}, \quad \{z\} = \begin{Bmatrix} s\{\psi\} \\ \{\psi\} \end{Bmatrix}$$

$$(9.10a{-}c)$$

The j-th eigenvalue and the corresponding j-th eigenvector for the problem defined by Eq. (9.9) are denoted by $s_j, \{\psi_j\}$, respectively. The following quantities are defined.

$$M_j = \{\psi_j\}^T[M]\{\psi_j\}, \quad C_j = \{\psi_j\}^T[C]\{\psi_j\}, \quad L_j = \{\psi_j\}^T[M]\{1\},$$

$$B_j = L_j/(2s_jM_j + C_j) \qquad (9.11a\text{--}d)$$

In the case where eigenvalues are complex numbers, it is known that those eigenvalues appear as a pair of complex conjugates. Assume here that there are N_c pairs of complex conjugates and introduce the following vectors $\{b_n\}, \{d_n\}, \{a_n\}$.

$$\{b_n\} = 2\text{Re}[B_n\{\psi_n\}], \quad \{d_n\} = 2\,\text{Im}[B_n\{\psi_n\}],$$

$$\{a_n\} = h_n\{b_n\} - \sqrt{1 - h_n^2}\{d_n\} \qquad (9.12a\text{--}c)$$

In Eq. (9.12a–c), $h_n = -\text{Re}[s_n]/|s_n|$ is the so-called n-th damping ratio. $\text{Im}[\cdot]$ in Eq. (9.12b) indicates the imaginary part of a complex number. The so-called n-th circular eigenfrequency and the so-called n-th damped circular eigenfrequency may be defined by

$$\Omega_n = |s_n|, \quad \Omega_{Dn} = \Omega_n\sqrt{1 - h_n^2} \qquad (9.13a, b)$$

The unit impulse response function for under-damped vibration has been given in Eq. (9.6c) and is used now.

In the case where eigenvalues are real numbers, it is known that those eigenvalues appear as a pair. Assume now that there are N_r pairs of real eigenvalues.

Let s_j, s_k denote a pair of real eigenvalues of this model. Assume that $|s_j| \langle |s_k|$ and define the following modal quantities in terms of s_j, s_k.

$$\Omega_j^* = \sqrt{s_js_k}, \quad h_j^* = -(s_j + s_k)/(2\sqrt{s_js_k}),$$

$$\Omega_{Dj}^* = \Omega_j^*\sqrt{h_j^{*2} - 1} = (s_j - s_k)/2 \qquad (9.14a\text{--}c)$$

The vectors $\{b_j^r\}, \{d_j^r\}, \{a_j^r\}$ related to real eigenvalues are defined by

$$\{b_j^r\} = B_k\{\psi_k\} + B_j\{\psi_j\},$$

$$\{d_j^r\} = B_k\{\psi_k\} - B_j\{\psi_j\},$$

$$\{a_j^r\} = h_j^*\{b_j^r\} - \sqrt{h_j^{*2} - 1}\{d_j^r\} \qquad (9.15a\text{--}c)$$

The unit impulse response function for over-damped vibration may be written as

$$g_j^r(t) = \left(1/\Omega_{Dj}^*\right)\exp\left(-h_j^*\Omega_j^* t\right) \sinh \Omega_{Dj}^* t \qquad (9.16)$$

The response nodal displacements $\{x\}$ may then be expressed in terms of $\{a_n\}, \{b_n\}, \{a_j^r\}, \{b_j^r\}$, i.e. the combination of the under-damped modes and the over-damped modes.

$$
\{x\} = \sum_{n=1}^{N_c}\left\{\{a_n\}\Omega_n q_n(t) + \{b_n\}\dot{q}_n(t)\right\}
$$

$$
\hspace{3cm} (9.17)
$$

$$
+ \sum_{j=1}^{N_r}\left\{\{a_j^r\}\Omega_j^* q_j^r(t) + \{b_j^r\}\dot{q}_j^r(t)\right\}
$$

The quantities $q_n(t), \dot{q}_n(t), q_j^r(t), \dot{q}_j^r(t)$ in Eq. (9.17) are defined by

$$
q_n(t) = \int_0^t \left\{-\ddot{u}_g(\tau)\right\}g_n(t-\tau)\mathrm{d}\tau,
$$

$$
\hspace{3cm} (9.18a, b)
$$

$$
\dot{q}_n(t) = \int_0^t \left\{-\ddot{u}_g(\tau)\right\}\dot{g}_n(t-\tau)\mathrm{d}\tau
$$

$$
q_j^r(t) = \int_0^t \left\{-\ddot{u}_g(\tau)\right\}g_j^r(t-\tau)\mathrm{d}\tau,
$$

$$
\hspace{3cm} (9.19a, b)
$$

$$
\dot{q}_j^r(t) = \int_0^t \left\{-\ddot{u}_g(\tau)\right\}\dot{g}_j^r(t-\tau)\mathrm{d}\tau
$$

Let $[U_V]$ denote a matrix consisting of vectors $\{a_n\}\Omega_n$ (for all n) and $[U_A]$ denote a matrix consisting of vectors $\{b_n\}$ (for all n). Furthermore let $[U_V^r]$ denote a matrix consisting of vectors $\{a_j^r\}\Omega_j^*$ (for all j) and $[U_A^r]$ denote a matrix consisting of vectors $\{b_j^r\}$ (for all j) in view of the case for under-damped modes. After some manipulation, the Fourier transform $H_{Vj}^r(\omega)$ of $-\dot{g}_j^r(t)$ may be described as

$$H^r_{Vj}(\omega) = \frac{1}{2\Omega^*_{Dj}} \left\{ \frac{\omega^2 - i\omega(-h^*_j\Omega^*_j - \Omega^*_{Dj})}{\omega^2 + (-h^*_j\Omega^*_j - \Omega^*_{Dj})^2} \right.$$

$$\left. - \frac{\omega^2 - i\omega(-h^*_j\Omega^*_j + \Omega^*_{Dj})}{\omega^2 + (-h^*_j\Omega^*_j + \Omega^*_{Dj})^2} \right\} \tag{9.20}$$

$H^r_{Aj}(\omega)$ for acceleration can then be evaluated by $H^r_{Aj}(\omega) = i\omega H^r_{Vj}(\omega)$.
The time derivative of Eq. (9.17) provides

$$\{\dot{x}\} = \sum_{n=1}^{N_c} \left\{ \{a_n\}\Omega_n \dot{q}_n(t) + \{b_n\}\ddot{q}_n(t) \right\}$$

$$+ \sum_{j=1}^{N_r} \left\{ \{a^r_j\}\Omega^*_j \dot{q}^r_j(t) + \{b^r_j\}\ddot{q}^r_j(t) \right\} \tag{9.21}$$

This can also be expressed in matrix form as

$$\{\dot{x}\} = [U_V]\{\dot{q}\} + [U_A]\{\ddot{q}\} + [U^r_V]\{\dot{q}^r\} + [U^r_A]\{\ddot{q}^r\} \tag{9.22}$$

The input energy can be obtained by substituting Eq. (9.22), its Fourier transform $\{\dot{X}\} = [U_V]\{\dot{Q}\} + [U_A]\{\ddot{Q}\} + [U^r_V]\{\dot{Q}^r\} + [U^r_A]\{\ddot{Q}^r\}$, the velocity and acceleration transfer relations $\{\dot{Q}\} = \{H_V\}A$, $\{\ddot{Q}\} = \{H_A\}A$ in under-damped vibration, the velocity and acceleration transfer relations $\{\dot{Q}^r\} = \{H^r_V\}A$, $\{\ddot{Q}^r\} = \{H^r_A\}A$ in over-damped vibration into Eq. (9.4).

$$E_I = \int_0^\infty F_{MNP}(\omega)|A(\omega)|^2 d\omega \tag{9.23}$$

$F_{MNP}(\omega)$ in Eq. (9.23) is the energy transfer function for nonproportionally damped models and is expressed by

$$F_{MNP}(\omega) = -\frac{1}{\pi}\text{Re}[Z(\omega)] \tag{9.24a}$$

$$Z(\omega) = \{H_V(\omega\,;\Omega_i, h_i)\}^T[U_V]^T[M][1] + \{H_A(\omega\,;\Omega_i, h_i)\}^T[U_A]^T[M]\{1\}$$

$$+ \{H^r_{Vi}(\omega)\}^T[U^r_V]^T[M]\{1\} + \{H^r_{Ai}(\omega)\}^T[U^r_A]^T[M]\{1\} \tag{9.24b}$$

It should be remarked that, if the ground motion velocity $\dot{u}_g$ is recorded and available, the earthquake input energy can also be expressed in the frequency domain by

$$E_I = \int_0^\infty \omega^2 F_{MNP}(\omega) |V(\omega)|^2 d\omega \qquad (9.25)$$

$V(\omega)$ in Eq. (9.25) is the Fourier transform of the ground velocity $\dot{u}_g$ and is given by $V(\omega) = A(\omega)/(i\omega)$ in terms of the Fourier transform of ground acceleration.

9.4. EARTHQUAKE INPUT ENERGY WITHOUT MODAL DECOMPOSITION

In order to examine the accuracy of the modal superposition method in the frequency domain, another method has been developed without modal decomposition. This method is applicable to both proportionally damped and nonproportionally damped structures. Using the Fourier transformation of the equations of motion, the Fourier transform $\{\dot{X}(\omega)\}$ of the nodal velocities can be expressed by

$$\{\dot{X}(\omega)\} = -i\omega\left(-\omega^2[M] + i\omega[C] + [K]\right)^{-1}[M]\{1\}A(\omega) \qquad (9.26)$$

The input energy can then be computed by Eq. (9.4). The input energy without modal decomposition may be evaluated by

$$E_I = \int_0^\infty F_M(\omega) |A(\omega)|^2 d\omega \qquad (9.27)$$

$F_M(\omega)$ and $[Y(\omega)]$ are defined by

$$F_M(\omega) = \mathrm{Re}[i\omega\{1\}^T[M]^T[Y(\omega)][M]\{1\}]/\pi \qquad (9.28a)$$

$$[Y(\omega)] = \left(-\omega^2[M] + i\omega[C] + [K]\right)^{-1} \qquad (9.28b)$$

The computation of $[Y(\omega)] = (-\omega^2[M] + i\omega[C] + [K])^{-1}$ for many frequencies is quite time-consuming especially for structures with many degrees of freedom. Therefore careful attention is necessary in the method without modal decomposition.

9.5. EXAMPLES

Consider eight 6-degree-of-freedom shear building models shown in Fig. 9.2. Models A–F and BI (Base-Isolation) indicate nonproportionally damped models and Model PD represents a proportionally damped model. It is assumed that Models A–F represent models with an added viscous damping system and Model BI represents a model with a base-isolation system. The constant floor masses are given by $m_i = 32 \times 10^3 (\text{kg})(i = 1, \cdots, 6)$. The constant story stiffnesses are given by $k_i = 3.76 \times 10^7 (\text{N/m})(i = 1, \cdots, 6)$ except the 1st story of Model BI $(k_1 = 3.76 \times 10^5 (\text{N/m}))$. Models A–F possess concentrated viscous dampers in the 1st through 6th story. For example, as for Model A, $c_1 = 3.76 \times 10^6 (\text{N·s/m})$, $c_i = 3.76 \times 10^5 (\text{N·s/m})(i \neq 1)$. Model BI has a concentrated viscous damper $c_1 = 3.76 \times 10^6 (\text{N·s/m})$ in the 1st story and the other damping coefficients are $c_i = 3.76 \times 10^5 (\text{N·s/m})$ $(i \neq 1)$. Model PD has a uniform damping coefficient distribution $c_i = 3.76 \times 10^5 (\text{N·s/m})$.

The undamped natural circular frequencies of Models A–F and Model PD are the same and are given by 8.27(rad/s), 24.3(rad/s), 39.0(rad/s), 51.3(rad/s), 60.7(rad/s) and 66.6(rad/s). Those of Model BI are given by 1.39(rad/s), 17.9(rad/s), 34.3(rad/s), 48.5(rad/s), 59.4(rad/s) and 66.3(rad/s). The lowest-mode damping ratio of Model PD is about 0.04. All the modal damping ratios for Models A–F, BI and PD are shown in Table 9.1. It can be seen that only one over-damped mode exists in every model. The effective

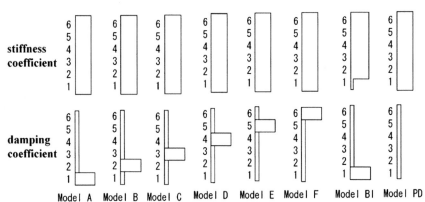

Figure 9.2 *6-degree-of-freedom proportionally damped and nonproportionally damped models.*

Table 9.1 Damping ratio for models A–F, BI and PD

Mode No.	1	2	3	4	5	6
Model A	0.126	0.192	**1.52**	0.248	0.298	0.331
Model B	0.116	0.140	0.198	0.274	**2.24**	0.325
Model C	0.0968	0.125	0.240	0.270	**2.29**	0.315
Model D	0.0752	0.168	0.206	**2.33**	0.292	0.307
Model E	0.0571	0.181	0.228	**2.37**	0.259	0.321
Model F	0.0454	0.142	0.222	**2.41**	0.286	0.328
Model BI	**17.3**	0.233	0.196	0.248	0.298	0.331
Model PD	0.0413	0.122	0.195	0.257	0.304	0.333

location of added viscous dampers can be found from this analysis which attains larger lowest-mode damping ratios. A more sophisticated mathematical treatment for the determination of effective added viscous damper location has been proposed by the present author (Takewaki 1997). A similar tendency, i.e. concentration of added viscous dampers in lower stories in the model with uniform story stiffnesses, has been observed in the previous study (Takewaki 1997).

Fig. 9.3 illustrates the function $F_{MNP}(\omega)$ for Models A–F, BI by use of Eq. (9.24) and $F_{MP}(\omega)$ for Model PD. The solid line shows the function evaluated from the complex and real modes. On the other hand, the dotted line illustrates the function evaluated from the complex modes only, i.e. the first two terms in Eq. (9.24b). It can be seen that the over-damped real modes play an important role in Models A, C and BI. For examining the accuracy of the present method via Eq. (9.24), the function by means of Eq. (9.28) has been plotted. It was confirmed that the present method via Eq. (9.24) is accurate enough to evaluate the function $F_{MNP}(\omega)$.

Fig. 9.4(a) shows the amplitude of the transfer function of the 1st-story drift to the base acceleration for Models A and BI. Fig. 9.4(b) illustrates the corresponding phase angle of the transfer function of the 1st-story drift to the base acceleration for Models A and BI. It can be understood that the over-damped real modes also play an important role in Models A and BI for evaluating the interstory drift accurately. In order to confirm the accuracy of Fig. 9.4, the same figure has been drawn by Eq. (9.28b) without modal decomposition. Sufficient accuracy has been confirmed by the method including over-damped real modes.

Fig. 9.5 illustrates the function $\omega^2 F_{MNP}(\omega)$ for Models A–F, BI is defined by Eqs. (9.25), (9.24a, b) and $\omega^2 F_{MP}(\omega)$ for Model PD given by

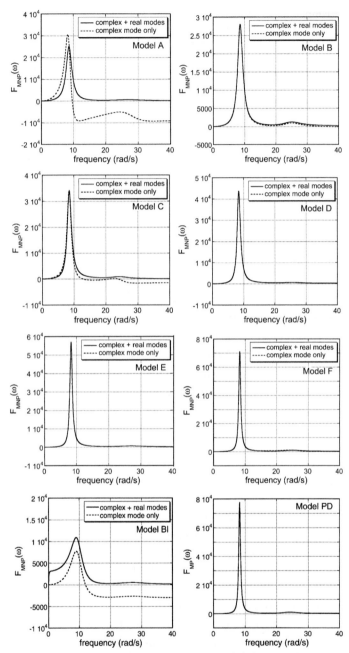

Figure 9.3 *Function* $F_{MNP}(\omega)$ *via Eq. (9.24) and* $F_{MP}(\omega)$ *for models A–F, BI and PD: complex plus real modes (solid line), complex mode only (dotted line).*

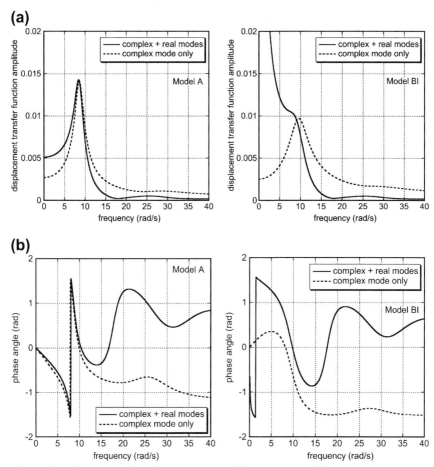

Figure 9.4 *(a) Transfer function amplitude of the 1st-story drift with respect to base acceleration for models A and BI: complex plus real modes (solid line), complex mode only (dotted line). (b) Phase angle of the transfer function of the 1st-story drift with respect to base acceleration for models A and BI: complex plus real modes (solid line), complex mode only (dotted line).*

Eq. (9.8). It can be seen that higher–mode effects become larger than $F_{MNP}(\omega)$ and $F_{MP}(\omega)$ because squared frequencies are multiplied on the energy transfer function.

It may be concluded from Figs. 9.3–9.5 that structures with passive control systems in appropriate and effective location, especially Model BI, have relatively wide-band energy transfer functions and are quite robust for various disturbances with different frequency contents. In other words, if the area (so-called acceleration power or Arias intensity) of the squared Fourier

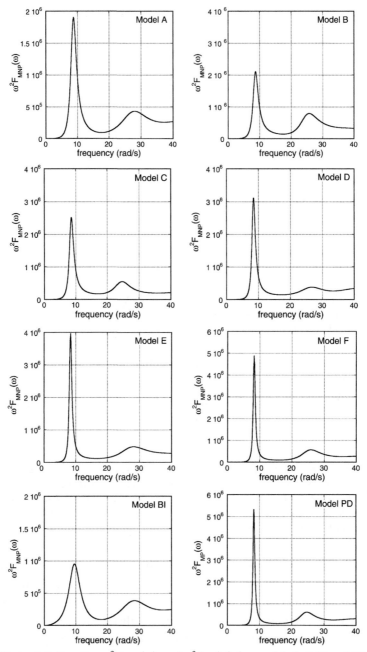

Figure 9.5 *Function $\omega^2 F_{MNP}(\omega)$ and $\omega^2 F_{MP}(\omega)$ for models A–F, BI and PD.*

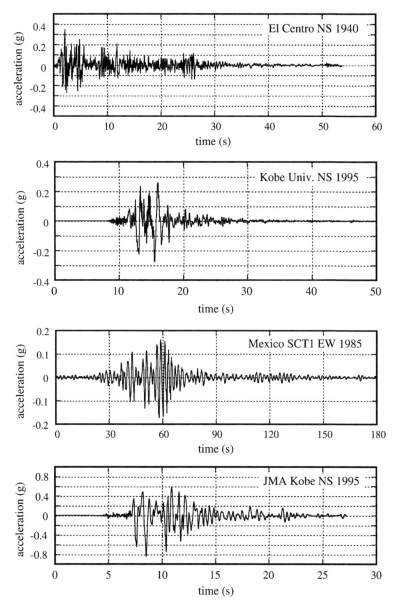

Figure 9.6 *Recorded ground motions: El Centro NS (Imperial Valley 1940), Kobe University NS (Hyogoken-Nanbu 1995), SCT1 EW (Mexico Michoacan 1985), JMA Kobe NS (Hyogoken-Nanbu 1995).*

amplitude spectrum of input ground accelerations is the same, the input energy to structures with passive control systems in appropriate and effective location is not affected much by the variability of the frequency contents of the input motion.

In order to examine the validity and accuracy of the present frequency-domain method, the earthquake input energy has been computed by both the frequency-domain method and the time-domain method to four recorded ground motions shown in Fig. 9.6. The corresponding Fourier amplitude spectra of these ground motions are shown in Fig. 9.7. The computed earthquake input energy is shown in Table 9.2. While several percent differences can be found in Models F and PD, reasonable accuracy

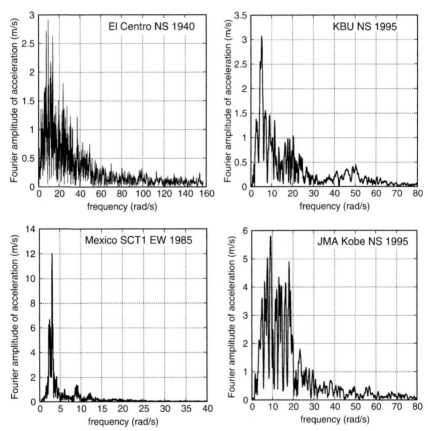

Figure 9.7 *Fourier amplitude spectra of recorded ground motions: El Centro NS (Imperial Valley 1940), Kobe University NS (Hyogoken-Nanbu 1995), SCT1 EW (Mexico Michoacan 1985), JMA Kobe NS (Hyogoken-Nanbu 1995).*

Table 9.2 Earthquake input energy by frequency and time domain analysis for models A, C, F, BI and PD to four recorded ground motions (unit in J)

		El Centro NS 1940		Kobe Univ. NS 1995		Mexico SCT1 NS 1985		JMA Kobe NS 1995	
Model A	F	1.37E+05	F	8.59E+04	F	4.21E+04	F	9.84E+05
	T	1.37E+05	T	8.52E+04	T	4.16E+04	T	9.99E+05
Model C	F	1.46E+05	F	8.54E+04	F	3.89E+04	F	1.09E+06
	T	1.47E+05	T	8.57E+04	T	3.89E+04	T	1.13E+06
Model F	F	1.18E+05	F	6.31E+04	F	2.91E+04	F	8.64E+05
	T	1.16E+05	T	6.56E+04	T	3.00E+04	T	9.25E+05
Model BI	F	1.18E+05	F	1.09E+05	F	1.80E+05	F	7.22E+05
	T	1.17E+05	T	1.09E+05	T	1.79E+05	T	7.19E+05
Model PD	F	1.16E+05	F	5.96E+04	F	2.74E+04	F	8.24E+05
(Prop. Damp.)	T	1.13E+05	T	6.22E+04	T	2.83E+04	T	8.73E+05

(**F**: frequency domain, **T**: time domain)

has been observed in other models. The cause of these discrepancies may be the difference in the integration procedures.

The computation of the Fourier amplitude spectra of ground motions is necessary only once even for different structural models. By using numerical integration in the frequency domain, the input energy can be evaluated by combining those with the energy transfer function. The energy transfer function is computed for each model via efficient modal analysis explained in this chapter. The structural designers can estimate approximate input energies from the relation of the Fourier amplitude spectra of ground motions with the energy transfer functions both of which are expressed in the frequency domain. With these data, it is not difficult to avoid resonance.

9.6. CRITICAL EXCITATION FOR EARTHQUAKE ENERGY INPUT IN MDOF SYSTEM

The following critical excitation problems may be stated for MDOF models.

Find $|A(\omega)|$

that maximizes $E_I = \int_0^\infty F_{MNP}(\omega)|A(\omega)|^2 d\omega$

Find $|V(\omega)|$

that maximizes $E_I = \int_0^\infty \omega^2 F_{MNP}(\omega)|V(\omega)|^2 d\omega$

The solution procedure developed for SDOF models in Chapter 8 can be applied to these problems without difficulty. It suffices to replace the energy transfer function $F(\omega)$ for SDOF models by $F_{MNP}(\omega)$ for MDOF models.

9.7. CONCLUSIONS

The conclusions may be stated as follows:

(1) A method based on complex modal analysis can be developed in the frequency domain for efficient computation of earthquake input energy to linear elastic passive control structures with high-level nonproportional damping in general. It requires only the computation of the Fourier amplitude spectrum of input ground motions and the derivation of velocity and acceleration-based energy transfer functions.

(2) Over-damped modes should not be neglected in highly damped passive control structures. Neglect of the over-damped modes may cause significant errors in some nonclassically damped models.

(3) When the upper bound of the Fourier amplitude spectrum of input ground accelerations is available, the formulation of earthquake input energy in the frequency domain is appropriate for deriving its bound. This is because the earthquake input energy in the frequency domain can be derived as the frequency-domain integration of the product of a non-negative time-invariant transfer function and the squared Fourier amplitude spectrum of the ground acceleration.

In this chapter, a structure with high-level damping has been treated in order to demonstrate the importance of over-damped modes in the computation of earthquake input energy. Such over-damped modes are often neglected in conventional modal analysis. The present formulation is certainly applicable to general linear elastic structures including lightly damped structures.

REFERENCES

Kuwamura, H., Kirino, Y., Akiyama, H., 1994. Prediction of earthquake energy input from smoothed Fourier amplitude spectrum. Earthquake Engrg. and Struct. Dyn. 23 (10), 1125–1137.

Lyon, R.H., 1975. Statistical Energy Analysis of Dynamical Systems. The MIT Press, Cambridge, MA.

Ohi, K., Takanashi, K., Tanaka, H., 1985. A simple method to estimate the statistical parameters of energy input to structures during earthquakes. Journal of Structural and Construction Eng., Archi. Inst. of Japan 347, 47–55 (in Japanese).

Ordaz, M., Huerta, B., Reinoso, E., 2003. Exact computation of input-energy spectra from Fourier amplitude spectra. Earthquake Engrg. and Struct. Dyn. 32 (4), 597–605.

Page, C.H., 1952. Instantaneous power spectra. Journal of Applied Physics 23 (1), 103–106.

Shinozuka, M., 1970. Maximum structural response to seismic excitations. Journal of Engineering Mechanics Division 96 (5), 729–738.

Takewaki, I., 1997. Optimal damper placement for minimum transfer functions. Earthquake Engrg. and Struct. Dyn. 26 (11), 1113–1124.

Takewaki, I., 2001a. A new method for non-stationary random critical excitation. Earthquake Engrg. and Struct. Dyn. 30 (4), 519–535.

Takewaki, I., 2001b. Probabilistic critical excitation for MDOF elastic–plastic structures on compliant ground. Earthquake Engrg. and Struct. Dyn. 30 (9), 1345–1360.

Takewaki, I., 2002a. Seismic critical excitation method for robust design: A review. J. Struct. Eng. 128 (5), 665–672.

Takewaki, I., 2002b. Robust building stiffness design for variable critical excitations. J. Struct. Eng. 128 (12), 1565–1574.

Takizawa, H., 1977. Energy-response spectrum of earthquake ground motions. In: Proc. of the 14th Natural Disaster Science Symposium. Hokkaido, pp. 359–362 (in Japanese).

Uang, C.-M., Bertero, V.V., 1990. Evaluation of seismic energy in structures. Earthquake Engrg. and Struct. Dyn. 19 (1), 77–90.

Veletsos, A.S., Ventura, C.E., 1986. Modal analysis of non-classically damped linear systems. Earthquake Engrg. and Struct. Dyn. 14 (2), 217–243.

CHAPTER TEN

Critical Excitation for Earthquake Energy Input in Soil-Structure Interaction System

Contents

10.1. INTRODUCTION

Most previous devastating earthquakes indicate that while intensive ground motions are recorded around building structures, the damage to those structures are not always serious. The selection of appropriate intensity measures (peak acceleration, peak velocity, input energy, Housner spectral intensity, Arias intensity (Arias 1970) etc.) of ground motions and soil-structure interaction (SSI) effects may be key factors to these arguments. The analysis of SSI effects is well established (e.g. Luco 1980; Cakmak et al. 1982; Wolf 1985, 1988; Gupta and Trifunac 1991; Meek and Wolf 1994; Wolf 1994) and some computer programs can be used for SSI analysis of

© 2013 Elsevier Ltd.
All rights reserved.
217

complicated models. It is also true that while the analysis of SSI effects has been focused on the investigation in terms of deformation and force, much attention has never been directed to the investigation in terms of earthquake input energy to the SSI system. There are a few investigations on this subject (Yang and Akiyama 2000; Trifunac et al. 2001). The works by Yang and Akiyama (2000) and Trifunac et al. (2001) are based on an approach in the time domain.

The purpose of this chapter is to explain a new evaluation method of earthquake input energy to SSI systems. The method is an approach in the frequency domain. Because the inertial interaction (foundation impedance) and the kinematic interaction (effective input motion) are well described by frequency-dependent functions, the present approach based on the frequency domain analysis is appropriate and effective. In particular, it is demonstrated that even SSI systems including embedded foundations can be treated in a simple way and the effects of the foundation embedment on the earthquake input energies to the super-structure and to the structure-foundation-soil system can be clarified systematically.

Earthquake ground motions involve a lot of uncertain factors in the modeling of various aspects and it does not appear easy to predict forth-coming events precisely at a specific site both in time and frequency (for example, Abrahamson et al. 1998). Some of the uncertainties may result from lack of information due to the low occurrence rate of large earthquakes and it does not seem that this problem can be resolved in the near future. In particular, the modeling of near-fault ground motions involves various uncertain factors in contrast to far-fault ground motions (Singh 1984).

It is therefore strongly desirable to develop a *robust* structural design method taking into account these uncertainties, and with limited informa-tion enabling the design of safer structures for a broader class of design earthquake (Drenick 1970; Shinozuka 1970; Takewaki 2001a, b, 2002a, b). It will be shown briefly in this chapter that the formulation of earthquake input energy in the frequency domain is effective for solving a critical excitation problem and deriving a bound on the earthquake input energy for a class of properly scaled ground motions.

Fig. 10.1 shows a schematic diagram of energy flow in SSI systems. It will be discussed in this chapter that the work by boundary forces on their corresponding displacements defined for various boundaries are utilized in evaluating the energy flow in SSI systems. Fig. 10.2 indicates (a) the free-body including a structure and its surrounding ground, (b) the free-body including the structure alone. These free-body diagrams are very useful in understanding the energy flow in SSI systems. The time histories of earthquake input energies to these subassemblages are shown in Fig. 10.3 for El Centro NS, Imperial Valley 1940 and Kobe University NS, Hyogoken-Nanbu 1995.

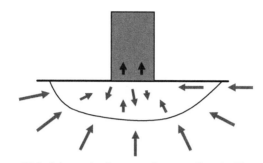

Figure 10.1 *Schematic diagram of energy flow in SSI system.*

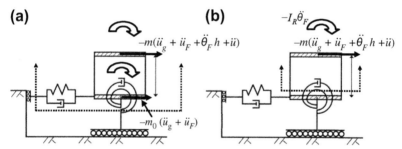

Figure 10.2 *(a) Free-body including structure and surrounding ground; (b) free-body including structure alone.*

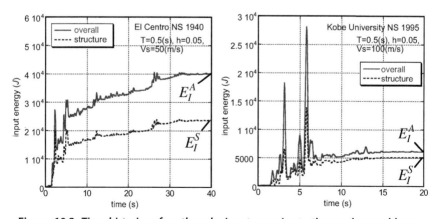

Figure 10.3 *Time histories of earthquake input energies to these subassemblages.*

10.2. EARTHQUAKE INPUT ENERGY TO FIXED-BASE SINGLE-DEGREE-OF-FREEDOM (SDOF) SYSTEM

A lot of work has been accumulated on the topics of earthquake input energy (for example, Tanabashi 1935; Housner 1956, 1959; Berg and Thomaides 1960; Housner and Jennings 1975; Kato and Akiyama 1975; Takizawa 1977; Mahin and Lin 1983; Zahrah and Hall 1984; Akiyama 1985; Ohi et al. 1985; Uang and Bertero 1990; Leger and Dussault 1992; Kuwamura et al. 1994; Fajfar and Vidic 1994; Riddell and Garcia 2001; Ordaz et al. 2003). In contrast to most of the previous work, the earthquake input energy is formulated here in the frequency domain (Page 1952; Lyon 1975, Takizawa 1977; Ohi et al. 1985; Ordaz et al. 2003) to facilitate both the derivation of bound of earthquake input energy and the formulation of dynamic soil-structure interaction in terms of kinematic and inertial effects. For later comparison between the fixed-base model and the SSI system, the formulation for the fixed-base model is presented first.

Consider a damped linear elastic SDOF system of mass m, stiffness k and damping coefficient c. Let $\Omega = \sqrt{k/m}$, and x denote the undamped natural circular frequency of the SDOF system, the damping ratio and the displacement of the mass relative to the ground, respectively. Time derivative is denoted by an over-dot. The input energy to the SDOF system by a unidirectional horizontal ground acceleration $\ddot{u}_g(t) = a(t)$ from $t = 0$ to $t = t_0$ (end of input) can be defined by the work on the ground made to the structural system and is expressed by

$$E_I = \int_0^{t_0} m(\ddot{u}_g + \ddot{x})\dot{u}_g dt \tag{10.1}$$

The term $m(\ddot{u}_g + \ddot{x})$ indicates the inertial force (although $-m(\ddot{u}_g + \ddot{x})$ is the exact inertial force) and is equal to the sum of the restoring force kx and the damping force $c\dot{x}$ in the system. Integration by parts of Eq. (10.1) provides

$$E_I = \int_0^{t_0} m(\ddot{x} + \ddot{u}_g)\dot{u}_g dt = \int_0^{t_0} m\ddot{x}\dot{u}_g dt + [(1/2)m\dot{u}_g^2]_0^{t_0}$$

$$= [m\dot{x}\dot{u}_g]_0^{t_0} - \int_0^{t_0} m\dot{x}\ddot{u}_g dt + [(1/2)m\dot{u}_g^2]_0^{t_0} \tag{10.2}$$

If $\dot{x} = 0$ at $t = 0$ and $\dot{u}_g = 0$ at $t = 0$ and $t = t_0$, the input energy can be reduced to the following form.

$$E_I = -\int_0^{t_0} m\ddot{u}_g \dot{x}\,dt \tag{10.3}$$

It is known (Page 1952; Lyon 1975; Takizawa 1977; Ohi et al. 1985; Ordaz et al. 2003) that the input energy per unit mass can also be expressed in the frequency domain by use of Fourier transformation so far as the system is linear elastic.

$$E_I/m = -\int_{-\infty}^{\infty} \dot{x}a\,dt = -\int_{-\infty}^{\infty} \left[(1/2\pi)\int_{-\infty}^{\infty} \dot{X}e^{i\omega t}\,d\omega \right] a\,dt$$

$$= -(1/2\pi)\int_{-\infty}^{\infty} A(-\omega)\{H_V(\omega\,;\Omega,\xi)A(\omega)\}\,d\omega$$

$$= \int_0^{\infty} |A(\omega)|^2 \{-\mathrm{Re}[H_V(\omega\,;\Omega,\xi)]/\pi\}\,d\omega \tag{10.4}$$

$$\equiv \int_0^{\infty} |A(\omega)|^2 F(\omega)\,d\omega$$

where $H_V(\omega\,;\Omega,\xi)$ is the transfer function defined by $\dot{X}(\omega) = H_V(\omega\,;\Omega,\xi)A(\omega)$ and $F(\omega) = -\mathrm{Re}[H_V(\omega\,;\Omega,\xi)]/\pi$. $\mathrm{Re}[\cdot]$ denotes a real part of a complex number. $\dot{X}$ and $A(\omega)$ are the Fourier transforms of $\dot{x}$ and $\ddot{u}_g(t) = a(t)$, respectively. The symbol i denotes the imaginary unit. The velocity transfer function $H_V(\omega\,;\Omega,\xi)$ can be expressed by

$$H_V(\omega\,;\Omega,\xi) = -i\omega/(\Omega^2 - \omega^2 + 2i\xi\Omega\omega) \tag{10.5}$$

Eq. (10.4) indicates that the earthquake input energy to damped linear elastic SDOF systems does not depend on the phase of input motions and this fact is well known (Page 1952; Lyon 1975; Takizawa 1977; Ohi et al. 1985; Kuwamura et al. 1994; Ordaz et al. 2003).

10.3. EARTHQUAKE INPUT ENERGY TO SSI SYSTEMS

Consider a linear elastic SDOF super-structure of story stiffness k and story damping coefficient c, as shown in Fig. 10.4, with a cylindrical rigid

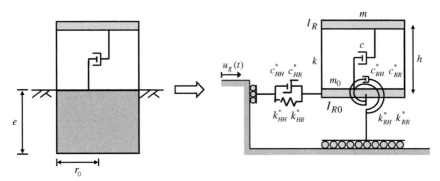

Figure 10.4 SDOF structure with embedded foundation and its modeling for analysis.

foundation embedded in the uniform half-space ground. Let r_0 and e denote the radius and the depth of the foundation, respectively. Let m and I_R denote the mass of the super-structure and the mass moment of inertia of the super-structure and let m_0 and I_{R0} denote the mass of the embedded foundation and the mass moment of inertia of the embedded foundation "around its top center node." The height of the super-structure mass from the ground surface is denoted by h.

U_0^* and Θ_0^* are the horizontal and rotational effective input motions in the frequency domain at the top center of the foundation. The corresponding effective input motions in the time domain may be expressed by

$$\ddot{u}_0^*(t) = \frac{1}{2\pi} \int_{-\infty}^{\infty} \ddot{U}_0^*(\omega) e^{i\omega t} d\omega = \frac{1}{2\pi} \int_{-\infty}^{\infty} S_{HT}(\omega) \ddot{U}_g(\omega) e^{i\omega t} d\omega \quad (10.6a)$$

$$\ddot{\theta}_0^*(t) = \frac{1}{2\pi} \int_{-\infty}^{\infty} \ddot{\Theta}_0^*(\omega) e^{i\omega t} d\omega = \frac{1}{2\pi} \int_{-\infty}^{\infty} S_{RT}(\omega) \ddot{U}_g(\omega) e^{i\omega t} d\omega$$

$$(10.6b)$$

$S_{HT}(\omega)$ and $S_{RT}(\omega)$ are the ratios of the horizontal and rotational effective input motions, U_0^* and Θ_0^*, in the frequency domain at the top center of the foundation to the Fourier transform $U_g(\omega)$ of the free-field horizontal ground-surface displacement. Assume that a vertically incident shear wave (SH wave) is considered. $S_{HT}(\omega)$ and $S_{RT}(\omega)$ are expressed in terms of the ratios $S_{HB}(\omega)$ and $S_{RB}(\omega)$, given in Meek and Wolf (1994) and Wolf (1994), of the horizontal and rotational ($\times r_0$) effective input motions in the frequency domain at the bottom center of the foundation to $U_g(\omega)$.

$$S_{HT}(\omega) = S_{HB}(\omega) + (e/r_0)S_{RB}(\omega) \qquad (10.7a)$$

$$S_{RT}(\omega) = (1/r_0)S_{RB}(\omega) \qquad (10.7b)$$

Let U, U_T, Θ_T denote the Fourier transform of the horizontal displacement u of the super-structural mass relative to the foundation without rocking component, the Fourier transform of the horizontal displacement u_T of the top center node of the foundation relative to u_0^* and the Fourier transform of the angle of rotation θ_T of the foundation relative to θ_0^*, respectively (see Fig. 10.5). The set of these components is denoted by $\mathbf{U} = \{\, U \quad U_T \quad \Theta_T \,\}^T$. The Fourier transforms of the force and moment corresponding to U_T and Θ_T are expressed by P_T and M_T. $\ddot{U}_g(\omega)$ denotes the Fourier transform of the free-field horizontal ground-surface acceleration.

The equations of motion in the frequency domain of the SDOF super-structure supported by the embedded rigid foundation and subjected to the effective input motions $\ddot{U}_0^*(= -\omega^2 U_0^* = S_{HT}(\omega)\ddot{U}_g(\omega))$ and $\ddot{\Theta}_0^*(= -\omega^2\Theta_0^* = S_{RT}(\omega)\ddot{U}_g(\omega))$ may be written as

$$\{-\omega^2\mathbf{M} + i\omega[\mathbf{C}_S + \mathbf{C}_F(\omega)] + [\mathbf{K}_S + \mathbf{K}_F(\omega)]\}\mathbf{U}(\omega)$$
$$= -\mathbf{M}\{\mathbf{r}_1 S_{HT}(\omega) + \mathbf{r}_2 S_{RT}(\omega)\}\ddot{U}_g(\omega) \qquad (10.8)$$

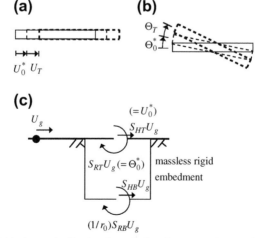

Figure 10.5 *(a) Horizontal effective foundation input motion and relative displacement; (b) rotational effective foundation input motion and relative angle of rotation; (c) effective foundation input motion at the top center and bottom center of the foundation.*

where $\mathbf{M}$ is the system mass matrix and $\mathbf{K}_S, \mathbf{K}_F, \mathbf{C}_S, \mathbf{C}_F$ are the element stiffness and damping matrices related to the structure and the foundation. These quantities are expressed by

$$\mathbf{M} = \begin{bmatrix} m & m & mh \\ m & m_0 + m & mh \\ mh & mh & mh^2 + I_{R0} + I_R \end{bmatrix}, \quad \mathbf{K}_S = \begin{bmatrix} k & 0 & 0 \\ 0 & 0 & 0 \\ 0 & 0 & 0 \end{bmatrix},$$

$$\mathbf{K}_F = \begin{bmatrix} 0 & 0 & 0 \\ 0 & k_{HH}^* & k_{HR}^* \\ 0 & k_{RH}^* & k_{RR}^* \end{bmatrix}, \quad \mathbf{C}_S = \begin{bmatrix} c & 0 & 0 \\ 0 & 0 & 0 \\ 0 & 0 & 0 \end{bmatrix}, \quad \mathbf{C}_F = \begin{bmatrix} 0 & 0 & 0 \\ 0 & c_{HH}^* & c_{HR}^* \\ 0 & c_{RH}^* & c_{RR}^* \end{bmatrix}$$

$$(10.9a-e)$$

Furthermore the quantities $\mathbf{U}(\omega), \mathbf{r}_1, \mathbf{r}_2$ in Eq. (10.8) are defined by

$$\mathbf{U}(\omega) = \{ U(\omega) \quad U_T(\omega) \quad \Theta_T(\omega) \}^T, \mathbf{r}_1 = \{0 \ 1 \ 0\}^T,$$

$$\mathbf{r}_2 = \{0 \ 0 \ 1\}^T \qquad (10.10a-c)$$

The coefficients k_{HH}^*, k_{HR}^*, k_{RH}^*, k_{RR}^* are the frequency-dependent foundation stiffnesses at the top center node of the foundation and c_{HH}^*, c_{HR}^*, c_{RH}^*, c_{RR}^* are the frequency-dependent foundation damping coefficients at the top center node of the foundation. These coefficients will be derived later.

Let U_B and Θ_B denote the Fourier transform of the horizontal displacement of the bottom center node of the foundation and the Fourier transform of the angle of rotation of the foundation, respectively. The Fourier transforms of the force and moment corresponding to U_B and Θ_B are expressed by P_B and M_B. The foundation impedances at the bottom center of the foundation may be expressed as

$$\begin{Bmatrix} P_B \\ M_B \end{Bmatrix} = \left(\begin{bmatrix} k_{HH} & k_{HR} \\ k_{RH} & k_{RR} \end{bmatrix} + i\omega \begin{bmatrix} c_{HH} & c_{HR} \\ c_{RH} & c_{RR} \end{bmatrix} \right) \begin{Bmatrix} U_B \\ \Theta_B \end{Bmatrix}$$

$$\equiv (\mathbf{K}_{FF} + i\omega \, \mathbf{C}_{FF}) \begin{Bmatrix} U_B \\ \Theta_B \end{Bmatrix} \qquad (10.11)$$

$\mathbf{K}_{FF}$ and $\mathbf{C}_{FF}$ are given in Meek and Wolf (1994) and Wolf (1994). Their components are expressed as

$$k_{HH}(\omega) + i\omega\, c_{HH}(\omega) = \frac{8\rho V_S^2 r_0}{2-\nu}\left(1+\frac{e}{r_0}\right)\left\{k_H(a_0) + ia_0 c_H(a_0)\right\}$$

(10.12a)

$$k_{RR}(\omega) + i\omega\, c_{RR}(\omega) = \frac{8\rho V_S^2 r_0^{\,3}}{3(1-\nu)}\left\{1 + 2.3\left(\frac{e}{r_0}\right) + 0.58\left(\frac{e}{r_0}\right)^3\right\}$$

$$\times\left\{k_R(a_0) + ia_0 c_R(a_0)\right\}$$

(10.12b)

$$k_{HR}(\omega) = k_{RH}(\omega) = \frac{e}{3}k_{HH}(\omega)$$

(10.12c)

$$c_{HR}(\omega) = c_{RH}(\omega) = \frac{e}{3}c_{HH}(\omega)$$

(10.12d)

where ρ, V_S, ν are the soil mass density, the ground shear wave velocity and the Poisson's ratio of soil and $a_0 = \omega r_0/V_S$ is the nondimensional frequency. The nondimensional parameters in Eqs. (10.12a, b) are given approximately as follows for various nondimensional depths of embedment: $k_H(a_0) = 1.0$ for $e/r_0 = 0.0$, 0.5, 1.0, 2.0; $c_H(a_0) = 0.6$, 1.05, 1.3, 1.75 for $e/r_0 = 0.0$, 0.5, 1.0, 2.0, respectively; $k_R(a_0)$ and $c_R(a_0)$ are shown in Takewaki et al. (2003).

The stiffness and damping coefficients in Eq. (10.9) may be derived as follows. The displacements of the node at the top center of the foundation are related to those at the bottom center (see Fig. 10.6). This relation may be expressed by

$$\begin{Bmatrix} U_T \\ \Theta_T \end{Bmatrix} = \begin{bmatrix} 1 & e \\ 0 & 1 \end{bmatrix}\begin{Bmatrix} U_B \\ \Theta_B \end{Bmatrix} \equiv \mathbf{T}\begin{Bmatrix} U_B \\ \Theta_B \end{Bmatrix}, \quad \begin{Bmatrix} U_B \\ \Theta_B \end{Bmatrix} = \mathbf{T}^{-1}\begin{Bmatrix} U_T \\ \Theta_T \end{Bmatrix}$$

(10.13a, b)

The inverse of the coefficient matrix $\mathbf{T}$ can be expressed as

$$\mathbf{T}^{-1} = \begin{bmatrix} 1 & -e \\ 0 & 1 \end{bmatrix} \equiv \mathbf{T}^*$$

(10.14)

The nodal force and moment at the top center of the foundation can be expressed in terms of those at the bottom center.

$$\begin{Bmatrix} P_T \\ M_T \end{Bmatrix} = \begin{bmatrix} 1 & 0 \\ -e & 1 \end{bmatrix}\begin{Bmatrix} P_B \\ M_B \end{Bmatrix} \equiv \mathbf{T}^{*T}\begin{Bmatrix} P_B \\ M_B \end{Bmatrix}$$

(10.15)

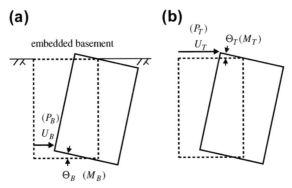

Figure 10.6 *(a) Horizontal displacement and angle of rotation of the bottom of the foundation; (b) horizontal displacement and angle of rotation of the top of the foundation.*

Substitution of Eqs. (10.11), (10.13b), (10.14) into Eq. (10.15) provides

$$\left\{ \begin{array}{c} P_T \\ M_T \end{array} \right\} = (\mathbf{T}^{*T}\mathbf{K}_{FF}\mathbf{T}^* + i\omega\mathbf{T}^{*T}\mathbf{C}_{FF}\mathbf{T}^*)\left\{ \begin{array}{c} U_T \\ \Theta_T \end{array} \right\}$$

$$\equiv \left(\begin{bmatrix} k^*_{HH} & k^*_{HR} \\ k^*_{RH} & k^*_{RR} \end{bmatrix} + i\omega \begin{bmatrix} c^*_{HH} & c^*_{HR} \\ c^*_{RH} & c^*_{RR} \end{bmatrix} \right)\left\{ \begin{array}{c} U_T \\ \Theta_T \end{array} \right\}$$

(10.16)

The solution of Eq. (10.8) leads to the following expression.

$$\frac{\mathbf{U}(\omega)}{\ddot{U}_g(\omega)} = -\{-\omega^2\mathbf{M} + i\omega\,(\mathbf{C}_S + \mathbf{C}_F) + (\mathbf{K}_S + \mathbf{K}_F)\}^{-1}\mathbf{M}\{\mathbf{r}_1 S_{HT}(\omega)$$

$$+ \mathbf{r}_2 S_{RT}(\omega)\}$$

(10.17)

Consider the earthquake input energy to the present model. The earthquake input energy E_I^A to the structure including the foundation mass may be obtained by considering the work by the forces acting just below the foundation mass on the foundation displacements (see Fig. 10.7(a)).

$$E_I^A = \int_0^\infty [m_0(\ddot{u}_0^* + \ddot{u}_T) + m\{\ddot{u}_0^* + \ddot{u}_T + h(\ddot{\theta}_0^* + \ddot{\theta}_T) + \ddot{u}\}](\dot{u}_0^* + \dot{u}_T)dt$$

$$+ \int_0^\infty [(I_{R0} + I_R)(\ddot{\theta}_0^* + \ddot{\theta}_T) + mh\{\ddot{u}_0^* + \ddot{u}_T + h(\ddot{\theta}_0^* + \ddot{\theta}_T) + \ddot{u}\}]$$

$$\times \, (\dot{\theta}_0^* + \dot{\theta}_T)\mathrm{d}t = \frac{1}{2\pi} \int_{-\infty}^{\infty} \frac{\mathrm{i}}{\omega}\bigg[m_0\{S_{HT}(\omega) + H_{\ddot{U}T}(\omega)\}$$

$$+ \, m\{S_{HT}(\omega) + H_{\ddot{U}T}(\omega) + h\{S_{RT}(\omega) + H_{\ddot{\Theta}T}(\omega)\} + H_{\ddot{U}}(\omega)\}\bigg] \cdot$$

$$\{S_{HT}(-\omega) + H_{\ddot{U}T}(-\omega)\}\big|\ddot{U}_g(\omega)\big|^2 \mathrm{d}\omega$$

$$+ \frac{1}{2\pi} \int_{-\infty}^{\infty} \frac{\mathrm{i}}{\omega}[(I_{R0} + I_R)\{S_{RT}(\omega) + H_{\ddot{\Theta}T}(\omega)\} \, + mh\{S_{HT}(\omega)$$

$$+ \, H_{\ddot{U}T}(\omega) + h\{S_{RT}(\omega) + H_{\ddot{\Theta}T}(\omega)\} + H_{\ddot{U}}(\omega)\}] \cdot$$

$$\{S_{RT}(-\omega) + H_{\ddot{\Theta}T}(-\omega)\}\big|\ddot{U}_g(\omega)\big|^2 \mathrm{d}\omega$$

$$(10.18)$$

where the transfer functions $H_{\ddot{U}}(\omega), H_{\ddot{U}T}(\omega), H_{\ddot{\Theta}T}(\omega)$ are the ratios of $\ddot{U}(\omega), \ddot{U}_T(\omega), \ddot{\Theta}_T(\omega)$ to $\ddot{U}_g(\omega)$ obtained from Eq. (10.17). In Eq. (10.18), the Fourier inverse transformation of accelerations and velocities are used. The terms $S_{HT}(-\omega)$, $H_{\ddot{U}T}(-\omega)$, $S_{RT}(-\omega)$ and $H_{\ddot{\Theta}T}(-\omega)$ in Eq. (10.18) result from the Fourier transformation of the corresponding time–domain quantities. The treatment of these quantities will be discussed after Eq. (10.20). When the term $\dot{u}_0^* + \dot{u}_T$ in the first line of Eq. (10.18) is replaced by $\dot{u}_g$ and the second line term is neglected, the resulting energy indicates the earthquake input energy E_I^{A*} to the structure-foundation system including the soil springs and dashpots (see Fig. 10.7(b)).

$$E_I^{A*} = \int_0^{\infty} [m_0(\ddot{u}_0^* + \ddot{u}_T) + m\{\ddot{u}_0^* + \ddot{u}_T + h(\ddot{\theta}_0^* + \ddot{\theta}_T) + \ddot{u}\}]\dot{u}_g \mathrm{d}t$$

$$= \frac{1}{2\pi} \int_{-\infty}^{\infty} \frac{\mathrm{i}}{\omega}[m_0\{S_{HT}(\omega) + H_{\ddot{U}T}(\omega)\}$$

$$+ \, m\{S_{HT}(\omega) + H_{\ddot{U}T}(\omega) + h\{S_{RT}(\omega) + H_{\ddot{\Theta}T}(\omega)\} + H_{\ddot{U}}(\omega)\}] \cdot$$

$$\{S_{HT}(-\omega) + H_{\ddot{U}T}(-\omega)\}\big|\ddot{U}_g(\omega)\big|^2 \mathrm{d}\omega$$

$$(10.19)$$

In a similar way, the earthquake input energy E_I^S to the structure may be obtained by considering the work by the forces acting just above the foundation mass on the foundation displacements (see Fig. 10.7(c)).

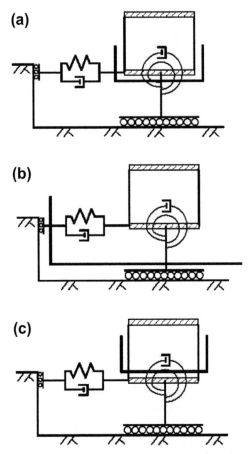

Figure 10.7 *Earthquake input energy (a) to structure-foundation system, (b) to structure-foundation-soil system, (c) to structure alone.*

$$E_I^S = \int_0^\infty m\{\ddot{u}_0^* + \ddot{u}_T + h(\ddot{\theta}_0^* + \ddot{\theta}_T) + \ddot{u}\}(\dot{u}_0^* + \dot{u}_T)\mathrm{d}t$$

$$+ \int_0^\infty [I_R(\ddot{\theta}_0^* + \ddot{\theta}_T) + mh\{\ddot{u}_0^* + \ddot{u}_T + h(\ddot{\theta}_0^* + \ddot{\theta}_T) + \ddot{u}\}](\dot{\theta}_0^* + \dot{\theta}_T)\mathrm{d}t$$

$$= \frac{1}{2\pi} \int_{-\infty}^\infty \frac{\mathrm{i}}{\omega} m[S_{HT}(\omega) + H_{\ddot{U}T}(\omega) + h\{S_{RT}(\omega)$$

$$+ H_{\ddot{\Theta}T}(\omega)\} + H_{\ddot{U}}(\omega)].$$

$$\{S_{HT}(-\omega) + H_{\ddot{U}T}(-\omega)\}|\ddot{U}_g(\omega)|^2 d\omega$$

$$+ \frac{1}{2\pi} \int_{-\infty}^{\infty} \frac{i}{\omega}[I_R\{S_{RT}(\omega) + H_{\ddot{\Theta}T}(\omega)\}$$

$$+ mh\{S_{HT}(\omega) + H_{\ddot{U}T}(\omega) + h\{S_{RT}(\omega) + H_{\ddot{\Theta}T}(\omega)\} + H_{\ddot{U}}(\omega)\}]\cdot$$

$$\{S_{RT}(-\omega) + H_{\ddot{\Theta}T}(-\omega)\}|\ddot{U}_g(\omega)|^2 d\omega$$

$$(10.20)$$

It should be noted that, from Eqs. (10.6a, b) and the property of time-histories as sets of real numbers, $S_{HT}(-\omega)$ and $S_{RT}(-\omega)$ can be computed as the complex conjugate pairs $S_{HT}^*(\omega)$ and $S_{RT}^*(\omega)$ of $S_{HT}(\omega)$ and $S_{RT}(\omega)$, and $H_{\ddot{U}T}(-\omega)$ and $H_{\ddot{\Theta}T}(-\omega)$ can be computed as the complex conjugate pairs $H_{\ddot{U}T}^*(\omega)$ and $H_{\ddot{\Theta}T}^*(\omega)$ of $H_{\ddot{U}T}(\omega)$ and $H_{\ddot{\Theta}T}(\omega)$. It should also be pointed out that E_I^A is equal to E_I^S exactly because, as the time approaches infinity, the kinetic energy of the foundation mass becomes zero due to the existence of damping in the system. This can be proved by the following manipulation.

$$E_I^A = \int_0^{\infty} [m_0(\ddot{u}_0^* + \ddot{u}_T) + m\{\ddot{u}_0^* + \ddot{u}_T + h(\ddot{\theta}_0^* + \ddot{\theta}_T) + \ddot{u}\}](\dot{u}_0^* + \dot{u}_T)dt$$

$$+ \int_0^{\infty} [(I_{R0} + I_R)(\ddot{\theta}_0^* + \ddot{\theta}_T) + mh\{\ddot{u}_0^* + \ddot{u}_T + h(\ddot{\theta}_0^* + \ddot{\theta}_T) + \ddot{u}\}]\cdot$$

$$(\dot{\theta}_0^* + \dot{\theta}_T)dt$$

$$= [(1/2)m_0(\dot{u}_0^* + \dot{u}_T)^2]_0^{\infty} + [(1/2)I_{R0}(\dot{\theta}_0^* + \dot{\theta}_T)^2]_0^{\infty}$$

$$+ \int_0^{\infty} m\{\ddot{u}_0^* + \ddot{u}_T + h(\ddot{\theta}_0^* + \ddot{\theta}_T) + \ddot{u}\}(\dot{u}_0^* + \dot{u}_T)dt$$

$$+ \int_0^{\infty} [I_R(\ddot{\theta}_0^* + \ddot{\theta}_T) + mh\{\ddot{u}_0^* + \ddot{u}_T + h(\ddot{\theta}_0^* + \ddot{\theta}_T) + \ddot{u}\}](\dot{\theta}_0^* + \dot{\theta}_T)dt$$

$$= E_I^S$$

$$(10.21)$$

This fact will be confirmed in the numerical examples shown later. Eqs. (10.18), (10.19) and (10.20) can also be expressed compactly by

$$E_I^A = \int_0^\infty F_A(\omega) |\ddot{U}_g(\omega)|^2 d\omega \qquad (10.22a)$$

$$E_I^{A*} = \int_0^\infty F_A^*(\omega) |\ddot{U}_g(\omega)|^2 d\omega \qquad (10.22b)$$

$$E_I^S = \int_0^\infty F_S(\omega) |\ddot{U}_g(\omega)|^2 d\omega \qquad (10.22c)$$

where $F_A(\omega)$ and $F_S(\omega)$ are the following energy transfer functions:

$$
\begin{aligned}
F_A(\omega) = &-\frac{1}{\pi\omega} \mathrm{Im}[[m_0\{S_{HT}(\omega) + H_{\ddot{U}T}(\omega)\} + m\{S_{HT}(\omega) + H_{\ddot{U}T}(\omega) \\
&+ h(S_{RT}(\omega) + H_{\ddot{\Theta}T}(\omega)) + H_{\ddot{U}}(\omega)\}] \cdot \{S_{HT}^*(\omega) + H_{\ddot{U}T}^*(\omega)\}] \\
&- \frac{1}{\pi\omega} \mathrm{Im}[[(I_{R0} + I_R)\{S_{RT}(\omega) + H_{\ddot{\Theta}T}(\omega)\} + mh\{S_{HT}(\omega) \\
&+ H_{\ddot{U}T}(\omega) + h(S_{RT}(\omega) + H_{\ddot{\Theta}T}(\omega)) + H_{\ddot{U}}(\omega)\}] \cdot \{S_{RT}^*(\omega) \\
&+ H_{\ddot{\Theta}T}^*(\omega)\}]
\end{aligned}
$$

$$(10.23a)$$

$$
\begin{aligned}
F_A^*(\omega) = &-\frac{1}{\pi\omega} \mathrm{Im}[m_0\{S_{HT}(\omega) + H_{\ddot{U}T}(\omega)\} + m\{S_{HT}(\omega) + H_{\ddot{U}T}(\omega) \\
&+ h(S_{RT}(\omega) + H_{\ddot{\Theta}T}(\omega)) + H_{\ddot{U}}(\omega)\}]
\end{aligned}
$$

$$(10.23b)$$

$$
\begin{aligned}
F_S(\omega) = &-\frac{1}{\pi\omega} \mathrm{Im}[m[S_{HT}(\omega) + H_{\ddot{U}T}(\omega) + h(S_{RT}(\omega) + H_{\ddot{\Theta}T}(\omega)) \\
&+ H_{\ddot{U}}(\omega)] \cdot \{S_{HT}^*(\omega) + H_{\ddot{U}T}^*(\omega)\}] - \frac{1}{\pi\omega} \mathrm{Im}[[I_R\{S_{RT}(\omega) \\
&+ H_{\ddot{\Theta}T}(\omega)\} + mh\{S_{HT}(\omega) + H_{\ddot{U}T}(\omega) + h(S_{RT}(\omega) \\
&+ H_{\ddot{\Theta}T}(\omega)) + H_{\ddot{U}}(\omega)\}] \cdot \{S_{RT}^*(\omega) + H_{\ddot{\Theta}T}^*(\omega)\}]
\end{aligned}
$$

$$(10.23c)$$

Eqs. (10.22a–c) imply that, if the energy transfer functions $F_A(\omega)$, $F_A^*(\omega)$ and $F_S(\omega)$ are calculated for every model as a function of excitation frequency, the corresponding critical excitation problem for the maximum input energy under the constraint $\int_{-\infty}^{\infty} \ddot{u}_g(t)^2 dt = (1/\pi) \int_0^{\infty} |\ddot{U}_g(\omega)|^2 d\omega = \overline{C}_A$ on acceleration can be stated and the Dirac delta function or the rectangular function of $|\ddot{U}_g(\omega)|$ can be found to be the critical input maximizing the input energy to the overall system and the structure (Takewaki 2002a, 2004).

Another constraint $\int_{-\infty}^{\infty} \dot{u}_g(t)^2 dt = (1/\pi) \int_0^{\infty} |\dot{U}_g(\omega)|^2 d\omega = \overline{C}_V$ on velocity may be an appropriate constraint in the longer natural period range of SSI systems (Tanabashi 1956; Housner and Jennings 1975; Trifunac et al. 2001). The corresponding critical excitation problem will be discussed elsewhere.

10.4. ACTUAL EARTHQUAKE INPUT ENERGY TO FIXED-BASE MODEL AND SSI SYSTEM

In order to examine the effect of degree of embedment of the foundation on the earthquake input energy to the structure under soil-structure interaction environment, the earthquake input energies to the corresponding fixed-base model and to the sway-rocking model have been computed by the proposed frequency domain method. Four recorded ground motions, shown in Fig. 10.8, are taken as the input motions. The corresponding Fourier amplitude spectra $|\ddot{U}_g(\omega)|$ of these ground motion accelerations are shown in Fig. 10.9.

Four cases of embedment $e/r_0 = 0.0, 0.5, 1.0, 2.0$ are considered and three cases of equivalent ground shear velocity $V_S = 50, 100, 200(\text{m/s})$ are treated. It should be noted that the rather small ground shear wave velocity $V_S = 50(\text{m/s})$ is not necessarily unrealistic if the dependence of ground shear wave velocity on the experienced shear strain level is considered. In fact, this low shear wave velocity represents an "equivalent quantity" evaluated for a rather high seismic input. The present author clarified that, under such intense seismic input, the equivalent shear wave velocity could become 10–20% of the initial one for a low strain level. It is also confirmed in some case studies that, even under such intense input, effective shear strains are smaller than 1% and no liquefaction occurs. It is the case in some areas near bays that a very soft ground exists and the investigation on the effect of such soft ground on the partially embedded structures is urgently required. It should be remarked that the analysis of

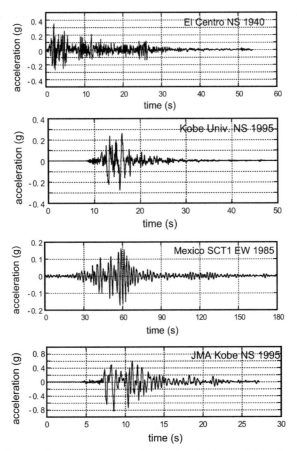

Figure 10.8 *Recorded ground motions: El Centro NS (Imperial Valley 1940), Kobe University NS (Hyogoken-Nanbu 1995), SCT1 EW (Mexico Michoacan 1985), JMA Kobe NS (Hyogoken-Nanbu 1995).*

extreme or limit states is inevitable for the reliable design of foundations and building structures.

    The other given parameters are as follows: $r_0 = 5(\text{m})$, $m = 62.8 \times 10^3 (\text{kg})$, $I_R = 5.23 \times 10^5 (\text{kg} \cdot \text{m}^2)$, $h = 4(\text{m})$, $k = \Omega^2 m$, $c = (2\xi/\Omega)k$, $\rho = 1.8 \times 10^3 (\text{kg/m}^3)$, $\nu = 0.25$, the damping ratio of the super-structure $\xi = 0.05$. The fundamental natural period of the fixed–base super-structure T is varied and $\Omega = 2\pi/T$. The foundation mass m_0 and mass moment of inertia I_{R0} are varied as the depth of embedment is changed; $m_0 = 1.88 \times 10^5 (\text{kg})$ for $e/r_0 = 0$, $m_0 = 2.52 \times 10^5 (\text{kg})$ for $e/r_0 = 0.5$, $m_0 = 3.60 \times 10^5 (\text{kg})$ for $e/r_0 = 1.0$, $m_0 = 6.48 \times 10^5 (\text{kg})$ for

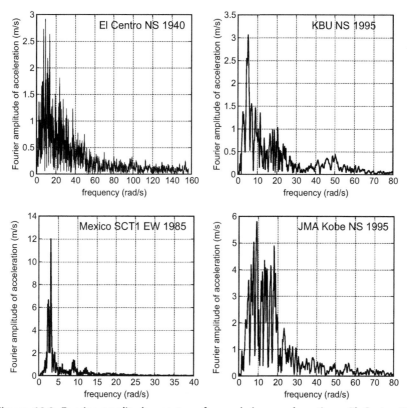

Figure 10.9 *Fourier amplitude spectra of recorded ground motions: El Centro NS (Imperial Valley 1940), Kobe University NS (Hyogoken-Nanbu 1995), SCT1 EW (Mexico Michoacan 1985), JMA Kobe NS (Hyogoken-Nanbu 1995).*

$e/r_0 = 2.0$; $I_{R0} = 1.57 \times 10^6 (\mathrm{kg \cdot m^2})$ for $e/r_0 = 0$, $I_{R0} = 2.62 \times 10^6$ $(\mathrm{kg \cdot m^2})$ for $e/r_0 = 0.5$, $I_{R0} = 6.00 \times 10^6 (\mathrm{kg \cdot m^2})$ for $e/r_0 = 1.0$, $I_{R0} = 2.70 \times 10^7 (\mathrm{kg \cdot m^2})$ for $e/r_0 = 2.0$. The foundation mass m_0 has been calculated by adding the mass of the basement to the ground floor mass. Therefore m_0 is not proportional to the degree of embedment e/r_0.

Fig. 10.10 illustrates the earthquake input energies to the fixed-base model E_I and the sway-rocking models E_I^S (structure only) with various degrees of embedment for El Centro NS (Imperial Valley 1940) plotted with respect to the natural period T of the super-structure. It can be observed that the input energy to the sway-rocking model without embedment may be almost the same as that for the fixed-base model. The rotational degree is considered in the sway-rocking model, but the

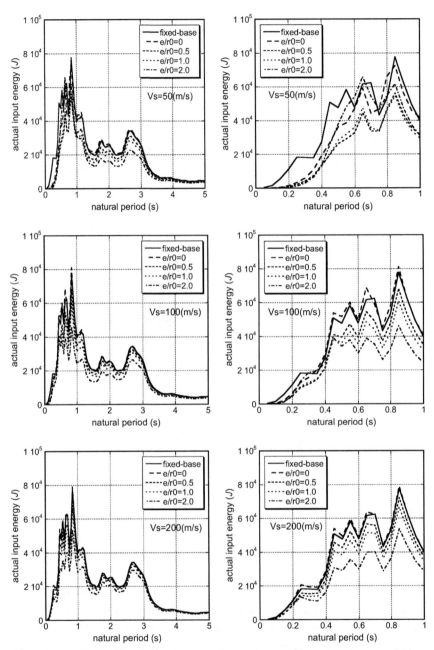

Figure 10.10 *Earthquake input energy to the structure with various degrees of foundation embedment for ground equivalent shear velocities 50, 100, 200(m/s) to El Centro NS of Imperial Valley 1940 (magnified figures are shown on the right-hand side).*

soil-structure interaction effect on the input energy is rather small in this case. On the other hand, as the degree of embedment becomes larger, the input energy is decreased regardless of the natural period range. This phenomenon may be one of the reasons that, while intensive ground motions are recorded around building structures, the damage to those structures is not always serious.

Figs. 10.11–10.13 show the earthquake input energies to the fixed-base model E_I and the sway-rocking models E_I^S with various degrees of embedment for Kobe University NS (Hyogoken-Nanbu 1995), SCT1 EW (Mexico Michoacan 1985) and JMA Kobe NS (Hyogoken-Nanbu 1995), respectively. As seen in the case for El Centro NS (Imperial Valley 1940), it can be observed that, as the degree of embedment becomes larger, the input energy is decreased regardless of the natural period range. It is also observed from Figs. 10.10 and 10.13 that remarkable reduction of input energy occurs in the shorter natural period range, smaller than 0.6(s), for rather soft ground (equivalent shear wave velocity = 50(m/s)). This corresponds well to the known fact that the soil-structure interaction effect is remarkable in rigid structures on soft grounds.

Fig. 10.14 shows the earthquake input energies to the structure-foundation-soil system E_I^{A*} and the structure only E_I^S with various degrees of foundation embedment $e/r_0 = 0.0, 0.5, 1.0, 2.0$ for the ground equivalent shear wave velocity = 100(m/s) to El Centro NS of Imperial Valley 1940. It has been confirmed that E_I^A is equal to E_I^S as proved in Eq. (10.21). It can be observed from Fig. 10.14 that, while the input energy to the structure alone is smaller than that to the structure-foundation-soil system in all the natural period range up to 2.0(s) for $e/r_0 = 0.0, 2.0$, that relation does not exist for $e/r_0 = 0.5, 1.0$. More detailed examination for a broader range of parameters will be necessary to clarify the effect of degree of foundation embedment on the input energies to a structure and to the corresponding structure-foundation-soil system.

It should be remarked that computation of the Fourier amplitude spectra of ground motion accelerations is necessary only once and the input energy can be evaluated by combining those, through numerical integration in the frequency domain, with the energy transfer function. The structural designers can easily understand approximate input energies from the relation of the Fourier amplitude spectra of ground motion accelerations with the energy transfer functions, both of which are expressed in the frequency domain.

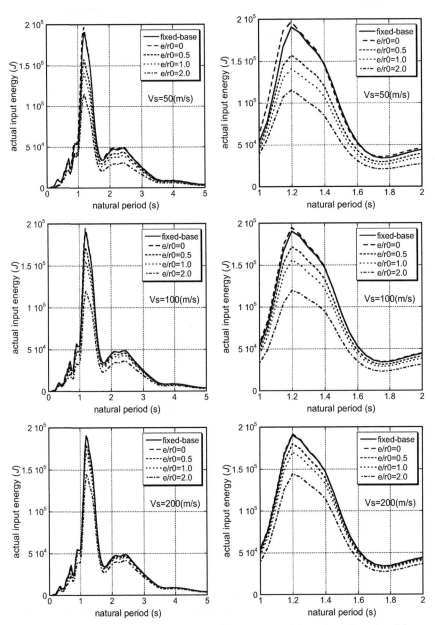

Figure 10.11 *Earthquake input energy to the structure with various degrees of foundation embedment for ground equivalent shear velocities 50, 100, 200(m/s) to Kobe University NS of Hyogoken-Nanbu 1995 (magnified figures are shown on the right-hand side).*

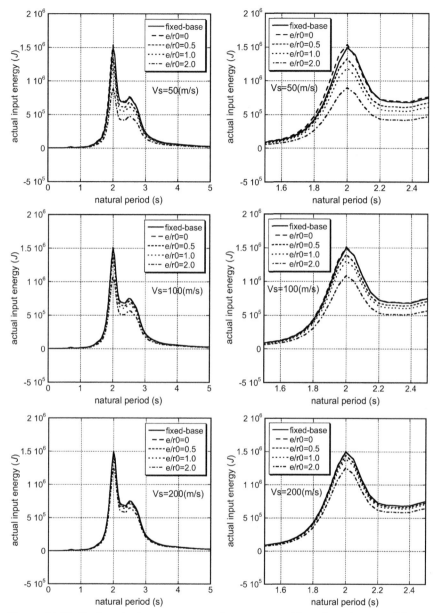

Figure 10.12 *Earthquake input energy to the structure with various degrees of foundation embedment for ground equivalent shear velocities 50, 100, 200(m/s) to SCT1 EW of Mexico Michoacan 1985 (magnified figures are shown on the right-hand side).*

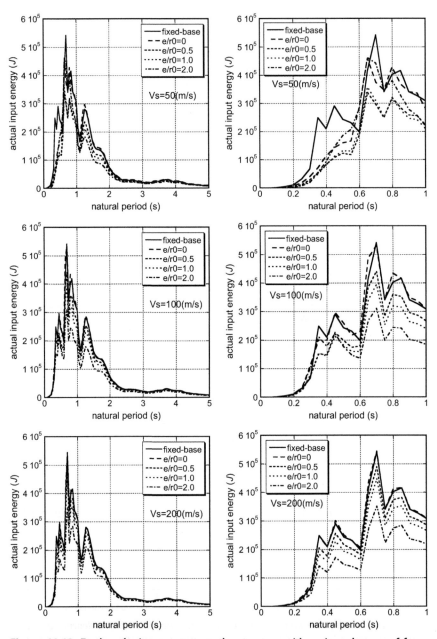

Figure 10.13 *Earthquake input energy to the structure with various degrees of foundation embedment for ground equivalent shear velocities 50, 100, 200(m/s) to JMA Kobe NS of Hyogoken-Nanbu 1995 (magnified figures are shown on the right-hand side).*

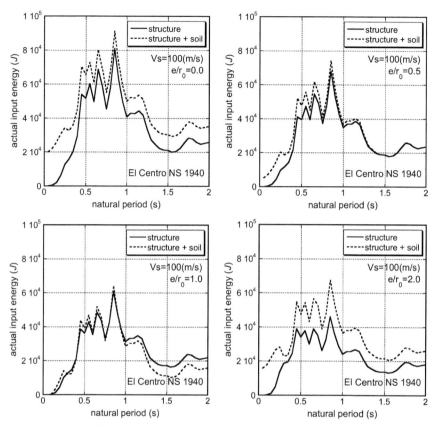

Figure 10.14 *Earthquake input energies to the structure-soil system and the structure only with various degrees of foundation embedment for ground equivalent shear velocity 100 (m/s) to El Centro NS of Imperial Valley 1940.*

10.5. CRITICAL EXCITATION FOR EARTHQUAKE ENERGY INPUT IN SSI SYSTEM

A new critical excitation method is explained for soil–structure interaction systems. In contrast to previous studies considering amplitude non-stationarity only (Takewaki 2001a, b, 2002a, b), no special constraint of input motions is needed on nonstationarity. The input energy to the soil–structure interaction system during an earthquake is introduced as a new measure of criticality (Takewaki 2003). In order to clarify the energy dissipation mechanism in the soil–structure interaction system, two kinds of input energy are defined, one to the overall soil–structure interaction system and the other to the super-structure only. The difference between these two energies indicates the energy dissipated in the soil or that radiating into the ground. The criticality of the input ground motions is expressed in terms of degree of concentration of input motion components on the maximum portion of the characteristic function defining the earthquake input energy. It is remarkable that no mathematical programming technique is required in the solution procedure. It is demonstrated that the input energy expression can be of a compact form via the frequency integration of the product between the input component (Fourier amplitude spectrum) and the structural model component (so-called energy transfer function). With the help of this compact form, it is shown that the formulation of earthquake input energy in the frequency domain is essential for solving the critical excitation problem and deriving a bound on the earthquake input energy for a class of ground motions. The extension of the concept to multi-degree-of-freedom (MDOF) systems is also presented.

Consider a one-story shear building model (mass m, stiffness k, damping coefficient c), as shown in Fig. 10.15, supported by swaying and rocking springs k_H, k_R and dashpots c_H, c_R. Let m_0, I_{R0}, h denote the foundation mass, its mass moment of inertia and the height of the structural mass from the base. The moment of inertia of structural mass is I_R. This model is subjected to a horizontal acceleration $\ddot{u}_g(t) = a(t)$ at the free-field ground surface. Let u_F, θ_F denote the foundation horizontal displacement and its angle of rotation. The horizontal displacement of the super-mass relative to the foundation is denoted by u.

The equations of motion of the model may be expressed as

$$\mathbf{M\ddot{u} + C\dot{u} + Ku} = -\mathbf{Mr}\ddot{u}_g \qquad (10.24)$$

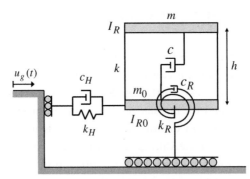

Figure 10.15 *One-story shear building model supported by swaying and rocking springs and dashpots.*

where $\mathbf{u} = \{u \quad u_F \quad \theta_F\}^T$. $\mathbf{M}, \mathbf{K}, \mathbf{C}, \mathbf{r}$ are the following mass, stiffness and damping matrices and influence coefficient vector, respectively.

$$\mathbf{M} = \begin{bmatrix} m & m & mh \\ m & m_0 + m & mh \\ mh & mh & mh^2 + I_{R0} + I_R \end{bmatrix} \qquad (10.25a)$$

$$\mathbf{K} = \text{diag}(k \quad k_H \quad k_R) \qquad (10.25b)$$

$$\mathbf{C} = \text{diag}(c \quad c_H \quad c_R) \qquad (10.25c)$$

$$\mathbf{r} = \{0 \quad 1 \quad 0\}^T \qquad (10.25d)$$

Consider the earthquake input energy (Housner 1959; Akiyama 1985; Uang and Bertero 1990) to the present model. Premultiplication of $\dot{\mathbf{u}}^T$ on Eq. (10.24) and integration of the resulting equation from 0 to t_0 lead to

$$[(1/2)\dot{\mathbf{u}}^T\mathbf{M}\dot{\mathbf{u}}]_0^{t_0} + \int_0^{t_0} \dot{\mathbf{u}}^T\mathbf{C}\dot{\mathbf{u}}\,dt + [(1/2)\mathbf{u}^T\mathbf{K}\mathbf{u}]_0^{t_0} = -\int_0^{t_0} \dot{\mathbf{u}}^T\mathbf{M}\mathbf{r}\ddot{u}_g\,dt \equiv E_I^A$$

$$(10.26)$$

Integration by parts of E_I^A and its rearrangement by use of $\dot{u}_g(0) = \dot{u}_g(t_0) = 0$ provide

$$E_I^A = -\left[\dot{\mathbf{u}}^T\mathbf{M}\mathbf{r}\dot{u}_g\right]_0^{t_0} + \int_0^{t_0} \ddot{\mathbf{u}}^T\mathbf{M}\mathbf{r}\dot{u}_g\,dt$$

$$= \int_0^{t_0} \{m\ddot{u} + (m_0 + m)\ddot{u}_F + mh\ddot{\theta}_F\}\dot{u}_g\,dt$$

$$= \int_0^{t_0} \{m_0(\ddot{u}_g + \ddot{u}_F) + m(\ddot{u}_g + \ddot{u}_F + \ddot{\theta}_F h + \ddot{u}) - (m_0 + m)\ddot{u}_g\}\dot{u}_g\,dt$$

$$= \int_0^{t_0} \{m_0(\ddot{u}_g + \ddot{u}_F) + m(\ddot{u}_g + \ddot{u}_F + \ddot{\theta}_F h + \ddot{u})\}\dot{u}_g\,dt$$

$$- [(1/2)(m_0 + m)\dot{u}_g^2]_0^{t_0}$$

$$= \int_0^{t_0} \{m_0(\ddot{u}_g + \ddot{u}_F) + m(\ddot{u}_g + \ddot{u}_F + \ddot{\theta}_F h + \ddot{u})\}\dot{u}_g\,dt$$

$$(10.27)$$

The expression in the braces in the last equation indicates the sum of inertial forces acting on the foundation and structural mass. Eq. (10.27) implies that the work by the ground on the swaying-rocking (SR) model is equal to E_I^A.

It is known (Takewaki 2003; Lyon 1975; Ohi et al. 1985; Kuwamura et al. 1994; Ordaz et al. 2003) that, in linear elastic structures, the earthquake input energy can also be expressed in the frequency domain. Let $\ddot{U}, \ddot{U}_F, \ddot{\Theta}_F, \ddot{U}_g$ denote the Fourier transforms of $\ddot{u}, \ddot{u}_F, \ddot{\theta}_F, \ddot{u}_g$, and $H_{\ddot{U}}(\omega), H_{\ddot{U}F}(\omega), H_{\ddot{\Theta}F}(\omega)$ denote the transfer functions of $\ddot{u}, \ddot{u}_F, \ddot{\theta}_F$ to $\ddot{u}_g$.

$$\ddot{U}/\ddot{U}_g = H_{\ddot{U}}(\omega) \tag{10.28a}$$

$$\ddot{U}_F/\ddot{U}_g = H_{\ddot{U}F}(\omega) \tag{10.28b}$$

$$\ddot{\Theta}_F/\ddot{U}_g = H_{\ddot{\Theta}F}(\omega) \tag{10.28c}$$

These quantities can be derived from the Fourier transformed equations of Eq. (10.24).

The Fourier inverse transformation of Eq. (10.27) after the extension of lower and upper limits from $(0, t_0)$ to $(-\infty, \infty)$ and use of Eq. (10.28) lead to

$$E_I^A = \frac{1}{2\pi} \int_{-\infty}^{\infty} \int_{-\infty}^{\infty} \{m_0(\ddot{U}_g + \ddot{U}_F) + m(\ddot{U}_g + \ddot{U}_F + \ddot{\Theta}_F h$$

$$+ \ddot{U})\} e^{i\omega t} \ddot{u}_g dt d\omega$$

$$= -\frac{1}{\pi} \int_{0}^{\infty} \frac{1}{\omega} \text{Im}[m_0 H_{\ddot{U}F}(\omega) + m\{H_{\ddot{U}F}(\omega) + H_{\ddot{\Theta}F}(\omega)h$$

$$+ H_{\ddot{U}}(\omega)\}]\left|\ddot{U}_g(\omega)\right|^2 d\omega \tag{10.29}$$

It is also possible to re-express Eq. (10.29) in terms of the velocity transfer functions $H_{\dot{U}}(\omega), H_{\dot{U}F}(\omega), H_{\dot{\Theta}F}(\omega)$ of $\dot{u}, \dot{u}_F, \dot{\theta}_F$ to $\ddot{u}_g$.

$$E_I^A = -\frac{1}{\pi} \int_{0}^{\infty} \text{Re}[m_0 H_{\dot{U}F}(\omega) + m\{H_{\dot{U}F}(\omega) + H_{\dot{\Theta}F}(\omega)h$$

$$+ H_{\dot{U}}(\omega)\}]\left|\ddot{U}_g(\omega)\right|^2 d\omega \tag{10.30}$$

It is also known that the earthquake input energy to a linear elastic structure or a linear elastic system does not depend on the phase characteristics of input motions (Takewaki 2003; Lyon 1975; Ohi et al. 1985; Kuwamura et al. 1994; Ordaz et al. 2003). Eq. (10.30) clearly supports this fact.

Consider next a work by the foundation on the structure alone. This quantity indicates the input energy to the structure alone and is expressed by

$$
E_I^S = \int_0^\infty m(\ddot{u}_g + \ddot{u}_F + \ddot{\theta}_F h + \ddot{u})(\dot{u}_g + \dot{u}_F)dt + \int_0^\infty \{m(\ddot{u}_g + \ddot{u}_F + \ddot{\theta}_F h
$$

$$
+ \ddot{u})h + I_R \ddot{\theta}_F \} \dot{\theta}_F dt
$$

$$(10.31)$$

The internal forces are in equilibrium with the inertial force $-m(\ddot{u}_g + \ddot{u}_F + \ddot{\theta}_F h + \ddot{u})$ and moment $-\{m(\ddot{u}_g + \ddot{u}_F + \ddot{\theta}_F h + \ddot{u})h + I_R \ddot{\theta}_F\}$ and do the work on $-(\dot{u}_g + \dot{u}_F)dt$ and $-\dot{\theta}_F dt$. The Fourier inverse transformation of Eq. (10.31) after the extension of the lower limit from 0 to $-\infty$ and use of the transfer functions defined in (10.28) provide

$$
E_I^S = -\frac{1}{\pi} \int_0^\infty \frac{1}{\omega} \mathrm{Im}[m\{1 + H_{\ddot{U}F}(\omega) + H_{\ddot{\Theta}F}(\omega)h + H_{\ddot{U}}(\omega)\}
$$

$$
\times \{1 + H_{\ddot{U}F}(-\omega)\} + (mh\{1 + H_{\ddot{U}F}(\omega) + H_{\ddot{\Theta}F}(\omega)h
$$

$$
+ H_{\ddot{U}}(\omega)\} + I_R H_{\ddot{\Theta}F}(\omega))H_{\ddot{\Theta}F}(-\omega)] \cdot |\ddot{U}_g(\omega)|^2 d\omega
$$

$$(10.32)$$

Let us define the following functions $F_A(\omega)$, $F_S(\omega)$ for Eqs. (10.29) and (10.32) to derive a compact form for a unified input energy expression.

$$
E_I^A = \int_0^\infty F_A(\omega)|\ddot{U}_g(\omega)|^2 d\omega
$$

$$(10.33a)$$

$$
E_I^S = \int_0^\infty F_S(\omega)|\ddot{U}_g(\omega)|^2 d\omega
$$

$$(10.33b)$$

Examples of time histories of earthquake input energies and their final values E_I^A, E_I^S for El Centro NS (Imperial Valley 1940) and Kobe University

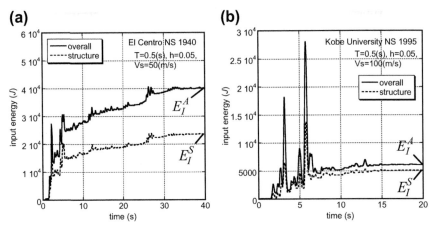

Figure 10.16 *Energy time histories to overall system and structure alone: (a) El Centro NS 1940, (b) Kobe University NS 1995.*

NS (Hyogoken-Nanbu 1995) are shown in Figs. 10.16(a) and (b) for natural period of the structure $T = 0.5$(s) and ground shear wave velocity $Vs = 50$, 100(m/s). The model parameters in Section 10.11 have been used. Figs. 10.17(a)–(c) show $F_A(\omega)$, $F_S(\omega)$ for $T = 0.2, 0.5$(s), $Vs = 50, 100, 200$(m/s).

10.6. CRITICAL EXCITATION PROBLEM

Consider the following critical excitation problem for modeling uncertainties in the input ground motions.

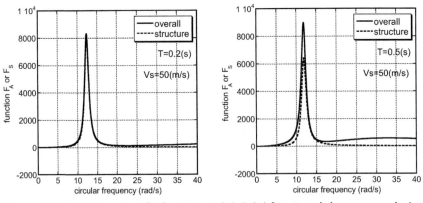

Figure 10.17(a) *Energy transfer functions $F_A(\omega)$, $F_S(\omega)$ for ground shear wave velocity $Vs = 50$ (m/s) and natural period $T = 0.2, 0.5$(s) of super-structure.*

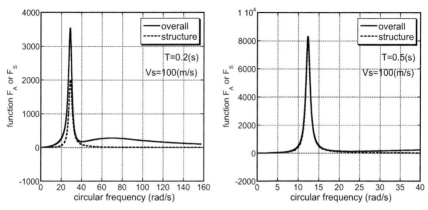

Figure 10.17(b) *Energy transfer functions $F_A(\omega)$, $F_S(\omega)$ for ground shear wave velocity Vs = 100 (m/s) and natural period T = 0.2, 0.5(s) of super-structure.*

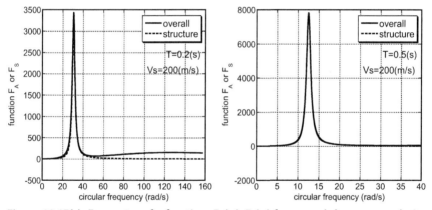

Figure 10.17(c) *Energy transfer functions $F_A(\omega)$, $F_S(\omega)$ for ground shear wave velocity Vs = 200(m/s) and natural period T = 0.2, 0.5(s) of super-structure.*

[Critical Excitation Problem 1a]

Find the Fourier amplitude spectrum $|A(\omega)| = |\ddot{U}_g(\omega)|$ of the free ground surface motion acceleration $\ddot{u}_g(t) = a(t)$ that maximizes the earthquake input energy (10.33a) to the overall SR model and satisfies the constraints

$$\int_{-\infty}^{\infty} a(t)^2 dt = (1/\pi) \int_0^{\infty} |A(\omega)|^2 d\omega = \overline{C} \tag{10.34}$$

$$|A(\omega)| \leq \overline{A} \tag{10.35}$$

Parseval's theorem is used in Eq. (10.34). Another critical excitation problem for another input energy index may be stated as follows:

[Critical Excitation Problem 1s]
Find the Fourier amplitude spectrum $|A(\omega)|$ of the free ground surface motion acceleration that maximizes the earthquake input energy (10.33b) to the structure alone and satisfies the constraints (10.34) and (10.35).

10.7. UPPER BOUND OF FOURIER AMPLITUDE SPECTRUM OF INPUT

It may be open to argument how to specify the upper bound of the Fourier amplitude spectrum of the free ground surface motion acceleration. One possibility is to introduce the relation between the duration and the magnitude of the Fourier amplitude spectrum. Fig. 10.18(a) shows the sine wave with the duration of 5 seconds. Its Fourier amplitude spectrum is shown in Fig. 10.18(b). On the other hand, Fig. 10.19(a) presents the sine wave with the duration of 20 sec. which has the same power $\int_{-\infty}^{\infty} a(t)^2 dt$ as Fig. 10.18(a). Its Fourier amplitude spectrum is shown in Fig. 10.19(b). It can be observed that, if the power $\int_{-\infty}^{\infty} a(t)^2 dt$ is the same, the long-duration motion has a larger Fourier amplitude spectrum. Because the duration of the ground motion can be bounded approximately by its characteristics, the upper bound of the Fourier amplitude spectrum may be able to be specified from this point of view.

(a)   **(b)**

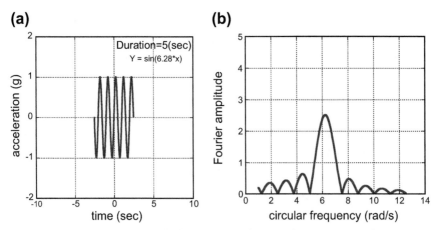

Figure 10.18 *(a) Sine wave with the duration of 5 sec.; (b) its corresponding Fourier amplitude spectrum.*

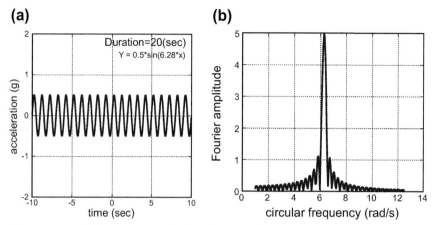

Figure 10.19 *(a) Sine wave with the duration of 20 sec. having the same power $\int_{-\infty}^{\infty} a(t)^2 dt$ as Fig. 10.18(a); (b) its corresponding Fourier amplitude spectrum.*

10.8. SOLUTION PROCEDURE AND UPPER BOUND OF INPUT ENERGY

When $\overline{A} \to \infty$, the solution to Problem 1A and 1S is the Dirac delta function (Takewaki 2003) (see Fig. 10.20). When $\overline{A}$ is finite, the solution is found to be a rectangular function that maximizes the definite integral of $F_A(\omega), F_S(\omega)$ (see Fig. 10.20). The frequency band width is given by $\Delta\omega = \pi\overline{C}/\overline{A}^2$ from Eq. (10.34). By the assumption that $F_A(\omega), F_S(\omega)$ attain their maximum at $\omega = \omega_1 = \Omega$ (ω_1: undamped fundamental natural circular

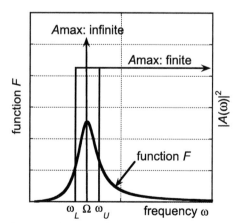

Figure 10.20 *Schematic diagram of solution procedure ($A_{max} = \overline{A}, \Omega = \omega_1$).*

frequency of the SR model). The absolute bound of the input energy for $\overline{A} \to \infty$ can be obtained approximately from

$$E_I^{Aabs} = \pi \overline{C} F_A(\omega_1), \quad E_I^{Sabs} = \pi \overline{C} F_S(\omega_1) \quad (10.36\text{a, b})$$

When $\overline{A}$ is finite, the credible bounds of the input energy for Problems 1A and 1S may be obtained from

credible bound for Problem 1A $= \overline{A}^2 \displaystyle\int_{\omega_L}^{\omega_U} F_A(\omega)d\omega$

$$(\omega_U - \omega_L = \Delta\omega, \ (\omega_U + \omega_L)/2 = \omega_1)$$
$$(10.37\text{a})$$

credible bound for Problem 1S $= \overline{A}^2 \displaystyle\int_{\omega_L}^{\omega_U} F_S(\omega)d\omega$

$$(\omega_U - \omega_L = \Delta\omega, \ (\omega_U + \omega_L)/2 = \omega_1)$$
$$(10.37\text{b})$$

There remain uncertainties in the setting of $\overline{A}$ (one possibility is shown in Section 10.7). The term "credible" is used in the sense that a finite credible value for $\overline{A}$ is employed.

Table 10.1 shows the maximum Fourier amplitude spectrum of ground motion acceleration, time integral of squared ground motion acceleration and frequency bandwidth of the critical rectangular Fourier amplitude spectrum of ground motion acceleration for four recorded ground motions.

10.9. CRITICAL EXCITATION PROBLEM FOR VELOCITY CONSTRAINTS

It can be shown that the upper bound of the input energy is quite large in the rather long natural period range. This is because the constraint on

Table 10.1 Characteristic values of acceleration constraint

	$A_{\max}(\text{m/s})$	$\overline{C}(\text{m}^2/\text{s}^3)$	$\Delta\omega \ (\text{rad/s})$
El Centro NS 1940	2.91	11.4	4.21
Kobe Univ. NS 1995	3.06	7.59	2.55
Mexico SCT1 EW 1985	12.0	15.2	0.332
JMA Kobe NS 1995	5.81	52.3	4.87

acceleration (10.34) is used (Takewaki 2003). The constraint on velocity is treated in the following section (Tanabashi 1956; Housner and Jennings 1975). The following new critical excitation problem can be stated:

[Critical Excitation Problem 2a]
Find the Fourier amplitude spectrum $|V(\omega)|$ of the free ground surface motion velocity $v(t)$ that maximizes the earthquake input energy

$$E_I^A = \int_0^\infty \omega^2 F_A(\omega)|V(\omega)|^2 d\omega \tag{10.38}$$

to the overall SR model and satisfies the constraints

$$\int_{-\infty}^\infty v(t)^2 dt = (1/\pi)\int_0^\infty |V(\omega)|^2 d\omega = \overline{C}_V \tag{10.39}$$

$$|V(\omega)| \le \overline{V} \tag{10.40}$$

The relation $|A(\omega)|^2 = \omega^2|V(\omega)|^2$ is used in Eq. (10.38) (see Eq. (10.33a)). It is noted that the constraint (10.39) is proportional to the energy of a traveling wave into a unit area of ground (Trifunac et al. 2001). Another critical excitation problem for another input energy index may be stated as follows:

[Critical Excitation Problem 2s]
Find the Fourier amplitude spectrum $|V(\omega)|$ of the free ground surface motion velocity that maximizes the earthquake input energy (see Eq. (10.33b) and $|A(\omega)|^2 = \omega^2|V(\omega)|^2$)

$$E_I^S = \int_0^\infty \omega^2 F_S(\omega)|V(\omega)|^2 d\omega \tag{10.41}$$

to the structure alone and satisfies the constraints (10.39) and (10.40).

10.10. SOLUTION PROCEDURE FOR VELOCITY CONSTRAINT PROBLEMS

The solution procedure to Problem 2A can be devised by replacing the function $F_A(\omega)$ in Problem 1A by $\omega^2 F_A(\omega)$. The absolute bound for

$\overline{V} \to \infty$ can be derived as follows by assuming that $\omega^2 F_A(\omega)$ attains the maximum at $\omega = \omega_1$.

$$E_I^{Aabs} = \pi \overline{C}_V \omega_1^2 F_A(\omega_1) \tag{10.42}$$

When $\overline{V}$ is finite, the corresponding credible bound can be obtained from

credible bound for Problem 2A $= \overline{V}^2 \int_{\omega_L}^{\omega_U} \omega^2 F_A(\omega) d\omega$

$$(\omega_U - \omega_L = \Delta \omega, \ (\omega_U + \omega_L)/2 = \omega_1) \tag{10.43}$$

The frequency band width is given by

$$\Delta \omega = \pi \overline{C}_V / \overline{V}^2 \tag{10.44}$$

The solution procedure to Problem 2S can be devised by replacing the function $F_S(\omega)$ in Problem 1S by $\omega^2 F_S(\omega)$. The absolute bound for $\overline{V} \to \infty$ can be derived as follows by assuming that $\omega^2 F_S(\omega)$ attains the maximum at $\omega = \omega_1$.

$$E_I^{Sabs} = \pi \overline{C}_V \omega_1^2 F_S(\omega_1) \tag{10.45}$$

When $\overline{V}$ is finite, the corresponding credible bound can be obtained from

credible bound for Problem 2S $= \overline{V}^2 \int_{\omega_L}^{\omega_U} \omega^2 F_S(\omega) d\omega$

$$\left(\omega_U - \omega_L = \Delta \omega, \ (\omega_U + \omega_L)/2 = \omega_1\right) \tag{10.46}$$

The frequency band width is given by

$$\Delta \omega = \pi \overline{C}_V / \overline{V}^2 \tag{10.47}$$

Table 10.2 shows the maximum Fourier amplitude spectrum of ground motion velocity, the time integral of squared ground motion velocity and the frequency bandwidth of the critical rectangular Fourier amplitude spectrum of ground motion velocity for four recorded ground motions.

Table 10.2 Characteristic values of velocity constraint

	V_{max} (m)	$\bar{C}_V$ (m²/s)	$\Delta\omega$ (rad/s)
El Centro NS 1940	0.696	0.332	2.15
Kobe Univ. NS 1995	0.597	0.307	2.70
Mexico SCT1 EW 1985	3.913	1.88	0.386
JMA Kobe NS 1995	0.746	0.854	4.82

10.11. NUMERICAL EXAMPLES-1 (ONE-STORY MODEL)

Numerical examples for one-story shear building models are pre-sented first for four recorded ground motions: El Centro NS (Imperial Valley 1940), Kobe University NS (Hyogoken-Nanbu 1995), SCT1 EW (Mexico Michoacan 1985), JMA Kobe NS (Hyogoken-Nanbu 1995). The actual maximum value of the Fourier amplitude spectrum has been adopted as $\bar{A}$. The floor masses and floor mass moments of inertia are as follows: $m = 30 \times 10^3 (\text{kg})$, $I_R = 1.6 \times 10^5 (\text{kg·m}^2)$, $m_0 = 90 \times 10^3 (\text{kg})$, $I_{R0} = 4.8 \times 10^5 (\text{kg·m}^2)$. The story height is $h = 3.5 (\text{m})$. The stiffnesses of swaying and rocking springs and the damping coefficients of dashpots are evaluated by the formula due to Parmelee (1970) as

$$k_H = (6.77/(1.79 - \nu))Gr$$

$$k_R = (2.52/(1.00 - \nu))Gr^3$$

$$c_H = (6.21/(2.54 - \nu))\rho V_S r^2$$

$$c_R = (0.136/(1.13 - \nu))\rho V_S r^4$$

The soil mass density and Poisson's ratio are $\rho = 1.8 \times 10^3 (\text{kg/m}^3)$ and $\nu = 0.35$. The radius of the equivalent circular foundation is $r = 4 (\text{m})$. The equivalent shear wave velocities are given by $Vs = 50, 100, 200 (\text{m/s})$. It should be noted that a rather small shear wave velocity of ground is considered because the "equivalent shear wave velocity" can be rather small in the large amplitude of shear strain of surface ground under earthquake loading.

The stiffnesses of swaying and rocking springs and the damping coeffi-cients of dashpots presented above are evaluated by the frequency-independent ones. In order to investigate the accuracy of this model, the comparison with the model including the frequency-dependent ones is made. The so-called cone model (Wolf 1994) is used to represent the frequency-dependent springs and dashpots.

Fig. 10.21 shows the comparison between two models (present frequency-independent model and frequency-dependent cone model) for overall earthquake input energy to structure-foundation system including soil springs and dashpots and earthquake input energy to structure. It can be seen that, up to the natural period of 1.0(s) of the super-structure, the present frequency-independent model can represent the input energy within an allowable accuracy regardless of the soil stiffness. As for the case of V_S= 200(m/s), there is no clear difference between the overall earthquake input energy to structure-foundation system and earthquake input energy to structure. Therefore such a case is not shown here. Furthermore, although a one-story structure model is used and the foundation size is relatively small in the present model, it has been confirmed that a similar tendency can be observed in models of several stories and larger foundation size.

The solid lines in Fig. 10.22 show the plots of earthquake input energies by the ground motion of El Centro NS to the overall system (structure plus surrounding soil) and the structure alone for $Vs = 50, 100, 200$(m/s) with respect to the natural period of the fixed-base structure. The damping ratio of the super-structure is 0.05. It can be observed from Fig. 10.22 that the input energy to stiff structures with short natural periods is governed primarily by the energy dissipated by the ground (surrounding soil) and the input energy to flexible structures with intermediate natural periods (around 1(s)) is governed mainly by the energy dissipated by the damping of super-structures. This phenomenon corresponds well to the well-known fact that the soil-structure

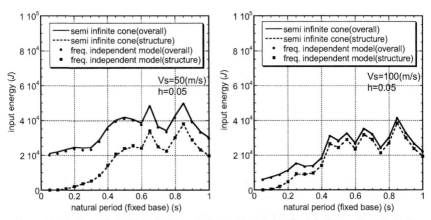

Figure 10.21 *Overall earthquake input energy to structure-foundation system including soil springs and dashpots and earthquake input energy to structure: Comparison of model with frequency-independent soil springs and dashpots (present model) with model including frequency-dependent soil springs and dashpots (cone model).*

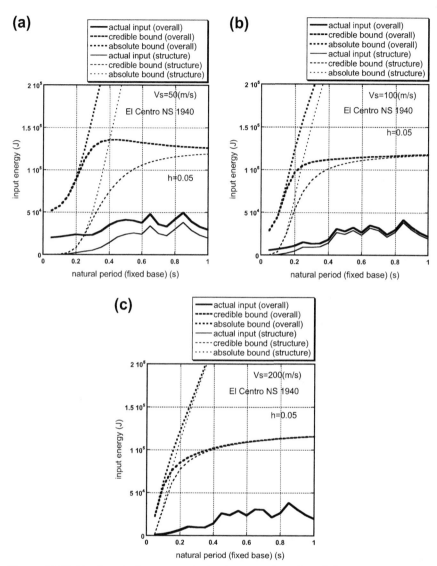

Figure 10.22 *Earthquake input energies by the ground motion of El Centro NS 1940 to overall system and structure alone: (a) Vs = 50(m/s), (b) Vs = 100(m/s), (c) Vs = 200(m/s).*

interaction effect is notable in the stiff structures on flexible ground. The dotted lines in Fig. 10.22 show the credible and absolute bounds, given by Eqs. (10.37) and (10.36), of earthquake input energies by the ground motion of El Centro NS to the overall system and the structure alone. As the shear wave velocity of the ground becomes larger, the input energy is governed mainly by the energy dissipated by the damping of super-structures. This results are in almost perfect agreement between the input energy for the entire system and the super-structure, though only for higher values of V_S. It should also be pointed out that the ground motion of El Centro NS does not have a notable predominant period and the distance between the actual input energy and the credible bound is almost constant with respect to the natural period of the super-structure.

Fig. 10.23 shows the earthquake input energies by the ground motion of Kobe University NS. It can be observed that this ground motion has a predominant period of about 1.2(s) and the actual earthquake input energy is close to the corresponding credible bound around this predominant period range both for the energy to the overall system and that to the structure alone. Since the soil-structure interaction effect is small in this rather long predominant period range, the input energy to the super-structure alone is close to that of the overall system.

Fig. 10.24 presents the earthquake input energies by the ground motion of SCT1 EW (Mexico 1985). It can be observed that no conspicuous difference exists between the energy to the overall system and the structure alone in the rather long natural period range (SCT1 EW has a predominant period around 2.0(s)).

Fig. 10.25 shows the earthquake input energies by the ground motion of JMA Kobe NS. A tendency similar to El Centro NS can be observed, i.e. the input energy to stiff structures with short natural periods is governed primarily by the energy dissipated by the ground and the input energy to flexible structures with intermediate natural periods governed mainly by the energy dissipated by the damping of super-structures.

In order to investigate the effect of the acceleration constraint (Eq. 10.34) and the velocity constraint (Eq. 10.39), four classes of ground motions have been considered (Abrahamson et al. 1998): (a) near-fault rock motion (JMA Kobe NS, Hyogoken–Nanbu 1995), (b) near-fault soil motion (Rinaldi EW, Northridge 1994), (c) long-duration rock motion (Caleta de Campos NS, Michoacan 1985), (d) long-duration soil motion (Vina del Mar NS, Chile 1985). The Fourier amplitude spectra of these ground motions are shown in Fig. 10.26. Fig. 10.27 shows the earthquake input energy per unit mass to a fixed-base model by these four classes of ground motions and its credible bounds for the acceleration constraint and the velocity constraint. It can be observed that the earthquake input energy in the shorter natural period range can be bounded appropriately by the credible bound for the acceleration constraint and that in the intermediate/or long natural period range can be bounded appropriately by the credible bound for the velocity constraint.

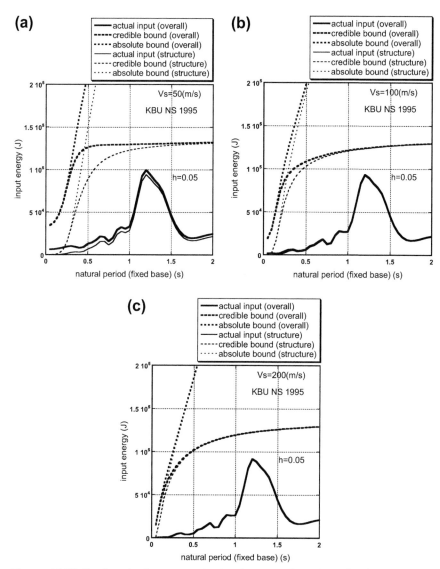

Figure 10.23 *Earthquake input energies by the ground motion of Kobe University NS 1995 to overall system and structure alone: (a) Vs = 50(m/s), (b) Vs = 100(m/s), (c) Vs = 200(m/s).*

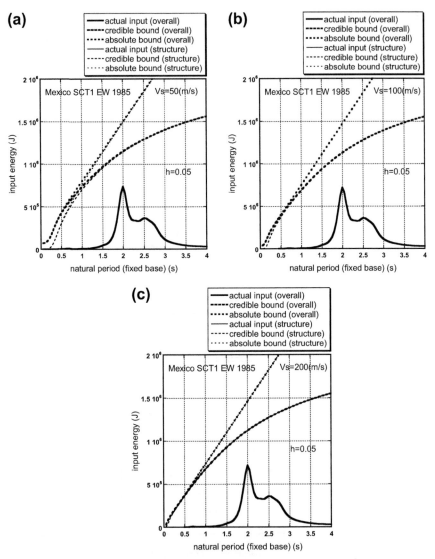

Figure 10.24 *Earthquake input energies by the ground motion of Mexico SCT1 EW 1985 to overall system and structure alone: (a) Vs = 50(m/s), (b) Vs = 100(m/s), (c) Vs = 200(m/s).*

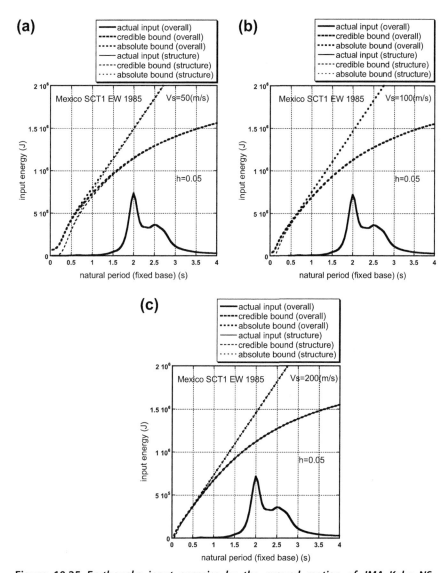

Figure 10.25 *Earthquake input energies by the ground motion of JMA Kobe NS 1995 to overall system and structure alone: (a) Vs = 50(m/s), (b) Vs = 100(m/s), (c) Vs = 200(m/s).*

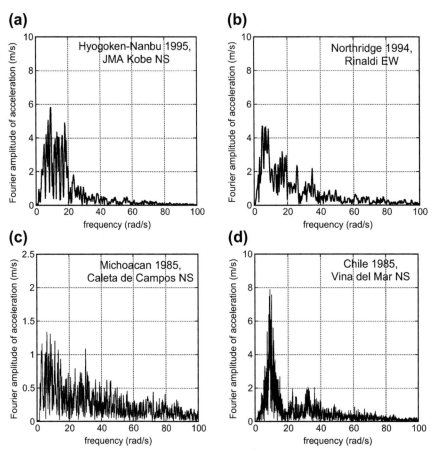

Figure 10.26 *Fourier amplitude spectrum of ground motion acceleration: (a) near-fault rock motion, (b) near-fault soil motion, (c) long-duration rock motion, (d) long-duration soil motion.*

10.12. NUMERICAL EXAMPLES-2 (3-STORY MODEL)

In order to present a more realistic example, 3-story shear building models supported by the swaying and rocking spring–dashpot system are considered. The input energies, Eq. (10.29) and Eq. (10.32), derived for a one-story model can be extended straightforwardly to an MDOF system by taking into account the inertial forces acting on the floor masses and by considering the work of the internal forces in equilibrium with the inertial forces on its corresponding displacements (see Eq. (10.27)).

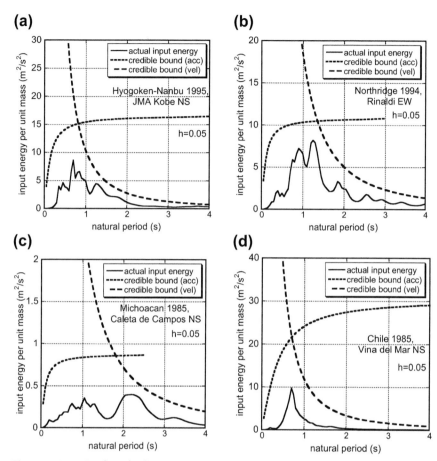

Figure 10.27 *Earthquake input energy per unit mass to fixed-base model by four classes of ground motions and its credible bounds for acceleration constraint and velocity constraint: (a) near-fault rock motion, (b) near-fault soil motion, (c) long-duration rock motion, (d) long-duration soil motion.*

The floor masses and floor mass moments of inertia are as follows: $m_i = 30 \times 10^3 (\text{kg})$, $I_{Ri} = 1.6 \times 10^5 (\text{kg} \cdot \text{m}^2)$, $m_0 = 90 \times 10^3 (\text{kg})$, $I_{R0} = 4.8 \times 10^5 (\text{kg} \cdot \text{m}^2)$. The story height is $h_i = 3.5 (\text{m})$. The lowest–mode damping ratio of the structure is 0.05. Poisson's ratio has been changed to $\nu = 0.45$. The radius of the equivalent circular foundation has also been changed to $r = 4.5 (\text{m})$. The properties of the swaying and rocking spring–dashpot system have been evaluated by the cone model (Wolf 1994) for a semi–infinite ground. Figs. 10.28(a)–(c) show the overall and structural input energies, the corresponding credible bounds and the corresponding absolute

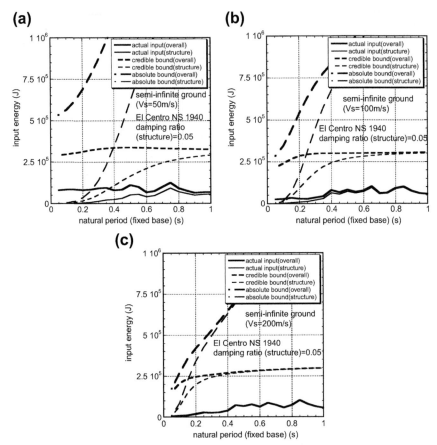

Figure 10.28 *Earthquake input energies by the ground motion of El Centro NS 1940 to overall system and structure alone (3-story model): (a) Vs = 50(m/s), (b) Vs = 100(m/s), (c) Vs = 200(m/s).*

bounds to the 3-story models on the ground Vs = 50, 100, 200(m/s), respectively, subjected to El Centro NS (Imperial Valley 1940). The horizontal axis indicates the fundamental natural period of the fixed-base model. The distribution of the story stiffnesses is determined so that the lowest mode of the fixed-base model has a straight-line mode. This ground motion does not have a sharp predominant period. Therefore the bound of the input energy is far from the actual one. Figs. 10.29(a)–(c) present these for the 3-story models on the ground Vs = 50, 100, 200(m/s), respectively, subjected to Kobe University NS (Hyogoken–Nanbu 1995). This ground motion has a sharp predominant period and the bound at the predominant period is very close to the actual one.

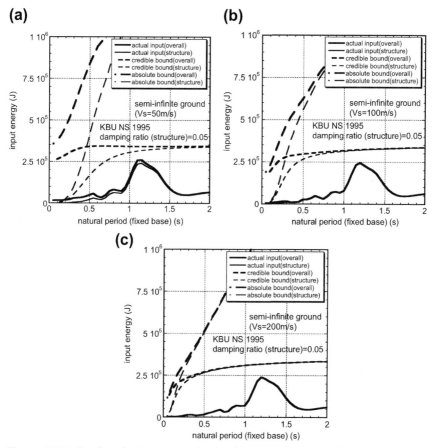

Figure 10.29 *Earthquake input energies by the ground motion of Kobe University NS 1995 to overall system and structure alone (3-story model): (a) Vs = 50(m/s), (b) Vs = 100(m/s), (c) Vs = 200(m/s).*

10.13. CONCLUSIONS

The conclusions may be stated as follows:

(1) A new evaluation method of earthquake input energy to soil–structure interaction systems has been developed. The method is an approach in the frequency domain. Because the inertial interaction (foundation impedance) and the kinematic interaction (effective input motion) are well described by frequency-dependent functions, the present approach based on the frequency domain analysis is appropriate and effective.

(2) Even soil–structure interaction systems including embedded foundations can be treated in a simple way and effects of the foundation embedment on the earthquake input energy to the super-structure can be clarified systematically by the proposed frequency domain formulation. It can be stated from a limited analysis that the input energy to the sway-rocking model without embedment is almost the same as that to the fixed-base model. As the degree of embedment becomes larger, the input energy is decreased regardless of the natural period range. The ratio of the input energy to the structure alone to that to the structure-foundation-soil system is affected in a complicated manner by the degree of embedment.

(3) The formulation of the earthquake input energy in the frequency domain is effective and essential for solving a critical excitation problem for the maximum earthquake input energy and deriving a bound on the earthquake input energy for a class of ground motions.

(4) A new critical excitation method has been developed for soil-structure interaction systems. The input energy to the soil-structure interaction systems during an earthquake has been introduced as a new measure of criticality. No mathematical programming technique is required in the solution procedure with the help of a compact and unified form of input energy expression. It has been shown that the formulation of earthquake input energy in the frequency domain is essential for solving the critical excitation problem and deriving a bound on the earthquake input energy for a class of ground motions.

(5) Definition of two input energies, one to the overall system (structure plus surrounding soil) and the other to the structure alone is very useful in understanding the mechanism of energy input and the effect of soil-structure interaction under various conditions of soil properties and natural period of structures. This advantage has been demonstrated through numerical examples for four representative recorded ground motions.

(6) Through another numerical examination for four classes of recorded ground motions: ((a) near-fault rock motions, (b) near-fault soil motions, (c) long-duration rock motions, (d) long-duration soil motions), the bounds under *acceleration and velocity constraints* (time integral of the squared base acceleration and time integral of the squared base velocity) are clarified to be meaningful in the short and interme-diate/long natural period ranges, respectively. More specifically, the solution with acceleration constraints can bound properly the

earthquake input energy in a shorter natural period range and that with velocity constraints can bound properly the earthquake input energy in an intermediate or longer natural period range. This fact implies that the measure of damage potential of earthquake ground motions should be discussed carefully (Housner and Jennings 1975).

(7) The present critical excitation method for SDOF superstructure models can be extended straightforwardly to MDOF superstructure models with the help of the aforementioned compact and unified form of input energy expression.

The evaluation of earthquake input energy in the time domain is suitable for the evaluation of the time history of input energy, especially for nonlinear systems. Dual use of the frequency-domain and time-domain techniques may be preferable in the advanced seismic analysis for more robust design.

The bound $\overline{A}$ of the Fourier amplitude spectrum of a ground motion has an uncertain characteristic. A treatment of this uncertainty has been proposed by the present author and Dr. Ben-Haim recently (Takewaki and Ben-Haim 2005) in terms of the unified formulation of this uncertainty and the other structural parameter uncertainty. Further investigation is necessary on the formulation of the bound $\overline{A}$.

In this section, a foundation is assumed to be on the ground surface and "the effective input motion" is not considered explicitly. In the case where the foundation is embedded, the effective input motion plays an important role (Wolf 1985; Takewaki et al. 2003; Takewaki and Fujimoto 2004). Even in such a case, almost the same formulation in the frequency domain as in the present chapter can be developed. This formulation will be presented elsewhere.

REFERENCES

Abrahamson, N., Ashford, S., Elgamal, A., Kramer, S., Seible, F., Somerville, P., 1998. Proc. of 1st PEER Workshop on Characterization of Special Source Effects, San Diego. Pacific Earthquake Engineering Research Center, Richmond.

Akiyama, H., 1985. Earthquake Resistant Limit-State Design for Buildings. University of Tokyo Press, Tokyo, Japan.

Arias, A., 1970. A measure of earthquake intensity. In: Hansen, R.J. (Ed.), Seismic Design for Nuclear Power Plants. The MIT Press, Cambridge, MA, pp. 438–469.

Berg, G.V., Thomaides, T.T., 1960. Energy consumption by structures in strong-motion earthquakes. Proc. of 2nd World Conf. on Earthquake Engineering. Gakujutsu Bunken Fukyu-kai, Tokyo and Kyoto. Tokyo, 681–696.

Cakmak, A.S., Abdel-Ghaffar, A.M., Brebbia, C.A. (Eds.), 1982. Soil dynamics and earthquake engineering, vol. I, II: Proc. of the Conf. on Soil Dynamics and Earthquake Engineering. A.A. Balkema.

Drenick, R.F., 1970. Model-free design of aseismic structures. Journal of Engineering Mechanics Division 96 (4), 483–493.

Fajfar, P., Vidic, T., 1994. Consistent inelastic design spectra: hysteretic and input energy. Earthquake Engineering and Structural Dynamics 23 (5), 523–537.

Gupta, V.K., Trifunac, M.D., 1991. Seismic response of multistoried buildings including the effects of soil-structure interaction. Soil Dyn. Earthquake Engng 10 (8), 414–422.

Housner, G.W., 1956. Limit design of structures to resist earthquakes. Proc. of the First World Conference on Earthquake Engineering. University of California, Berkeley, CA, Berkeley, 5:1–11.

Housner, G.W., 1959. Behavior of structures during earthquakes. Journal of the Engineering Mechanics Division 85 (EM4), 109–129.

Housner, G.W., Jennings, P.C., 1975. The capacity of extreme earthquake motions to damage structures. Prentice-Hall Englewood Cliff, NJ. In: Hall, W.J. (Ed.), Structural and Geotechnical Mechanics: A volume honoring N.M. Newmark, pp. 102–116.

Kato, B., Akiyama, H., 1975. Energy input and damages in structures subjected to severe earthquakes. Journal of Structural and Construction Eng., Archi. Inst. of Japan 235, 9–18 (in Japanese).

Kuwamura, H., Kirino, Y., Akiyama, H., 1994. Prediction of earthquake energy input from smoothed Fourier amplitude spectrum. Earthquake Engineering and Structural Dynamics 23 (10), 1125–1137.

Léger, P., Dussault, S., 1992. Seismic-energy dissipation in MDOF structures. J. Struct. Eng. 118 (5), 1251–1269.

Luco, J.E., 1980. Linear soil-structure interaction. UCRL-15272, Lawrence Livermore Laboratory. California, Livermore.

Lyon, R.H., 1975. Statistical energy analysis of dynamical systems. The MIT Press, Cambridge, MA.

Mahin, S.A., Lin, J., 1983. Construction of inelastic response spectrum for single-degree-of-freedom system. Report No. UCB/EERC-83/17, Earthquake Engineering Research Center. University of California, Berkeley, CA.

Meek, J., Wolf, J., 1994. Cone models for embedded foundation. J. Geotech. Engrg. 120 (1), 60–80.

Ohi, K., Takanashi, K., Tanaka, H., 1985. A simple method to estimate the statistical parameters of energy input to structures during earthquakes. Journal of Structural and Construction Eng., Archi. Inst. of Japan 347, 47–55 (in Japanese).

Ordaz, M., Huerta, B., Reinoso, E., 2003. Exact computation of input-energy spectra from Fourier amplitude spectra. Earthquake Engineering and Structural Dynamics 32 (4), 597–605.

Page, C.H., 1952. Instantaneous power spectra. Journal of Applied Physics 23 (1), 103–106.

Parmelee, R.A., 1970. The influence of foundation parameters on the seismic response of interaction systems. Proc. of the 3rd Japan Earthquake Engineering Symposium.

Riddell, R., Garcia, J.E., 2001. Hysteretic energy spectrum and damage control. Earthquake Engineering and Structural Dynamics 30 (12), 1791–1816.

Shinozuka, M., 1970. Maximum structural response to seismic excitations. Journal of Engineering Mechanics Division 96 (5), 729–738.

Singh, J.P., 1984. Characteristics of near-field ground motion and their importance in building design. ATC-10-1 Critical aspects of earthquake ground motion and building damage potential. ATC, 23–42.

Takewaki, I., 2001a. A new method for non-stationary random critical excitation. Earthquake Engineering and Structural Dynamics 30 (4), 519–535.

Takewaki, I., 2001b. Probabilistic critical excitation for MDOF elastic–plastic structures on compliant ground. Earthquake Engineering and Structural Dynamics 30 (9), 1345–1360.

Takewaki, I., 2002a. Seismic critical excitation method for robust design: A review. J. Struct. Eng. 128 (5), 665–672.

Takewaki, I., 2002b. Robust building stiffness design for variable critical excitations. J. Struct. Eng. 128 (12), 1565–1574.

Takewaki, I., 2003. Bound of earthquake input energy to structures with various damping distributions. J. Struct. Construction Engng. (Transactions of AIJ) 572, 65–72 (in Japanese).

Takewaki, I., 2004. Bound of earthquake input energy. J. Struct. Eng. 130 (9), 1289–1297.

Takewaki, I., Ben-Haim, Y., 2005. Info-gap robust design with load and model uncertainties. Journal of Sound and Vibration 288 (3), 551–570.

Takewaki, I., Fujimoto, H., 2004. Earthquake input energy to soil-structure interaction systems: A frequency-domain approach. Int. J. Advances in Structural Engineering 7 (5), 399–414.

Takewaki, I., Takeda, N., Uetani, K., 2003. Fast practical evaluation of soil–structure interaction of embedded structures. Soil Dynamics and Earthquake Engineering 23 (3), 195–202.

Takizawa, H., 1977. Energy-response spectrum of earthquake ground motions. Proc. of the 14th Natural Disaster Science Symposium, Hokkaido, 359–362 (in Japanese).

Tanabashi, R., 1935. Personal view on destructiveness of earthquake ground motions and building seismic resistance. Journal of Architecture and Building Science. Archi. Inst. of Japan, 48(599) (in Japanese).

Tanabashi, R., 1956. Studies on the nonlinear vibrations of structures subjected to destructive earthquakes. Proc. of the First World Conference on Earthquake Engineering. University of California, Berkeley, CA. Berkeley, 6:1–16.

Trifunac, M.D., Hao, T.Y., Todorovska, M.I., 2001. On energy flow in earthquake response. Report CE 01–03, July. University of Southern Calif.

Uang, C.-M., Bertero, V.V., 1990. Evaluation of seismic energy in structures. Earthquake Engineering and Structural Dynamics 19 (1), 77–90.

Wolf, J.P., 1985. Dynamic Soil-Structure Interaction. Prentice-Hall, Englewood Cliffs, NJ.

Wolf, J.P., 1988. Soil-Structure Interaction Analysis in time Domain. Prentice-Hall, Englewood Cliffs, NJ.

Wolf, J.P., 1994. Foundation Vibration Analysis using Simple Physical Models. Prentice-Hall, Englewood Cliffs, NJ.

Yang, Z., Akiyama, H., 2000. Evaluation of soil-structure interaction in terms of energy input. Journal of Structural and Construction Eng. (Transactions of AIJ) 536, 39–45 (in Japanese).

Zahrah, T., Hall, W., 1984. Earthquake energy absorption in SDOF structures. J. Struct. Eng. 110 (8), 1757–1772.

CHAPTER ELEVEN

Critical Excitation for Earthquake Energy Input in Structure-Pile-Soil System

Contents

11.1. INTRODUCTION

The purpose of this chapter is to explain a new method in the frequency domain (Takewaki 2004a; Takewaki and Fujimoto 2004) for the computation of earthquake input energies both to a structure-pile system and a structure only. In Takewaki (2004a) and Takewaki and Fujimoto (2004), a frequency-domain formulation has been developed for a simple sway-rocking model. In investigating the energy flow in the structure-pile system, many difficulties arise resulting from the dynamic interaction between the pile and the surrounding soil. It can be shown that the formulation of the earthquake input energy in the frequency domain is effective and efficient for deriving the earthquake input energy both to a structure-pile system and a structure only. This is because, in contrast to the

Critical Excitation Methods in Earthquake Engineering   © 2013 Elsevier Ltd.
All rights reserved.

formulation in the time domain which requires time–history response analysis, the formulation in the frequency domain only requires the computation of the Fourier amplitude spectrum of the input motion acceleration and the use of time–invariant transfer functions related to the structure-pile system. This means that the input and the structural model can be treated independently.

An efficient continuum model consisting of a dynamic Winkler-type soil element and a pile is used to express the dynamic behavior of the structure-pile system accurately. It is shown that the formulation of the earthquake input energy in the frequency domain is appropriate for introducing the frequency-dependent vibration property of the surface ground in terms of the wave propagation theory and the frequency-dependent property of the dynamic Winkler-type soil element around the pile. It is also demonstrated that (1) the present formulation is effective for various input levels and various ground properties and (2) the energy input mechanism or energy flow in the building structure-pile system can be well described by the newly introduced energy transfer function (F-function). A new concept, called input energy densities at various underground levels, is further introduced to disclose the energy input mechanism or energy flow in the building structure-pile system.

11.2. TRANSFER FUNCTION TO BEDROCK ACCELERATION INPUT

Consider a building-pile system as shown in Fig. 11.1. Let m_i and m_{bb} denote the mass in the i-th floor of the building and the mass in the foundation mat, respectively. It is assumed here that the surface ground consists of horizontal soil layers. The building-pile system is connected with the free-field ground through the Winkler-type soil element (Gazetas and Dobry 1984; Kavvadas and Gazetas 1993; Nikolaou et al. 2001). A single pile is considered and the building is modeled by a shear building model. If necessary, the effect of pile groups may be included by introducing a frequency–dependent pile-group coefficient. The pile head is assumed to be fixed to the foundation beam, i.e. the nodal rotation of the pile head is zero. The aspect ratio of the building is assumed to be so small that the rocking motion is negligible. The stiffness of the Winkler-type soil element is taken from the value corresponding to the fixed pile head proposed in Gazetas and Dobry (1984) and Kavvadas and Gazetas (1993). The damping of the Winkler-type soil element is assumed to

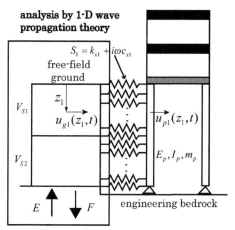

Figure 11.1 *Building-pile system supported by the free-field ground through Winkler-type soil element.*

be a combination of the radiation damping into the horizontal direction and the linear hysteretic damping (Gazetas and Dobry 1984; Kavvadas and Gazetas 1993). This leads to the introduction of a frequency-dependent damping coefficient.

[Free-Field Ground]

The numbering of the soil layer starts from the ground surface, i.e. the top soil layer is the first soil layer, and the coordinate z_1 in the first soil layer is directed downward from the ground surface. Let L_1 denote the thickness of the first soil layer and let V_{s1} denote the complex shear wave velocity including the linear hysteretic damping ratio in the first soil layer. The one-dimensional wave propagation theory for the free-field ground provides the horizontal displacement in the first soil layer.

$$u_{g1}(z_1, t) = \hat{E}_1(z_1)e^{i\omega t} \tag{11.1}$$

where i is the imaginary unit. $\hat{E}_1(z_1)$ can be expressed in terms of the amplitude E_1 at the ground surface of the upward propagating wave.

$$\hat{E}_1(z_1) = \{e^{i\omega z_1/V_{s1}} + e^{-i\omega z_1/V_{s1}}\}E_1 \tag{11.2}$$

[Pile-Soil System]

Let E_p, I_p, m_p denote Young's modulus, the second moment of area and the mass per unit length of the pile. The impedance of the Winkler-type soil element in the first soil layer is denoted by $S_1 = k_{x1} + i\omega c_{x1}$ where k_{x1} and c_{x1} are the stiffness and damping coefficients of the Winkler–type soil element. The damping of the pile is assumed to be negligible. Let u_{p1} denote the horizontal displacement of the pile in the first soil layer. The equation of motion of the pile in the first soil layer can be expressed by

$$E_p I_p \frac{\partial^4 u_{p1}}{\partial z_1^4} + m_p \frac{\partial^2 u_{p1}}{\partial t^2} = S_1(u_{g1} - u_{p1}) \qquad (11.3)$$

Let us introduce u_{p1} as

$$u_{p1}(z_1, t) = \hat{U}_{p1}(z_1) e^{i\omega t} \qquad (11.4)$$

Substitution of Eqs. (11.1) and (11.4) into Eq. (11.3) provides

$$(\partial^4 \hat{U}_{p1}/\partial z_1^4) - \lambda_1^4 \hat{U}_{p1} = \alpha_1 \hat{E}_1 \qquad (11.5)$$

where

$$\lambda_1 = ((m_p \omega^2 - S_1)/E_p I_p)^{0.25}, \quad \alpha_1 = S_1/E_p I_p \qquad (11.6a,b)$$

The general solution of Eq. (11.5) may be expressed as

$$\hat{U}_{p1}(z_1) = D_1^{(1)} e^{-\lambda_1 z_1} + D_2^{(1)} e^{\lambda_1 z_1} + D_3^{(1)} e^{-i\lambda_1 z_1} + D_4^{(1)} e^{i\lambda_1 z_1} + s_1 \hat{E}_1(z_1) \qquad (11.7)$$

where

$$s_1 = \alpha_1/\{(\omega/V_{s1})^4 - \lambda_1^4\} \qquad (11.8)$$

Similarly let us introduce the pile horizontal displacement in the second soil layer as $u_{p2}(z_2, t) = \hat{U}_{p2}(z_2) e^{i\omega t}$. The general solution of the pile horizontal displacement in the second soil layer may be derived as

$$\hat{U}_{p2}(z_2) = D_1^{(2)} e^{-\lambda_2 z_2} + D_2^{(2)} e^{\lambda_2 z_2} + D_3^{(2)} e^{-i\lambda_2 z_2} + D_4^{(2)} e^{i\lambda_2 z_2} + s_2 \hat{E}_2(z_2) \qquad (11.9)$$

where

$$s_2 = \alpha_2/\{(\omega/V_{s2})^4 - \lambda_2^4\},$$

$$\lambda_2 = ((m_p\omega^2 - S_2)/E_p I_p)^{0.25},$$

$$\alpha_2 = S_2/E_p I_p,$$

$$S_2 = k_{x2} + i\omega c_{x2},$$

$$\hat{E}_2(z_2) = \{E_2 e^{i\omega z_2/V_{s2}} + F_2 e^{-i\omega z_2/V_{s2}}\} \qquad (11.10\text{a}-\text{e})$$

V_{s2} is the complex shear wave velocity in the second soil layer and E_2, F_2 are the amplitudes of the incident wave (upward propagating wave) and the reflected wave (downward propagating wave), respectively, at the top of the second soil layer. As for the relation of E_2, F_2 with E_1, see Appendix 11.1. The expression, Eq. (11.9), has been shown here because a detailed expression for determining undetermined parameters is shown in the following.

Since four undetermined coefficients exist in every soil layer, a set of $4 \times n + n_b$ simultaneous linear equations has to be solved for every excitation frequency ω where n and n_b are the number of the soil layers and the number of the stories of the building, respectively. This set of $4 \times n + n_b$ simultaneous linear equations can be constructed from the boundary conditions, the continuity conditions and the equilibrium equations in the building-pile-soil system. The explicit expression of these simultaneous linear equations for $n = n_b = 2$ can be found in Appendix 11.1.

The validity of the proposed method has been demonstrated through the comparison with the results by the finite-element method and the earthquake record observation (Nikolaou et al. 2001). The comparison with recorded data during an earthquake is shown in Appendix 11.2. Part of the comparison with the results by the finite-element method will be shown later in Figs. 11.8 and 11.9.

For later formulation, let us consider the case where the building-pile system is subjected to the outcropping horizontal acceleration $\ddot{u}_g(t) = a(t)$ at the bedrock. The super-dot indicates the time derivative. Let $A(\omega)$ denote the Fourier transform of $\ddot{u}_g(t) = a(t)$. The displacement transfer functions $H_{g1}(z_1, \omega) = U_{g1}(z_1, \omega)/A(\omega)$ and $H_{g2}(z_2, \omega) = U_{g2}(z_2, \omega)/A(\omega)$ of the free-field ground to $A(\omega)$ can be obtained from the one-dimensional wave propagation theory. On the other hand, the displacement transfer functions $H_{p1}(z_1, \omega) = U_{p1}(z_1, \omega)/A(\omega)$ and $H_{p2}(z_2, \omega) = U_{p2}(z_2, \omega)/A(\omega)$ of the pile to $A(\omega)$ can be derived from the set of $4 \times n + n_b$ simultaneous linear equations stated above.

11.3. EARTHQUAKE INPUT ENERGY TO STRUCTURE-PILE SYSTEM

Consider free bodies as shown in Fig. 11.2(a) and (b). For simplicity of expression, it is assumed again that $n = n_b = 2$. Let us define the work by the forces at the side of the surrounding soil and at the pile tip on the corresponding displacements as the input energy into the building-pile system. This input energy can be expressed in the time domain as

$$E_I^A = \sum_{i=1}^{2} \int_0^{L_i} \int_{-\infty}^{\infty} \{k_{xi}(u_{gi} - u_{pi}) + c_{xi}(\dot{u}_{gi} - \dot{u}_{pi})\}\dot{u}_{gi} dt dz_i$$
$$+ \int_{-\infty}^{\infty} \{-E_p I_p u_{p2}'''(L_2)\}\dot{u}_{p2}(L_2) dt$$

(11.11)

where k_{xi} and c_{xi} are the stiffness and damping coefficients of the Winkler-type soil element in the i-th soil layer. $(\quad)'$ denotes the differentiation with respect to the space coordinate. The first term in Eq. (11.11) indicates the input energy from the side of the surrounding soil and the second term does that from the pile tip. L_i is the thickness of the i-th soil layer. Eq. (11.11) can be transformed to the following form via the Fourier inverse transformation of the free-field ground displacement and the pile displacement.

$$E_I^A = \sum_{i=1}^{2} \int_0^{L_i} \int_{-\infty}^{\infty} [(1/2\pi) \int_{-\infty}^{\infty} S_i(U_{gi} - U_{pi})e^{i\omega t}d\omega] \dot{u}_{gi} dt dz_i$$
$$- \int_{-\infty}^{\infty} E_p I_p [(1/2\pi) \int_{-\infty}^{\infty} U_{p2}'''(L_2, \omega)e^{i\omega t}d\omega]\dot{u}_{p2}(L_2) dt$$

(11.12)

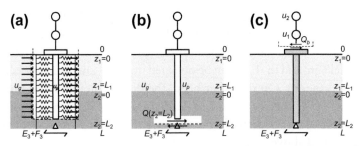

Figure 11.2 *Free-body diagrams: (a) Forces at the side of the surrounding soil, (b) force at the pile tip, (c) force at the virtual plane at the bottom of the building.*

where $U_{p2}'''(L_2, \omega)$ is the Fourier transform of $u_{p2}''(L_2)$. Application of the Fourier transformation of $\ddot{u}_{gi}$ and $\ddot{u}_{p2}(L_2)$, the relations $\dot{U}_{gi}(-\omega) = -i\omega U_{gi}(-\omega)$, $\dot{U}_{p2}(-\omega) = -i\omega U_{p2}(-\omega)$ and the utilization of the transfer functions $H_{gi}(z_i, \omega), H_{pi}(z_i, \omega)$ to Eq. (11.12) may provide

$$E_I^A = \sum_{i=1}^{2} \int_0^{L_i} \int_{-\infty}^{\infty} \left[(1/2\pi) S_i (U_{gi} - U_{pi}) \left(\int_{-\infty}^{\infty} \ddot{u}_{gi} e^{i\omega t} dt \right) \right] d\omega dz_i$$

$$- \int_{-\infty}^{\infty} E_p I_p \left[(1/2\pi) U_{p2}'''(L_2, \omega) \left(\int_{-\infty}^{\infty} \ddot{u}_{p2}(L_2) e^{i\omega t} dt \right) \right] d\omega$$

$$= -(1/2\pi) \sum_{i=1}^{2} \int_{-\infty}^{\infty} \int_0^{L_i} (i\omega) S_i \{ H_{gi}(z_i, \omega)$$

$$- H_{pi}(z_i, \omega) \} H_{gi}(z_i, -\omega) |A(\omega)|^2 dz_i d\omega$$

$$+ (E_p I_p / 2\pi) \int_{-\infty}^{\infty} (i\omega) H_{p2}'''(L_2, \omega) H_{p2}(L_2, -\omega) |A(\omega)|^2 d\omega$$

(11.13)

Let us express Eq. (11.13) compactly as

$$E_I^A = \int_{-\infty}^{\infty} F_A(\omega) |A(\omega)|^2 d\omega$$

(11.14)

where $F_A(\omega)$ is called the energy transfer function for the building-pile system and characterizes the energy input to the building-pile system. $F_A(\omega)$ is described as

$$F_A(\omega) = -(1/2\pi)(i\omega) \sum_{i=1}^{2} \int_0^{L_i} S_i \{ H_{gi}(z_i, \omega) - H_{pi}(z_i, \omega) \} H_{gi}(z_i, -\omega) dz_i$$

$$+ (E_p I_p / 2\pi)(i\omega) H_{p2}'''(L_2, \omega) H_{p2}(L_2, -\omega)$$

(11.15)

It should be noted that $\text{sgn}(\omega)$ has to be introduced in the imaginary part of the complex stiffness of the Winkler-type soil element in dealing with the linear hysteretic damping in the negative frequency range.

11.4. EARTHQUAKE INPUT ENERGY TO STRUCTURE

Consider a free body as shown in Fig. 11.2(c). Let u_i denote the absolute horizontal displacement of the i-th floor mass. U_i and $\ddot{U}_i$ are the Fourier transforms of u_i and $\ddot{u}_i$, respectively. Let us define the work by the horizontal force at the virtual section at the bottom of the building on the corresponding displacement as the input energy into the building. The input energy may be expressed in the time domain as

$$E_I^S = \int_{-\infty}^{\infty} \left\{ \sum_{j=1}^{2} m_j \ddot{u}_j \right\} \dot{u}_{p1}(0) dt \qquad (11.16)$$

Let us introduce the following transfer functions of the displacements of the floor masses.

$$H_1(\omega) = U_1(\omega)/A(\omega), \quad H_2(\omega) = U_2(\omega)/A(\omega) \qquad (11.17a, b)$$

These transfer functions can be derived from the the set of $4 \times n + n_b$ simultaneous linear equations stated above. Application of the inverse Fourier transformation of the displacements of the floor masses to Eq. (11.16) provides

$$E_I^S = \int_{-\infty}^{\infty} (1/2\pi) \int_{-\infty}^{\infty} \left\{ \sum_{j=1}^{2} m_j \ddot{U}_j \right\} e^{i\omega t} d\omega \, \dot{u}_{p1}(0) dt \qquad (11.18)$$

Application of the Fourier transformation of $\dot{u}_{p1}(0)$, the relations $\dot{U}_{p1}(0, -\omega) = -i\omega U_{p1}(0, -\omega)$, $\ddot{U}_j = (i\omega)^2 U_j$ and the utilization of the transfer functions $H_j(\omega)$ in Eq. (11.17) to Eq. (11.18) may provide

$$E_I^S = \int_{-\infty}^{\infty} (1/2\pi) \left\{ \sum_{j=1}^{2} m_j \ddot{U}_j \right\} \left(\int_{-\infty}^{\infty} \dot{u}_{p1}(0) e^{i\omega t} dt \right) d\omega$$

$$= (1/2\pi) \int_{-\infty}^{\infty} (i\omega^3) \left\{ \sum_{j=1}^{2} m_j H_j(\omega) \right\} H_{p1}(0, -\omega) |A(\omega)|^2 d\omega$$

$$(11.19)$$

Let us express Eq. (11.19) compactly as

$$E_I^S = \int_{-\infty}^{\infty} F_S(\omega) |A(\omega)|^2 d\omega \qquad (11.20)$$

$F_S(\omega)$ in Eq. (11.20) is called the energy transfer function of the structure and characterizes the energy input to the structure.

11.5. INPUT ENERGIES BY DAMAGE-LIMIT LEVEL EARTHQUAKE AND SAFETY-LIMIT LEVEL EARTHQUAKE

[Energy Transfer Functions for Two-Level Design Earthquakes]
Consider two ground models, called Ground A and Ground B. Ground A is a soft ground and Ground B is a harder ground. The soil profiles (soil properties and layer thicknesses) of both grounds are shown in Fig. 11.3. In addition to the shear wave velocity, the depth of the surface ground is also an important factor for evaluating the property of the ground, e.g. the natural frequency of the surface ground. The depth of the surface ground of Ground A is larger than that of Ground B and Ground A is much softer than Ground B. A soil layer of the shear wave velocity around or larger than 400(m/s) is called "engineering bedrock." In order to describe the nonlinear behavior of the soil, a well-known equivalent linear model is used (Schnabel et al. 1972). The dependency of the soil properties on the strain level is shown in Fig. 11.4. Instead of the time-history evaluation of the maximum soil shear strain in the equivalent linear model, a response spectrum method (Takewaki 2004b) is utilized. In that response spectrum method, a method based on complex eigenvalue analysis is used to take into account the non-proportional damping characteristics of the soil model. The damping model (Gazetas and Dobry 1984; Kavvadas and Gazetas 1993) of the Winkler-type

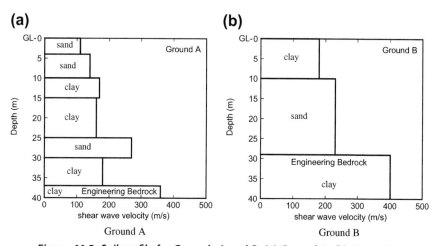

Figure 11.3 *Soil profile for Grounds A and B. (a) Ground A, (b) Ground B.*

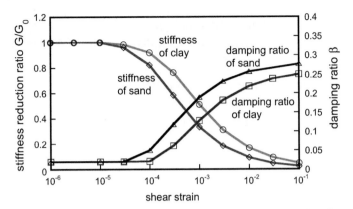

Figure 11.4 *Stiffness reduction and damping ratio of clay and sand with respect to shear strain.*

soil element is shown in Fig. 11.5. The damping consists of hysteretic damping and radiation damping. The damping ratio evaluated in the equivalent linearization of the free-field ground is substituted into the hysteretic damping term in Fig. 11.5. The effectiveness and validity of this response spectrum method have been demonstrated in Takewaki 2004b.

Let us consider two-level earthquake input motions, one is the damage-limit level earthquake and the other is the safety-limit level earthquake. These earthquake motions are defined at the engineering bedrock surface as outcropping motions. The acceleration response spectra for 5% damping ratio for these two-level earthquake input motions are shown in Fig. 11.6.

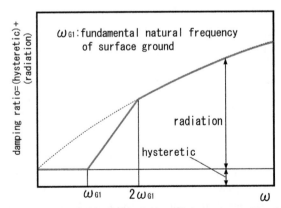

Figure 11.5 *Damping ratio of the Winkler-type soil element: combination of hysteretic and radiation dampings.*

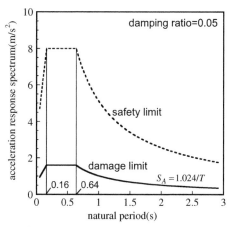

Figure 11.6 *Design acceleration response spectra at bedrock for damage and safety limits (damping ratio = 0.05).*

The fundamental natural periods of Ground A for the damage-limit and safety-limit levels, respectively, are 1.0(s) and 1.6(s). The fundamental natural periods of Ground B for the damage-limit and safety-limit levels, respectively, are 0.6(s) and 1.2(s).

For simple presentation of the energy input mechanism, the building is assumed to be modeled by a 2-story shear building model as shown in Fig. 11.7. The story stiffness distribution is determined so that the lowest eigenmode of the model with fixed base is a straight line. The magnitude of

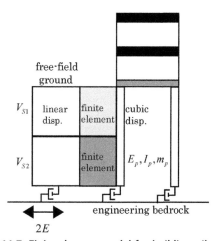

Figure 11.7 *Finite-element model for building-pile system.*

the story stiffnesses is controlled by modifying the fundamental natural period of the model with fixed base. The model floor masses are $m_i = 10 \times 10^3 (\text{kg})(i = 1, 2)$ and $m_{bb} = 30 \times 10^3 (\text{kg})$. The story heights are $h_i = 3.5 (\text{m})(i = 1, 2)$. The soil mass density and Poisson's ratio are $\rho_i = 1.6 \times 10^3 (\text{kg/m}^3)$ and $\nu = 0.45$. The pile diameter is 1.5(m) and the pile Young's modulus and mass density are $2.1 \times 10^{10} (\text{N/m}^2)$ and $2.4 \times 10^3 (\text{kg/m}^3)$, respectively. The damping of the building is assumed to be stiffness-proportional and the damping ratio for the lowest eigenvibration is 0.05.

The solid line in Fig. 11.8 shows the real parts of the energy transfer function $F_A(\omega)$, Eq. (11.15), for the building-pile system and that $F_S(\omega)$, Eq. (11.20), for the building only under the damage-limit level earthquake. The proposed model is called here "a continuum model." Figs. 11.8(a) and

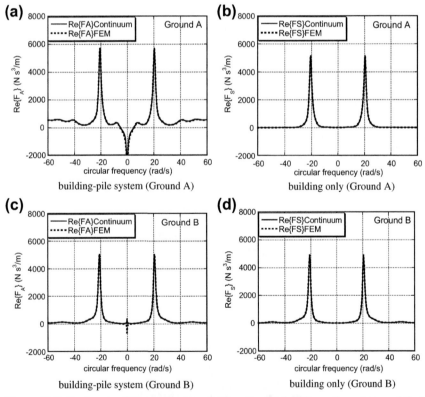

Figure 11.8 *Real parts of the energy transfer function for building-pile system and that for building only by the proposed continuum model and the finite-element model (damage-limit level earthquake). (a) Ground A, building-pile system, (b) Ground A, building only, (c) Ground B, building-pile system, (d) Ground B, building only.*

(b) are for Ground A and Figs. 11.8(c) and (d) are for Ground B. The stiffness and damping ratio of the free-field ground and the Winkler-type soil element have been evaluated to the damage-limit level earthquake input. The fundamental natural period of the building with fixed base is 0.3(s). The imaginary part of the energy transfer function $F_A(\omega)$ becomes an odd function of frequency that does not contribute to the input energy. The solid line in Fig. 11.9 shows the real parts of the energy transfer function $F_A(\omega)$, Eq. (11.15), for the building-pile system and that $F_S(\omega)$, Eq. (11.20), for the building only under the safety-limit level earthquake. It can be observed from Figs. 11.8 and 11.9 that, while the energy transfer functions $F_A(\omega)$ and $F_S(\omega)$ for both Grounds A and B under the damage-limit level earthquake exhibit a large value around the fundamental natural circular frequency (about 20 rad/s) of the building, $F_A(\omega)$ for Ground A (soft ground) under the safety-limit level earthquake shows a large value around the fundamental natural

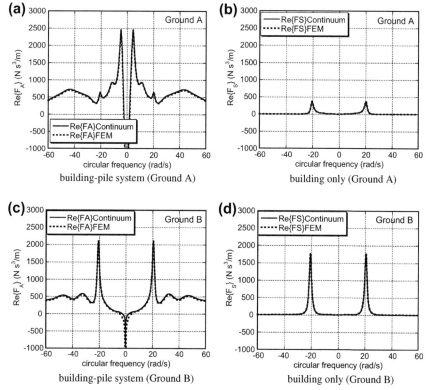

Figure 11.9 *Real parts of the energy transfer function for building-pile system and for building only by the proposed continuum model and the finite-element model (safety-limit level earthquake). (a) Ground A, building-pile system, (b) Ground A, building only, (c) Ground B, building-pile system, (d) Ground B, building only.*

circular frequency (about 4rad/s) of the surface ground. It can also be observed from Figs. 11.9(c) and (d) that the input energy to the building on Ground A (soft ground) under the safety-limit level earthquake is greatly smaller than that on Ground B (hard ground). This may result from the fact that a building with a rather short natural period is considered here.

In order to demonstrate the validity of the proposed model and method, the finite-element model (see Fig. 11.7) proposed in Nakamura et al. (1996), Takewaki (1999) has been used to evaluate the energy transfer functions $F_A(\omega)$ and $F_S(\omega)$. It should be noted that viscous boundaries have been introduced at the bottom of the surface ground and frequency-dependent characteristics of the Winkler-type soil element have been used. The dotted line in Figs. 11.8 and 11.9 shows the real parts of the energy transfer function $F_A(\omega)$ for the building-pile system and that $F_S(\omega)$ for the building only by the FEM model. A very good correspondence can be seen and it can be concluded that the proposed model and method are valid.

In order to clarify the characteristics in the case where the fundamental natural period of the building is resonant to that of the surface ground, the energy transfer functions for the building with the fundamental natural period 1.6(s) under the safety-limit level earthquake have been computed. It should be reminded that the fundamental natural period of the surface ground of Ground A under the safety-limit level earthquake is about 1.6(s). Fig. 11.10 shows the real parts of the energy transfer function for building-pile system and that for building only in case of the building with fundamental natural period = 1.6(s). It can be observed that a large energy input can be expected in such a resonant case.

[Input Energy by Actual Ground Motion]

For evaluating the characteristics of input energies by an actual ground motion, the ground motion of El Centro NS 1940 has been chosen. The maximum velocities are scaled to 0.1m/s and 0.5m/s for the damage-limit level and the safety-limit level, respectively, so as to be compatible with the Japanese code (see Fig. 11.6). Fig. 11.11 shows the earthquake input energy to the building-pile system and that to building only with respect to building fundamental natural period subjected to El Centro NS 1940 of the damage-limit level. It can be observed that, while the input energy to the building-pile system and that to building only differ greatly in case of Ground A (soft ground), those do not differ so much in case of Ground B (hard ground). This is because the energy dissipated in the ground or radiated into the ground is large in Ground A. Fig. 11.12 illustrates the earthquake input

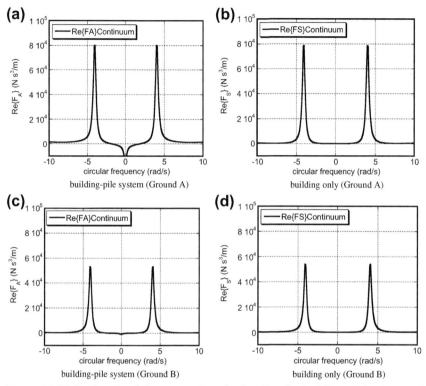

Figure 11.10 *Real parts of the energy transfer function for building-pile system and that for building only in case of the building with fundamental natural period = 1.6(s): resonance with Ground A (safety-limit level earthquake). (a) Ground A, building-pile system, (b) Ground A, building only, (c) Ground B, building-pile system, (d) Ground B, building only.*

energy to the building–pile system and that from the pile-tip with respect to building fundamental natural period subjected to El Centro NS 1940 of the damage-limit level. It can be seen that the energy from the pile-tip is almost constant irrespective of the building fundamental natural period. This is because the energy from the pile-tip does not depend on the building and depends mainly on the soil condition. It can also be observed that the input energy to the building–pile–soil system is larger than that from the pile-tip in the building fundamental natural period range of 0.5–1.5s in Ground A and 0.3–3.0s in Ground B. This means that the input energy from the side of the surrounding soil is positive in these fundamental natural period ranges. On the contrary, in the other fundamental natural period range, this relation is reversed. This means that the input energy from the side of the surrounding soil is negative in these fundamental natural period ranges.

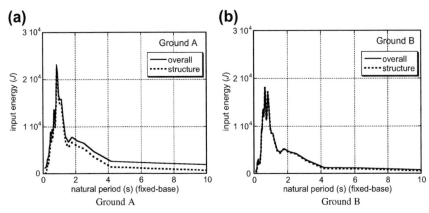

Figure 11.11 *Earthquake input energy to building-pile system and that to building only with respect to building fundamental natural period subjected to El Centro NS 1940 (damage-limit level). (a) Ground A, (b) Ground B.*

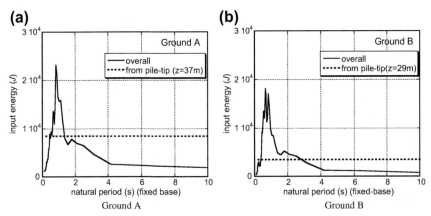

Figure 11.12 *Earthquake input energy to building-pile system and that from the pile-tip with respect to building fundamental natural period subjected to El Centro NS 1940 (damage-limit level). (a) Ground A, (b) Ground B.*

In order to confirm this fact, an additional computation has been made. Fig. 11.13 shows the real part of the energy transfer function corresponding to the energy input from the side of the surrounding soil in Ground A under the damage-limit level earthquake. The building fundamental natural periods have been chosen as 1.0s and 5.0s. In the building with the fundamental natural period 5.0s, the negative portion is relatively wide. This fact indicates the negative input energy from the side of the surrounding soil in this fundamental natural period range.

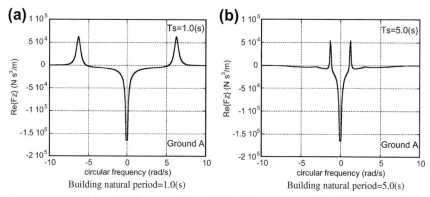

Figure 11.13 *Real part of the energy transfer function corresponding to the energy input from the side of the surrounding soil (damage-limit level earthquake). (a) Ts = 1.0 (s), (b) Ts = 5.0 (s).*

Fig. 11.14 indicates the earthquake input energy to the building-pile system and that to building only with respect to building fundamental natural period subjected to El Centro NS 1940 of the safety-limit level. It can be observed that the difference between the input energy to the building-pile system and that to the building only is larger than that to the input of the damage-limit level. This is because the energy dissipated in the ground or radiated into the ground is large under the input of the safety-limit level.

Fig. 11.15 shows the earthquake input energy to the building-pile system and that from the pile-tip with respect to building fundamental natural

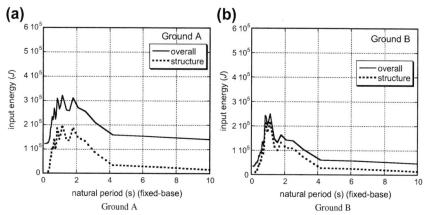

Figure 11.14 *Earthquake input energy to building-pile system and that to building only with respect to building fundamental natural period subjected to El Centro NS 1940 (damage-limit level). (a) Ground A, (b) Ground B.*

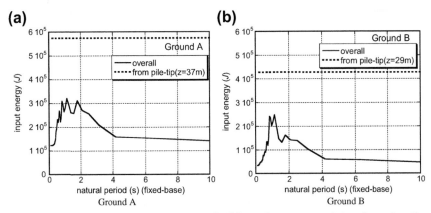

Figure 11.15 *Earthquake input energy to building-pile system and that from the pile-tip with respect to building fundamental natural period subjected to El Centro NS 1940 (safety-limit level). (a) Ground A, (b) Ground B.*

period subjected to El Centro NS 1940 of the safety-limit level. It can be seen that the input energy from the side of the surrounding soil is negative in all the fundamental natural period range. This means that, while the input energy from the pile-tip is always positive, the energy dissipated in the ground or radiated into the ground is extremely large under the input of the safety-limit level due to the reduction of soil stiffness.

Fig. 11.16 illustrates the underground distribution of the newly defined input energy density for Ground A under the damage-limit level and safety-limit level earthquakes. The integration of the input energy density over the pile represents the input energy from the side of the surrounding soil. It can be understood that there is a portion with a positive density and one with a negative density. This investigation is very useful in disclosing the energy input mechanism underground. Fig. 11.17 shows the underground distribution of the input energy density for Ground B under the damage-limit level and safety-limit level earthquakes.

Fig. 11.18 indicates the real part of the energy transfer function at the two different depths, GL-10m and GL-30m, corresponding to the energy input from the side of the surrounding soil for Ground A under the damage-limit level earthquake. The building natural period is 0.5s. It can be seen that, while the input energy from the side of the surrounding soil at GL-10m is positive, that at GL-30m is negative. It may be concluded that, since the present approach is based on the energy transfer function independent of input ground motions and being able to be defined at each

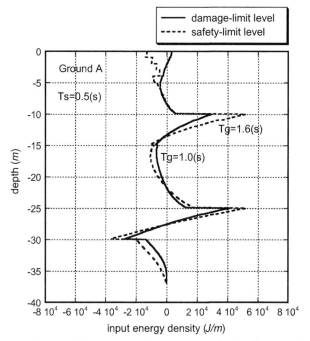

Figure 11.16 *Underground distribution of input energy density for Ground A (damage-limit level and safety-limit level earthquakes).*

point or area, it is possible to disclose the earthquake energy input mechanism in detail.

11.6. CRITICAL EXCITATION FOR EARTHQUAKE ENERGY INPUT IN STRUCTURE-PILE-SOIL SYSTEM

The following critical excitation method may be stated for structure-pile-soil systems.

Find $|A(\omega)|$

that maximizes $E_I^S = \int\limits_{-\infty}^{\infty} F_S(\omega)|A(\omega)|^2 d\omega$

The solution procedure developed for fixed-base single-degree-of-freedom (SDOF) models in Chapter 8, for fixed-base multi-degree-of-freedom (MDOF) models in Chapter 9 and for SSI models in Chapter 10 can be applied to this problem. It suffices to replace the energy transfer function $F(\omega)$ for the SDOF model by $F_S(\omega)$ for the present model.

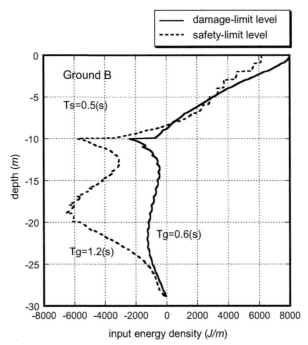

Figure 11.17 *Underground distribution of input energy density for Ground B (damage-limit level and safety-limit level earthquakes).*

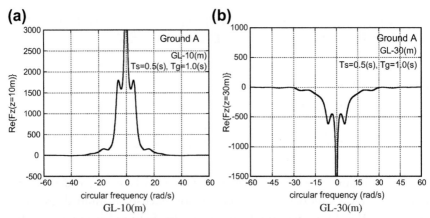

Figure 11.18 *Real part of the energy transfer function at the two different depths corresponding to the energy input from the side of the surrounding soil for Ground A (damage-limit level earthquake). (a) GL-10(m), (b) GL-30(m).*

11.7. CONCLUSIONS

The conclusions may be stated as follows:

(1) A new evaluation method of earthquake input energy to structure-pile interaction systems has been developed. The method is an approach in the frequency domain. The energy transfer function, which plays a key role in the input energy evaluation, is derived from the transfer functions of the structure-pile interaction system to the earthquake acceleration input at the bedrock.

(2) The present approach in the frequency domain has an advantage; to be able to include directly the frequency-dependent characteristics (stiffness and damping) of the Winkler-type soil element in addition to the treatment of the free-field ground via the wave propagation theory.

(3) The introduction of the definition of two input energies, one to the overall system (structure plus pile and surrounding soil) and the other to the structure alone, is very useful in understanding the mechanism of the earthquake energy input and the effect of soil-structure interaction under various conditions of soil properties and natural period of structures on the earthquake energy input.

(4) While the approach in the time domain requires time-series ground motions, the approach in the frequency domain requires only the Fourier spectrum of the acceleration at the bedrock and enables one to capture the general characteristics of the energy input through a general form of the Fourier spectrum of the input acceleration. This treatment is completely compatible with most of the seismic resistant design codes in which a response spectrum or the corresponding Fourier spectrum is provided.

(5) The proposed input energy density whose integration over the pile represents the input energy from the side of the surrounding soil is appropriate for disclosing the energy input mechanism or energy flow underground.

(6) The energy transfer function can be defined even at a point (including underground) and it is easy to understand general characteristics of the flow of the earthquake energy at the point or in the area. These characteristics underground depend on the soil property and its depth. It has been shown that these general characteristics can be characterized by the proposed energy transfer function.

The evaluation of earthquake input energy in the time domain is suitable for the evaluation of the time history of input energy, especially for nonlinear systems. Dual use of the frequency-domain and time-domain techniques may be preferable in advanced seismic analysis for more robust design.

The formulation of the earthquake input energy in the frequency domain is effective and essential for solving a critical excitation problem (Takewaki 2004a) for the maximum earthquake input energy and deriving a bound on the earthquake input energy for a class of ground motions.

APPENDIX 11.1 10 × 10 SIMULTANEOUS LINEAR EQUATIONS FOR DERIVING TRANSFER FUNCTIONS FOR THE 2-STORY BUILDING MODEL ON GROUND WITH TWO SOIL LAYERS

Let $\alpha_{i1} = \rho_1 V_{s1}/\rho_2 V_{s2}$ and $\alpha_{i2} = \rho_2 V_{s2}/\rho_3 V_{s3}$ denote the complex impedance ratios between the first and second soil layers and the second and third soil layers, respectively, where ρ_i and V_{si} are the mass density and the complex shear wave velocity in the i-th soil layer. The story stiffness and the corresponding damping coefficient of the i-th story are denoted by k_i and c_i, respectively. The transfer functions are derived from the boundary conditions, continuity conditions and equilibrium equations. Those equations may be described compactly from Eqs. (11.2), (11.7), (11.9) as

$$-\omega^2 \mathbf{X}\mathbf{Y} = \mathbf{Z} \qquad (A11.1)$$

where

$$\mathbf{X} = [\mathbf{X}_1 \ \cdots \ \mathbf{X}_{10}]$$

$$\mathbf{Y} = \frac{1}{-\omega^2(2E_3)}$$

$$\times \left\{ U_1 \quad U_2 \quad D_1^{(1)} \quad D_2^{(1)} \quad D_3^{(1)} \quad D_4^{(1)} \quad D_1^{(2)} \quad D_2^{(2)} \quad D_3^{(2)} \quad D_4^{(2)} \right\}^T$$

$$\mathbf{Z} = \frac{1}{A_1^*} \begin{Bmatrix} -(i\omega c_1 + k_1)s_1 \\ 0 \\ 0 \\ (m_{bb}\omega^2 - i\omega c_1 - k_1)s_1 \\ (-s_1 + s_2)\cos(\omega L_1/V_{s1}) \\ \{-s_1(\omega/V_{s1}) + s_2(\omega/V_{s2})\alpha_{i1}\}\sin(\omega L_1/V_{s1}) \\ \{s_1(\omega/V_{s1})^2 - s_2(\omega/V_{s2})^2\}\cos(\omega L_1/V_{s1}) \\ \{s_1(\omega/V_{s1})^3 - s_2(\omega/V_{s2})^3\alpha_{i1}\}\sin(\omega L_1/V_{s1}) \\ s_2 J_2(\omega/V_{s2})^2\cos\{(\omega L_2/V_{s2}) + \beta\} \\ -s_2 J_2\cos\{(\omega L_2/V_{s2}) + \beta\} + (A_1^* + A_2^*)/2 \end{Bmatrix} \qquad (A11.2a-c)$$

$$\mathbf{X}_1 = \left\{ \begin{array}{c} m_1\omega^2 - i\omega(c_1 + c_2) - (k_1 + k_2) \\ -(i\omega c_2 + k_2) \\ 0 \\ -(i\omega c_1 + k_1) \\ 0 \\ 0 \\ 0 \\ 0 \\ 0 \\ 0 \end{array} \right\}, \mathbf{X}_2 = \left\{ \begin{array}{c} i\omega c_2 + k_2 \\ -m_2\omega^2 + i\omega c_2 + k_2 \\ 0 \\ 0 \\ 0 \\ 0 \\ 0 \\ 0 \\ 0 \\ 0 \end{array} \right\},$$

$$\mathbf{X}_3 = \left\{ \begin{array}{c} i\omega c_1 + k_1 \\ 0 \\ \lambda_1 \\ -4E_pI_p\lambda_1^3 + i\omega c_1 + k_1 - m_{bb}\omega^2 \\ e^{-\lambda_1 L_1} \\ \lambda_1 e^{-\lambda_1 L_1} \\ \lambda_1^2 e^{-\lambda_1 L_1} \\ \lambda_1^3 e^{-\lambda_1 L_1} \\ 0 \\ 0 \end{array} \right\},$$

$$\mathbf{X}_4 = \left\{ \begin{array}{c} i\omega c_1 + k_1 \\ 0 \\ -\lambda_1 \\ 4E_pI_p\lambda_1^3 + i\omega c_1 + k_1 - m_{bb}\omega^2 \\ e^{\lambda_1 L_1} \\ -\lambda_1 e^{\lambda_1 L_1} \\ \lambda_1^2 e^{\lambda_1 L_1} \\ -\lambda_1^3 e^{\lambda_1 L_1} \\ 0 \\ 0 \end{array} \right\}, \mathbf{X}_5 = \left\{ \begin{array}{c} i\omega c_1 + k_1 \\ 0 \\ i\lambda_1 \\ 4E_pI_pi\lambda_1^3 + i\omega c_1 + k_1 - m_{bb}\omega^2 \\ e^{-i\lambda_1 L_1} \\ i\lambda_1 e^{-i\lambda_1 L_1} \\ -\lambda_1^2 e^{-i\lambda_1 L_1} \\ -i\lambda_1^3 e^{-i\lambda_1 L_1} \\ 0 \\ 0 \end{array} \right\}$$

$$\mathbf{X}_6 = \left\{ \begin{array}{c} i\omega c_1 + k_1 \\ 0 \\ -i\lambda_1 \\ -4E_p I_p i\lambda_1^3 + i\omega c_1 + k_1 - m_{bb}\omega^2 \\ e^{i\lambda_1 L_1} \\ -i\lambda_1 e^{i\lambda_1 L_1} \\ -\lambda_1^2 e^{i\lambda_1 L_1} \\ i\lambda_1^3 e^{i\lambda_1 L_1} \\ 0 \\ 0 \end{array} \right\}, \mathbf{X}_7 = \left\{ \begin{array}{c} 0 \\ 0 \\ 0 \\ 0 \\ -1 \\ -\lambda_2 \\ -\lambda_2^2 \\ -\lambda_2^3 \\ \lambda_2^2 e^{-\lambda_2 L_2} \\ e^{-\lambda_2 L_2} \end{array} \right\},$$

$$\mathbf{X}_8 = \left\{ \begin{array}{c} 0 \\ 0 \\ 0 \\ 0 \\ -1 \\ \lambda_2 \\ -\lambda_2^2 \\ \lambda_2^3 \\ \lambda_2^2 e^{\lambda_2 L_2} \\ e^{\lambda_2 L_2} \end{array} \right\}, \mathbf{X}_9 = \left\{ \begin{array}{c} 0 \\ 0 \\ 0 \\ 0 \\ -1 \\ -i\lambda_2 \\ \lambda_2^2 \\ i\lambda_2^3 \\ -\lambda_2^2 e^{-i\lambda_2 L_2} \\ e^{-i\lambda_2 L_2} \end{array} \right\}, \mathbf{X}_{10} = \left\{ \begin{array}{c} 0 \\ 0 \\ 0 \\ 0 \\ -1 \\ i\lambda_2 \\ \lambda_2^2 \\ -i\lambda_2^3 \\ -\lambda_2^2 e^{i\lambda_2 L_2} \\ e^{i\lambda_2 L_2} \end{array} \right\}$$

$$(A11.3a-j)$$

$$J_2 = \sqrt{\cos^2 \omega(L_1/V_{s1}) + \alpha_{i1}^2 \sin^2 \omega(L_1/V_{s1})},$$

$$\beta = \cos^{-1}\{[\cos \omega(L_1/V_{s1})]/J_2\} \qquad (A11.4a, b)$$

A_1^* and A_2^* in Eq. (A11.2c) are defined via the one-dimensional wave propagation theory as

$$
\begin{Bmatrix} E_3 \\ F_3 \end{Bmatrix} = [A_2][A_1] \begin{Bmatrix} 1 \\ 1 \end{Bmatrix} E_1 = \begin{bmatrix} A_{11}^{(2)} & A_{12}^{(2)} \\ A_{21}^{(2)} & A_{22}^{(2)} \end{bmatrix} \begin{bmatrix} A_{11}^{(1)} & A_{12}^{(1)} \\ A_{21}^{(1)} & A_{22}^{(1)} \end{bmatrix} \begin{Bmatrix} 1 \\ 1 \end{Bmatrix} E_1
$$

$$
= \begin{Bmatrix} A_{11}^{(2)} A_{11}^{(1)} + A_{12}^{(2)} A_{21}^{(1)} + A_{11}^{(2)} A_{12}^{(1)} + A_{12}^{(2)} A_{22}^{(1)} \\ A_{21}^{(2)} A_{11}^{(1)} + A_{22}^{(2)} A_{21}^{(1)} + A_{21}^{(2)} A_{12}^{(1)} + A_{22}^{(2)} A_{22}^{(1)} \end{Bmatrix} E_1
$$

$$
= \begin{Bmatrix} A_1^* \\ A_2^* \end{Bmatrix} E_1
$$

$$(A11.5a)$$

$$
\begin{Bmatrix} E_2 \\ F_2 \end{Bmatrix} = [A_1] \begin{Bmatrix} 1 \\ 1 \end{Bmatrix} E_1 \tag{A11.5b}
$$

where

$$
[A_m] = \begin{bmatrix} \frac{1}{2}(1 + \alpha_{im}) e^{i\omega(L_m/V_{sm})} & \frac{1}{2}(1 - \alpha_{im}) e^{i\omega(-L_m/V_{sm})} \\ \frac{1}{2}(1 - \alpha_{im}) e^{i\omega(L_m/V_{sm})} & \frac{1}{2}(1 + \alpha_{im}) e^{i\omega(-L_m/V_{sm})} \end{bmatrix} \tag{A11.6}
$$

The coefficients E_3 and F_3 in Eq. (A11.5a) are the amplitudes of the incident wave (upward propagating wave) and the reflected wave (downward propagating wave) at the top of the third soil layer. Similarly E_1 is the amplitude of the incident wave (upward propagating wave) at the top of the first soil layer.

Each equation in the set of simultaneous linear equations (A11.1) corresponds to the following equation or condition:

(1) Equilibrium equation at the 1st-floor mass.
(2) Equilibrium equation at the 2nd-floor mass.
(3) Nodal rotation of the pile top = 0.
(4) Shear force of the pile at the top is in equilibrium with the inertial force at the foundation and the 1st-story shear force.
(5) Continuity condition of the pile horizontal displacements at the interface of the soil layers.

(6) Continuity condition of the pile nodal rotations at the interface of the soil layers.

(7) Equilibrium condition of pile bending moments at the interface of the soil layers.

(8) Equilibrium condition of pile shear forces at the interface of the soil layers.

(9) Bending moment at the pile tip = 0 (pin-jointed).

(10) Horizontal displacement of the pile tip = horizontal displacement of the free-field ground at the same level.

APPENDIX 11.2 COMPARISON WITH RECORDED DATA DURING AN EARTHQUAKE

In order to verify the present Winkler-type soil element model, an analytical model has been constructed for an actual building with piles (Nikolaou et al. 2001). The overview of the building-pile system in Yokohama, Japan is shown in Fig. 11.19. This building consists of a steel frame of 12 stories and is supported by 20 cast-in-place reinforced concrete piles, 35m long and 1.7m in diameter. To compare the peak response of bending strains of piles, a finite-element model as shown in Fig. 11.7 has been used. This finite-element

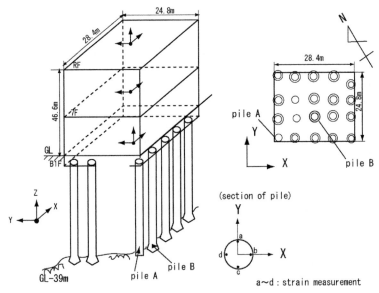

Figure 11.19 *12-story steel building with 20 piles at Yokohama in Japan.*

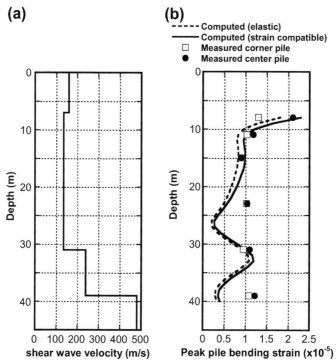

Figure 11.20 *(a) Shear wave velocity profile of ground; (b) comparison of peak pile bending strain computed by analytical model including the present Winkler-type soil element with that recorded during an earthquake in 1992.*

model includes the present Winkler-type soil element and the difference is the shape functions for the free-field ground and piles, i.e. a linear function for the free-field ground and a cubic function for the piles. Fig. 11.20(a) shows the shear wave velocity profile of the ground. Fig. 11.20(b) illustrates the comparison of the peak pile bending strain computed by the analytical model including the present Winkler-type soil element with that recorded during an earthquake in 1992. A good agreement can be observed near the pile head and this demonstrates the validity of the present continuum model with the Winkler-type soil element. This bending strain contains both the inertial effect and the kinematic effect. It has been confirmed from the analytical model that both the inertial effect and the kinematic effect are included almost in the same magnitude in this case. It may also be said that the pile-group effect is rather small in this case.

REFERENCES

Gazetas, G., Dobry, R., 1984. Horizontal response of piles in layered soils. J. Geotech. Engrg. 110 (1), 20–40.

Kavvadas, M., Gazetas, G., 1993. Kinematic seismic response and bending of free-head piles in layered soil. Geotechnique 43 (2), 207–222.

Nakamura, T., Takewaki, I., Asaoka, Y., 1996. Sequential Stiffness Design for Seismic Drift Ranges of a Shear Building–Pile–Soil System. Earthquake Engineering and Structural Dynamics 25 (12), 1405–1420.

Nikolaou, S., Mylonakis, G., Gazetas, G., Tazoh, T., 2001. Kinematic pile bending during earthquakes: analysis and field measurements. Geotechnique 51 (5), 425–440.

Schnabel, P.B., Lysmer, J., Seed, H.B., 1972. SHAKE: A computer program for earthquake response analysis of horizontally layered sites, A computer program distributed by NISEE/Computer Applications. Berkeley.

Takewaki, I., 1999. Inverse Stiffness Design of Shear-Flexural Building Models Including Soil–Structure Interaction. Engineering Structures 21 (12), 1045–1054.

Takewaki, I., 2004a. Bound of Earthquake Input Energy to Soil-Structure Interaction Systems, Proc. of The 11th International Conference on Soil Dynamics and Earthquake Engineering (ICSDEE) and the 3rd International Conference on Earthquake Geotechnical Engineering (ICEGE). Berkeley, January 7–9, (2), 734–741.

Takewaki, I., 2004b. Response spectrum method for nonlinear surface ground analysis. Advances in Structural Engineering 7 (6), 503–514.

Takewaki, I., Fujimoto, H., 2004. Earthquake Input Energy to Soil-Structure Interaction Systems: A Frequency-Domain Approach,. Advances in Structural Engineering 7 (5), 399–414.

CHAPTER TWELVE

Critical Excitation for Earthquake Energy Input Rate

Contents

12.1. INTRODUCTION

The purpose of this chapter is to explain a new probabilistic critical excitation method for identifying the critical frequency content of ground motions maximizing the mean earthquake energy input rate to structures. The critical excitation problem includes a double maximization procedure with respect to time and to the power spectral density (PSD) function. The key to finding the critical frequency content is the order interchange in the double maximization procedure. It should be remarked that no mathematical programming technique is required in the proposed method. It is shown that the proposed technique is systematic and the critical excitation can be found extremely efficiently within a reasonable accuracy. Extension of the proposed method to a more general ground motion model, i.e. nonuniformly modulated nonstationary models, and to a more general problem for variable envelope functions and variable frequency contents is

Critical Excitation Methods in Earthquake Engineering
© 2013 Elsevier Ltd.
All rights reserved.
295

discussed. The novel points of this chapter are (1) to derive a new expression on the probabilistic earthquake input energy and its rate in terms of uniformly modulated and nonuniformly modulated ground motion models, (2) to formulate a new critical excitation problem with the probabilistic earthquake energy input rate as the criticality measure and (3) to propose a systematic solution procedure to that problem.

A deterministic expression of earthquake energy input rate to a base-isolated building model is also presented in order to capture the properties of earthquake energy input rate in more detail.

12.2. NONSTATIONARY GROUND MOTION MODEL

In most of this chapter, it is assumed that the input horizontal base acceleration follows the uniformly modulated nonstationary random process.

$$\ddot{u}_g(t) = c(t)w(t) \tag{12.1}$$

where $c(t)$ denotes a given deterministic envelope function and $w(t)$ represents a stationary Gaussian process with zero mean. Let T_D denote the duration of $\ddot{u}_g(t)$.

The auto-correlation function $R_w(t_1, t_2)$ of $w(t)$ may be defined by

$$R_w(t_1, t_2) = E[w(t_1)w(t_2)] = \int_{-\infty}^{\infty} S_w(\omega)e^{i\omega(t_1 - t_2)}d\omega \tag{12.2}$$

In Eq. (12.2), $S_w(\omega)$ is the PSD function of $w(t)$ and $E[\cdot]$ indicates the ensemble mean. The time-dependent evolutionary PSD function of the input motion $\ddot{u}_g(t)$ can then be expressed by

$$S_g(t; \omega) = c(t)^2 S_w(\omega) \tag{12.3}$$

The constraint on the mean of the total energy (Drenick 1970; Shinozuka 1970) is considered here, which is described by

$$C \equiv E\left[\int_0^{T_D} \ddot{u}_g(t)^2 dt\right] = \overline{C} \tag{12.4}$$

$\overline{C}$ is a given value of the mean total energy. It can be shown (Housner and Jennings 1975) that this quantity C has a relationship with the input energy to the single-degree-of-freedom (SDOF) model. This quantity is also related to the response spectrum. These relationships may be useful in the specification of C. Substitution of Eq. (12.1) into Eq. (12.4) yields

$$\int_0^{T_D} c(t)^2 E[w(t)^2]dt = \overline{C} \tag{12.5}$$

Substitution of Eq. (12.2) $(t_1 = t_2 = t)$ into Eq. (12.5) leads to

$$\left(\int_{-\infty}^{\infty} S_w(\omega)d\omega\right)\left(\int_0^{T_D} c(t)^2 dt\right) = \overline{C} \tag{12.6}$$

Eq. (12.6) is reduced finally to

$$\int_{-\infty}^{\infty} S_w(\omega)d\omega = \overline{C}/\int_0^{T_D} c(t)^2 dt = \overline{S}_w \tag{12.7}$$

More elaborate nonuniformly modulated nonstationary models are available (Conte and Peng 1997; Fang and Sun 1997). Advanced critical excitation methods for such elaborate models will be briefly discussed later.

12.3. PROBABILISTIC EARTHQUAKE ENERGY INPUT RATE: A FREQUENCY-DOMAIN APPROACH

A lot of work has been conducted on the topics of earthquake input energy. For example, (see Housner 1956, 1959; Housner and Jennings 1975; Akiyama 1985; Uang and Bertero 1990; Trifunac et al. 2001; Austin and Lin 2004). In contrast to most of the previous works, the earthquake input energy is formulated and evaluated here in the frequency domain (Page 1952; Lyon 1975; Ordaz et al. 2003; Takewaki 2004a–c, 2005a, b; Takewaki and Fujimoto 2004) to facilitate the development of a unique critical excitation method and the corresponding efficient solution procedure.

Together with the earthquake input energy, the earthquake energy input rate has been focused on as an important measure of ground motion destructiveness. Page (1952) introduced the concept of instantaneous power spectrum and Ohi et al. (1991), Kuwamura et al. (1997a, 1997b), Iyama and Kuwamura (1999) developed several interesting theories on the earthquake energy input rate. It has been clarified that, while the earthquake input energy can quantify the energy absorbed in a structure during a ground motion input, the earthquake energy input rate can capture the load effect of ground motions for producing the maximum deformation in a structure. Actually many researchers pointed out that the earthquake energy input rate may be more crucial for structures than the earthquake input energy for near-fault ground motions (Hall et al. 1995; Kuwamura et al. 1997a, 1997b; Iyama and Kuwamura 1999; Bozorgnia and Bertero 2003). In this chapter, the earthquake energy input rate is taken as a measure of criticality of ground motions.

Fig. 12.1(a) shows the time history of the input energy per unit mass to a damped, linear elastic SDOF model with the fundamental natural period 1.0(s) and the damping ratio of 0.05 subjected to El Centro NS (Imperial Valley 1940). The solid line is computed from the relative velocity of the mass to the base and the dotted line indicates the time history of the input energy representing the work by the base on the structure (Uang and Bertero 1990). Next, Fig. 12.1(b) illustrates the time history of the energy input rate per unit mass corresponding to the solid line in Fig. 12.1(a). Finally, Fig. 12.1(c) indicates the time history of the story drift of the model. Figs. 12.2(a)–(c) show the corresponding figures of the same model subjected to Kobe University NS (Hyogoken-Nanbu 1995) and Figs. 12.3(a)–(c) illustrate the corresponding figures subjected to JMA Kobe NS (Hyogoken-Nanbu 1995). It can be seen that while the earthquake input energy exhibits a positive value, the energy input rate takes both positive and negative values. Furthermore, it can be found that the time producing the maximum story drift corresponds fairly well to the time attaining the maximum earthquake energy input rate. This fact implies that the earthquake energy input rate can be used as an alternative for measuring the power of a ground motion for producing the maximum deformation. It should also be remarked that the earthquake input energy rates of the ground motions from the same earthquake can vary greatly (see, for example, Bozorgnia and Bertero 2003) and JMA Kobe NS has a large earthquake input energy rate compared to Kobe University NS.

The probabilistic earthquake energy input rate is introduced in the following. While a lot of work has been conducted on the random vibration theory under nonstationary ground motions (see, for example, Crandall 1958, 1963; Lin 1967; Vanmarcke 1977; Der Kiureghian 1980; Nigam 1983), the probabilistic earthquake input energy and its rate have never been formulated explicitly.

For clear presentation of a new concept, consider a damped, linear elastic SDOF model with a mass m subjected to a base acceleration $\ddot{u}_g(t)$. Let ω_n, h, $\omega_D = \omega_n\sqrt{1 - h^2}$ denote the undamped natural circular frequency, the

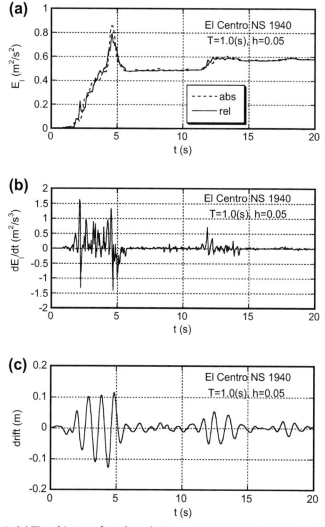

Figure 12.1 *(a) Time history of earthquake input energy per unit mass; (b) energy input rate; (c) story drift for El Centro NS (Imperial Valley 1940).*

damping ratio and the damped natural circular frequency of the model. The horizontal displacement $x(t)$ of the mass relative to the ground may then be expressed analytically by

$$x(t) = \int_{-\infty}^{t} g(t - \zeta)\ddot{u}_g(\zeta)d\zeta \qquad (12.8)$$

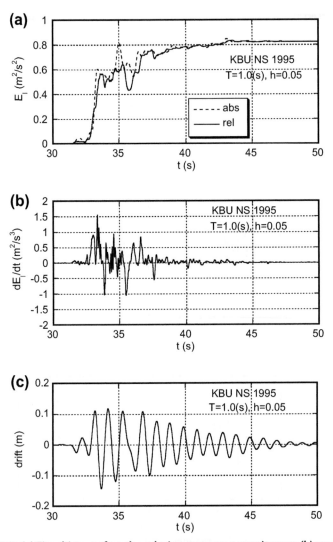

Figure 12.2 *(a) Time history of earthquake input energy per unit mass; (b) energy input rate; (c) story drift for KBU NS (Hyogoken-Nanbu 1995).*

where $g(t)$ is the unit impulse response function in the underdamped case and is described by

$$g(t) = -\frac{1}{\omega_D}e^{-h\omega_n t}\sin \omega_D t \qquad (12.9)$$

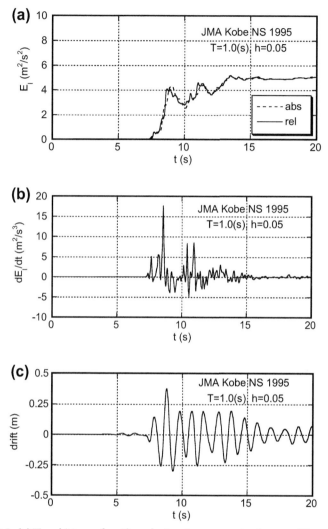

Figure 12.3 *(a) Time history of earthquake input energy per unit mass; (b) energy input rate; (c) story drift for JMA Kobe NS (Hyogoken-Nanbu 1995).*

The time derivative of $g(t)$ may be obtained as

$$\dot{g}(t) = \frac{h}{\sqrt{1 - h^2}} e^{-h\omega_n t} \sin \omega_D t - e^{-h\omega_n t} \cos \omega_D t \qquad (12.10)$$

It can be shown after some manipulations that the horizontal velocity $\dot{x}(t)$ of the mass relative to the ground may be expressed analytically by

$$\dot{x}(t) = \int_{-\infty}^{t} \dot{g}(t - \zeta)\ddot{u}_g(\zeta)d\zeta \qquad (12.11)$$

By using Eqs. (12.1) and (12.11), the earthquake input energy $E_I(t)$ (solid line in Figs. 12.1(a), Figs. 12.2(a), Figs. 12.3(a)) per unit mass to the SDOF model until time t may be expressed by

$$
\begin{aligned}
E_I(t) &= - \int_{-\infty}^{t} \dot{x}(\tau)\ddot{u}_g(\tau)d\tau \\
&= - \int_{-\infty}^{t} \int_{-\infty}^{\tau} \dot{g}(\tau - \zeta)\ddot{u}_g(\zeta)d\zeta\ddot{u}_g(\tau)d\tau \qquad (12.12) \\
&= - \int_{-\infty}^{t} \int_{-\infty}^{\tau} \dot{g}(\tau - \zeta)c(\zeta)w(\zeta)c(\tau)w(\tau)d\zeta d\tau
\end{aligned}
$$

Ensemble mean evaluation of Eq. (12.12) and substitution of Eq. (12.2) into the resulting equation lead to the following expression of mean of the earthquake input energy.

$$
\begin{aligned}
E[E_I(t)] &= - \int_{-\infty}^{\infty} \int_{-\infty}^{t} \int_{-\infty}^{\tau} \dot{g}(\tau - \zeta)c(\zeta)c(\tau)e^{i\omega\zeta}e^{-i\omega\tau}S_w(\omega)d\zeta d\tau d\omega \\
&= \int_{-\infty}^{\infty} G(t, \omega)S_w(\omega)d\omega
\end{aligned}
$$

$$(12.13)$$

where $G(t, \omega)$ in the integrand can be defined by

$$G(t, \omega) = - \int_{-\infty}^{t} \int_{-\infty}^{\tau} \dot{g}(\tau - \zeta)c(\zeta)c(\tau)e^{i\omega\zeta}e^{-i\omega\tau}d\zeta d\tau \qquad (12.14)$$

From Eq. (12.13), the mean of the energy input rate may be written as

$$E\left[\frac{d}{dt}E_I(t)\right] = \int_{-\infty}^{\infty} \frac{\partial}{\partial t}[G(t, \omega)]S_w(\omega)d\omega \qquad (12.15)$$

The rearrangement of Eq. (12.14) provides

$$G(t,\omega) = -\int_{-\infty}^{t} c(\tau)e^{-i\omega\tau}\left(\int_{-\infty}^{\tau} \dot{g}(\tau-\zeta)c(\zeta)e^{i\omega\zeta}d\zeta\right)d\tau \qquad (12.16)$$

The time derivative of the function $G(t,\omega)$ may then be expressed by

$$\frac{\partial}{\partial t}[G(t,\omega)] = -c(t)e^{-i\omega t}\left(\int_{-\infty}^{t} \dot{g}(t-\zeta)c(\zeta)e^{i\omega\zeta}d\zeta\right) \qquad (12.17)$$

Since the mean energy input rate $E[dE_I(t)/dt]$ and the PSD function $S_w(\omega)$ are real numbers, only the real part of $\partial[G(t,\omega)]/\partial t$ is meaningful in the evaluation of $E[dE_I(t)/dt]$. Therefore Eq. (12.15) may be reduced to

$$E\left[\frac{d}{dt}E_I(t)\right] = \int_{-\infty}^{\infty} \text{Re}[\partial G(t,\omega)/\partial t]S_w(\omega)d\omega \qquad (12.18)$$

To the best of the author's knowledge, the expressions (12.12)–(12.18) have never been presented in the past and will be used in the following.

12.4. CRITICAL EXCITATION PROBLEM FOR EARTHQUAKE ENERGY INPUT RATE

The critical excitation problem for energy input rate may be stated as:

Given the floor mass, story stiffness and viscous damping coefficient of an SDOF shear building model and the excitation envelope function $c(t)$, find the critical PSD function $S_w(\omega)$ to maximize the specific function $E[dE_I(t^*)/dt]$ (t^*: the time when the maximum value of the function $E[dE_I(t)/dt]$ to the input $S_w(\omega)$ is attained) subject to the excitation power limit (integral of the PSD function in the frequency range)

$$\int_{-\infty}^{\infty} S_w(\omega)d\omega \leq \overline{S}_w \qquad (12.19)$$

and to the PSD amplitude limit.

$$\sup S_w(\omega) \leq \overline{s}_w \qquad (12.20)$$

It should be remarked that the PSD amplitude is closely related to the duration of the motion. As an example, Fig. 12.4(a) shows a sine wave of duration $= 5$ seconds and Fig. 12.4(b) presents its Fourier amplitude

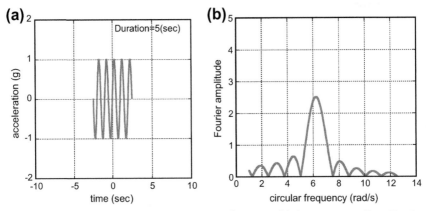

Figure 12.4 *(a) Sine wave with duration of 5 sec.; (b) its corresponding Fourier amplitude spectrum.*

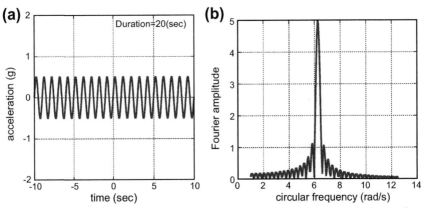

Figure 12.5 *(a) Sine wave with duration of 20 sec. having the same power $\int_{-\infty}^{\infty} \ddot{u}_g{}^2 dt$ as Fig. 12.4(a); (b) its corresponding Fourier amplitude spectrum.*

spectrum. Fig. 12.5(a) illustrates a sine wave of duration $= 20$ seconds having the same power $\int_{-\infty}^{\infty} \ddot{u}_g{}^2 dt$ as the sine wave in Fig. 12.4(a). Fig. 12.5(b) presents its Fourier amplitude spectrum. The amplitude of the PSD function should be specified appropriately based on the information on the duration of the ground motion.

12.5. SOLUTION PROCEDURE FOR DOUBLE MAXIMIZATION PROBLEM

This problem consists of the double maximization procedures. This may be described by

$$\max_{S_w(\omega)} \max_t \{E[dE_I(t)/dt]\}$$

The first maximization is performed with respect to time for a given PSD function $S_w(\omega)$ (see Fig. 12.6) and the second maximization is conducted with respect to the PSD function $S_w(\omega)$. In the first maximization process, the time t^* when $E[dE_I(t)/dt]$ attains its maximum must be obtained for each PSD function. This original problem is complex and needs much computation. To avoid this cumbersome computation, a new procedure

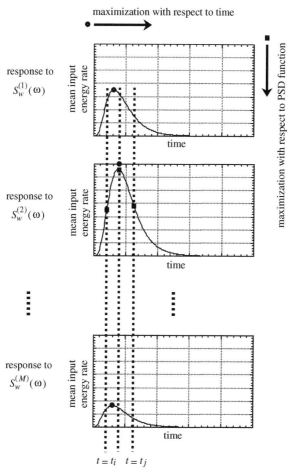

Figure 12.6 *Schematic diagram of proposed solution procedure (interchange of double maximization procedure with respect to time and to the power spectral density function).*

based on the interchange of the order of the maximization procedures is explained. This procedure is guaranteed because the function $E[dE_I(t)/dt]$ is a single-valued functional with respect to $S_w(\omega)$ and to time. The procedure explained here can be expressed by

$$\max_{t} \max_{S_w(\omega)} \{E[dE_I(t)/dt]\}$$

The first maximization with respect to the PSD function for a given time can be performed very efficiently by using the solution procedure similar to that for the critical excitation problem under stationary ground motions (Takewaki 2002) (see Fig. 12.7). If the time is fixed, the function $\mathrm{Re}[\partial G(t,\omega)/\partial t]$ can be regarded as a function of the frequency only and the previous theory for stationary input can be used. In the case where $\bar{s}_w$ is infinite, the critical PSD function is reduced to the Dirac delta function. On the other hand, when $\bar{s}_w$ is finite, the critical PSD function turns out to be a constant value of $\bar{s}_w$ in a finite interval $\Omega = \overline{S}_w/(2\bar{s}_w)$. This interval may consist of multiple intervals. The intervals, Ω_1, Ω_2, ..., which constitute Ω, can be obtained by controlling the level of $S_w(\omega)$ in the diagram of $\mathrm{Re}[\partial G(t,\omega)/\partial t]$ (see Fig. 12.7(b)). The tentative critical excitation obtained for a specific time has a rectangular PSD function or multiple rectangular functions as shown in Fig. 12.7(b). The second maximization with respect to time can be implemented without difficulty by sequentially changing the time and comparing the values at various times directly.

The algorithm used here may be summarized as:

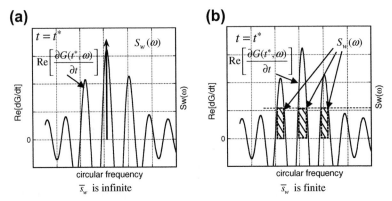

Figure 12.7 *(a) Critical power spectral density function for infinite $\bar{s}_w$; (b) critical power spectral density function for finite $\bar{s}_w$.*

(i) Compute $\mathrm{Re}[\partial G(t_i, \omega)/\partial t]$ in Eq. (12.18) at a specific time $t = t_i$.

(ii) Find the critical PSD function at time $t = t_i$ as the rectangular PSD function.

(iii) Compute $E[dE_I(t)/dt]$ to the rectangular PSD function obtained in step (ii) from Eq. (12.18).

(iv) Repeat steps (i)–(iii) for every time step and obtain $E[dE_I(t_m)/dt] = \max E[dE_I(t_i)/dt]$.

(v) The PSD function obtained at $t = t_m$ is taken as the PSD function of the critical excitation.

In this algorithm the global optimality is guaranteed because the global optimality in the maximization with respect to the shape of PSD functions is guaranteed by the property of a single-valued function of $\mathrm{Re}[\partial G(t_i, \omega)/\partial t]$ and that with respect to time is guaranteed by a sequential search algorithm in each time step. The present algorithm including the interchange of the order of the double maximization procedures is applicable to more elaborate nonuniformly modulated nonstationary excitation models although the expression of Eq. (12.18) must be modified and a new critical excitation problem must be stated. This will be discussed later.

12.6. MEAN ENERGY INPUT RATE FOR SPECIAL ENVELOPE FUNCTION

Assume that the envelope function $c(t)$ has the following form.

$$c(t) = \begin{cases} e^{-\alpha t} - e^{-\beta t} & (t \geq 0) \\ 0 & (t < 0) \end{cases} \qquad (12.21)$$

The time derivative of the function $G(t, \omega)$ defined by Eq. (12.14) can be expressed by

$$\frac{\partial}{\partial t}[G(t, \omega)] = -(e^{-\alpha t} - e^{-\beta t})e^{-i\omega t}$$

$$\times \int_0^t \left\{ \frac{h}{\sqrt{1 - h^2}} e^{-h\omega_n(t-\zeta)} \sin \omega_D(t - \zeta) \right.$$

$$\left. - e^{-h\omega_n(t-\zeta)} \cos \omega_D(t - \zeta) \right\} \cdot (e^{-\alpha\zeta} - e^{-\beta\zeta})e^{i\omega\zeta} d\zeta$$

$$(12.22)$$

Rearrangement of Eq. (12.22) may provide

$$\frac{\partial}{\partial t}[G(t,\omega)] = -(e^{-\alpha t} - e^{-\beta t})e^{-i\omega t}e^{-h\omega_n t}$$

$$\times \left[\frac{h}{\sqrt{1-h^2}} \left\{ \sin\omega_D t \int_0^t \cos\omega_D\zeta (e^{-\alpha\zeta} - e^{-\beta\zeta})e^{h\omega_n\zeta}e^{i\omega\zeta}d\zeta \right. \right.$$

$$\left. - \cos\omega_D t \int_0^t \sin\omega_D\zeta (e^{-\alpha\zeta} - e^{-\beta\zeta})e^{h\omega_n\zeta}e^{i\omega\zeta}d\zeta \right\}$$

$$+ \left\{ -\cos\omega_D t \int_0^t \cos\omega_D\zeta (e^{-\alpha\zeta} - e^{-\beta\zeta})e^{h\omega_n\zeta}e^{i\omega\zeta}d\zeta \right.$$

$$\left. \left. - \sin\omega_D t \int_0^t \sin\omega_D\zeta (e^{-\alpha\zeta} - e^{-\beta\zeta})e^{h\omega_n\zeta}e^{i\omega\zeta}d\zeta \right\} \right]$$

$$(12.23)$$

As stated before, since $E[dE_I(t)/dt]$ and $S_w(\omega)$ are real numbers, only the real part of $\partial[G(t,\omega)]/\partial t$ is meaningful in the evaluation of $E[dE_I(t)/dt]$. The real part of $\partial[G(t,\omega)]/\partial t$ may be expressed by

$$\text{Re}\left[\frac{\partial}{\partial t}[G(t,\omega)]\right] = -(e^{-\alpha t} - e^{-\beta t})$$

$$\times \left[\cos\omega t \int_0^t \left\{ \frac{h}{\sqrt{1-h^2}} e^{-h\omega_n(t-\zeta)} \sin\omega_D(t-\zeta) \right. \right.$$

$$\left. - e^{-h\omega_n(t-\zeta)} \cos\omega_D(t-\zeta) \right\} (e^{-\alpha\zeta} - e^{-\beta\zeta}) \cos\omega\zeta d\zeta$$

$$+ \sin\omega t \int_0^t \left\{ \frac{h}{\sqrt{1-h^2}} e^{-h\omega_n(t-\zeta)} \sin\omega_D(t-\zeta) \right.$$

$$\left. \left. - e^{-h\omega_n(t-\zeta)} \cos\omega_D(t-\zeta) \right\} (e^{-\alpha\zeta} - e^{-\beta\zeta}) \sin\omega\zeta d\zeta \right]$$

$$(12.24)$$

Rearrangement of Eq. (12.24) provides

$$
\mathrm{Re}\left[\frac{\partial}{\partial t}[G(t,\omega)]\right] = -(e^{-\alpha t} - e^{-\beta t})e^{-h\omega_n t}
$$

$$
\times \left[\sin\omega t \left\{\frac{h}{\sqrt{1-h^2}}(\sin\omega_D t G_{CS} - \cos\omega_D t G_{SS})\right.\right.
$$

$$
\left. - (\cos\omega_D t G_{CS} + \sin\omega_D t G_{SS})\right\}
$$

$$
+ \cos\omega t \left\{\frac{h}{\sqrt{1-h^2}}(\sin\omega_D t G_{CC} - \cos\omega_D t G_{SC})\right.
$$

$$
\left.\left. - (\cos\omega_D t G_{CC} + \sin\omega_D t G_{SC})\right\}\right]
$$

(12.25)

where the quantities G_{CS}, G_{SS}, G_{CC}, G_{SC} are defined in Takewaki (2001) (ω_n, ω_D in this chapter should be read as ω_j, ω_{jd}, respectively).

On the other hand, the imaginary part of $\partial[G(t,\omega)]/\partial t$ may be expressed by

$$
\mathrm{Im}\left[\frac{\partial}{\partial t}[G(t,\omega)]\right] = -(e^{-\alpha t} - e^{-\beta t})
$$

$$
\times \left[\cos\omega t \int_0^t \left\{\frac{h}{\sqrt{1-h^2}} e^{-h\omega_n(t-\zeta)}\sin\omega_D(t-\zeta)\right.\right.
$$

$$
\left. - e^{-h\omega_n(t-\zeta)}\cos\omega_D(t-\zeta)\right\}(e^{-\alpha\zeta} - e^{-\beta\zeta})\sin\omega\zeta d\zeta
$$

$$
- \sin\omega t \int_0^t \left\{\frac{h}{\sqrt{1-h^2}} e^{-h\omega_n(t-\zeta)}\sin\omega_D(t-\zeta)\right.
$$

$$
\left.\left. - e^{-h\omega_n(t-\zeta)}\cos\omega_D(t-\zeta)\right\}(e^{-\alpha\zeta} - e^{-\beta\zeta})\cos\omega\zeta d\zeta\right]
$$

(12.26)

Rearrangement of Eq. (12.26) provides

$$\text{Im}\left[\frac{\partial}{\partial t}[G(t,\omega)]\right] = -(e^{-\alpha t} - e^{-\beta t})e^{-h\omega_n t}$$

$$\times \left[\cos\omega t \left\{\frac{h}{\sqrt{1-h^2}}(\sin\omega_D t G_{CS} - \cos\omega_D t G_{SS})\right.\right.$$

$$\left.- (\cos\omega_D t G_{CS} + \sin\omega_D t G_{SS})\right\}$$

$$- \sin\omega t \left\{\frac{h}{\sqrt{1-h^2}}(\sin\omega_D t G_{CC} - \cos\omega_D t G_{SC})\right.$$

$$\left.\left.- (\cos\omega_D t G_{CC} + \sin\omega_D t G_{SC})\right\}\right]$$

$$(12.27)$$

It can be proved that the real part of $\partial[G(t,\omega)]/\partial t$ is an even function of ω and the imaginary part of $\partial[G(t,\omega)]/\partial t$ is an odd function of ω.

12.7. CRITICAL EXCITATION PROBLEM FOR NONUNIFORMLY MODULATED GROUND MOTION MODEL

Consider a nonuniformly modulated, nonstationary ground motion model (Conte and Peng 1997) where the ground motion acceleration can be expressed by

$$\ddot{u}_g(t) = \sum_i c_i(t)w_i(t) \qquad (12.28)$$

In Eq. (12.28) i denotes the i-th element of the nonuniformly modulated ground motion model. For example, the primary, secondary and surface waves are candidates for those elements. In Eq. (12.28) $c_i(t)$ is the envelope function of the i-th element of the ground motion and $w_i(t)$ is the stationary Gaussian process with zero mean in the i-th element. It is assumed that $w_i(t)$'s are statistically independent.

The ensemble mean of the total energy for this model may be expressed by

$$C = E\left[\int_0^{T_D} \ddot{u}_g(t)^2 dt\right] = \sum_i \int_0^{T_D} c_i(t)^2 E\left[w_i(t)^2\right] dt$$

$$= \sum_i \left(\int_0^{T_D} c_i(t)^2 dt\right)\left(\int_{-\infty}^{\infty} S_{w_i}(\omega) d\omega\right) \tag{12.29}$$

In Eq. (12.29) the cross terms vanish due to the statistical independence of $w_i(t)$'s. From Eqs. (12.12), (12.13) and (12.28), the mean of the earthquake input energy under the nonuniformly modulated ground motion may be expressed by

$$E[E_I(t)] = -\sum_j \int_{-\infty}^{\infty} \int_{-\infty}^{t} \int_{-\infty}^{\tau} \dot{g}(\tau-\zeta)c_j(\zeta)c_j(\tau)e^{i\omega\zeta}e^{-i\omega\tau}S_{w_j}(\omega)d\zeta d\tau d\omega$$

$$= \sum_j \left(\int_{-\infty}^{\infty} G_j(t,\omega)S_{w_j}(\omega)d\omega\right) \tag{12.30}$$

where $G_j(t,\omega)$ in the integrand can be defined by

$$G_j(t,\omega) = -\int_{-\infty}^{t} \int_{-\infty}^{\tau} \dot{g}(\tau-\zeta)c_j(\zeta)c_j(\tau)e^{i\omega\zeta}e^{-i\omega\tau}d\zeta d\tau \tag{12.31}$$

Following Eqs. (12.15)–(12.18), the mean of the energy input rate may then be written as

$$E\left[\frac{d}{dt}E_I(t)\right] = \sum_j \left(\int_{-\infty}^{\infty} \mathrm{Re}[\partial G_j(t,\omega)/\partial t]S_{w_j}(\omega)d\omega\right) \tag{12.32}$$

The critical excitation problem may be described as follows:

Given the envelope functions $c_i(t)$ for all i, find the critical PSD function $\tilde{S}_{w_i}(\omega)$ for all i to maximize the specific function $E[dE_I(t^*)/dt]$ (t^*: the time when the maximum value of the function $E[dE_I(t)/dt]$ to the input $\{S_{w_j}(\omega)\}$ is attained) subject to the excitation power limit

$$\int_{-\infty}^{\infty} S_{w_i}(\omega)d\omega = \overline{S}_{w_i} \text{ for all } i \tag{12.33}$$

and to the PSD amplitude limit

$$\sup S_{w_i}(\omega) \le \bar{s}_{w_i} \text{ for all } i \tag{12.34}$$

It should be remarked that $\overline{S}_{w_i}$ has to satisfy the following condition.

$$\sum_i \left(\int_0^{T_D} c_i(t)^2 \, dt \right) \overline{S}_{w_i} = \overline{C} \tag{12.35}$$

A procedure devised for the problem under the uniformly modulated model can be applied to this problem. The algorithm (i)–(v) stated for the problem under the uniformly modulated model can be used by replacing $\mathrm{Re}[\partial G(t,\omega)/\partial t]$ by $\mathrm{Re}[\partial G_j(t,\omega)/\partial t]$ and changing the time sequentially to find the time maximizing the specific function $E[dE_I(t)/dt]$. The algorithm (i) and (ii) has to be applied to each element j in Eq. (12.32).

12.8. GENERAL PROBLEM FOR VARIABLE ENVELOPE FUNCTION AND VARIABLE FREQUENCY CONTENT

The critical envelope function has been found by Takewaki (2004d) with "the maximum value in time of the mean–square drift" as the criticality measure for a fixed frequency content of $w(t)$. It has been confirmed by a lot of numerical simulations that the critical frequency content is the "resonant one" irrespective of the envelope functions.

Once the critical frequency content, i.e. resonant one, is found by the present method, that can be used as the fixed frequency content in the method (Takewaki, 2004d). It may therefore be concluded that the combination of the critical frequency content, i.e. resonant one, and the critical envelope function is the critical set for a more general problem in which both frequency contents and envelope functions are variable.

12.9. NUMERICAL EXAMPLES

Numerical examples are presented for a simple problem, i.e. the problem including the frequency content as a variable under uniformly modulated ground motions.

The parameters $\alpha = 0.13, \beta = 0.45$ in the envelope function $c(t)$, Eq. (12.21), are used (see Fig. 12.8). The natural periods T of the SDOF models are 0.2(s), 1.0(s) and the damping ratios h are 0.05, 0.20. The power

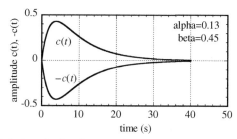

Figure 12.8 *Amplitude of nonstationary ground motion employed.*

of $S_w(\omega)$ is specified as $\overline{S}_w = 1.0(m^2/s^4)$. Because the present theory is linear with respect to the power of PSD functions, a specific value of the power is treated here.

Figs. 12.9(a), (b) show the function $\text{Re}[\partial G(t, \omega)/\partial t]$ defined by Eq. (12.25) of the model with $T = 1.0$(s) and $h = 0.05$ at every two seconds, $t = 2, 4, \cdots, 20$(s). The closed–form expressions $G_{CS}, G_{SS}, G_{CC}, G_{SC}$ in Takewaki (2001) have been substituted into Eq. (12.25). It can be seen that, as the time passes, the amplitude of $\text{Re}[\partial G(t, \omega)/\partial t]$ becomes larger until a specific time corresponding to the variation of the envelope function $c(t)$ and multiple peaks arise. It can also be observed that the peak value is attained around $t = 6$(s), which is somewhat later than $t = 4$(s) corresponding to the maximum value in the envelope function $c(t)$. This property corresponds to well-known fact. Fig. 12.10 illustrates the function $\text{Re}[\partial G(t, \omega)/\partial t]$ defined by Eq. (12.25) of the model with $T = 0.2$(s) and $h = 0.05$ at every two seconds, $t = 2, 4, \cdots, 10$(s) and Fig. 12.11 shows that of the model with $T = 1.0$(s) and $h = 0.20$ at every two seconds, $t = 2, 4, \cdots, 10$(s). It should be remarked that the maximum value of $\text{Re}[\partial G(t, \omega)/\partial t]$ occurs nearly at the natural frequency only in lightly damped structures.

For simple and clear presentation of the solution procedure, the amplitude $\overline{s}_w$ of $S_w(\omega)$ has been specified so as to be infinite in the first example. The solution algorithm explained in Fig. 12.7(a) has been applied. Fig. 12.12(a) illustrates the function $E[dE_I(t)/dt]$ for the model with $T = 1.0$(s) and $h = 0.05$. This can also be expressed by $\overline{S}_w \partial G(t, \omega_n)/\partial t$. It is assumed here that the maximum value of $\partial G(t, \omega)/\partial t$ at every time is attained at $\omega = \omega_n$. It can be seen that the maximum value of $E[dE_I(t)/dt]$ is attained around $t = 6$(s), which is somewhat later than $t = 4$(s) corresponding to the maximum value in the envelope function $c(t)$. Fig. 12.12(b) shows the function $E[dE_I(t)/dt]$ for the stiffer model with

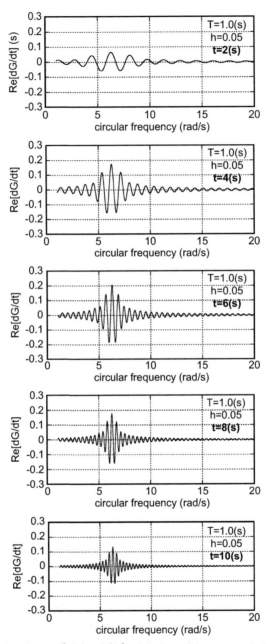

Figure 12.9(a) *Function* $\mathrm{Re}[\partial G(t, \omega)/\partial t]$ *at various times* t $= 2, 4, 6, 8, 10(s)$ *for the model* T $= 1.0(s), h = 0.05.$

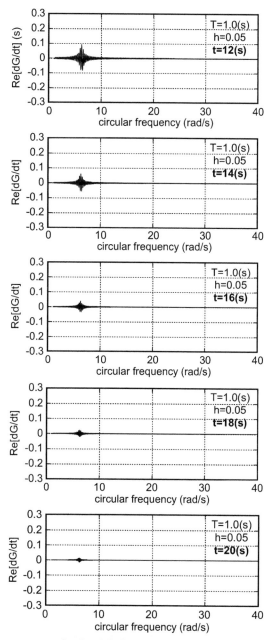

Figure 12.9(b) *Function* $\text{Re}[\partial G(t, \omega)/\partial t]$ *at various times* t $= 12, 14, 16, 18, 20(s)$ *for the model* T $= 1.0(s), h = 0.05.$

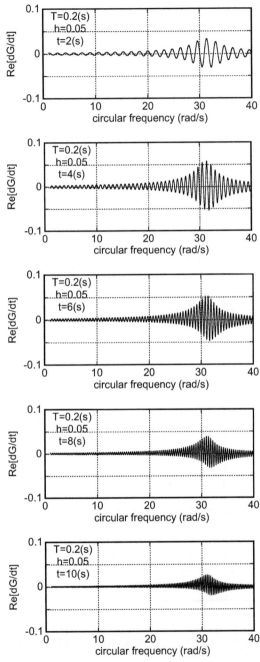

Figure 12.10 *Function* **Re**$[\partial G(t, \omega)/\partial t]$ *at various times* **t** $= 2, 4, 6, 8, 10(s)$ *for the model* **T** $= 0.2(s), h = 0.05.$

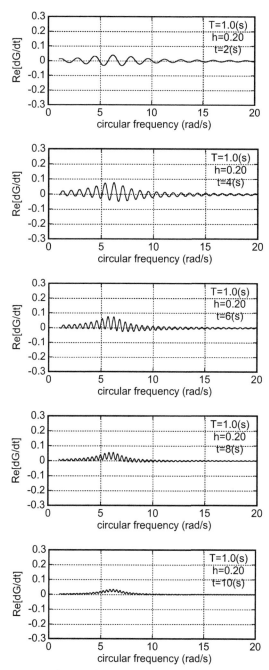

Figure 12.11 *Function* $\mathrm{Re}[\partial G(t, \omega)/\partial t]$ *at various times* $t = 2, 4, 6, 8, 10(s)$ *for the model* $T = 1.0(s), h = 0.20$.

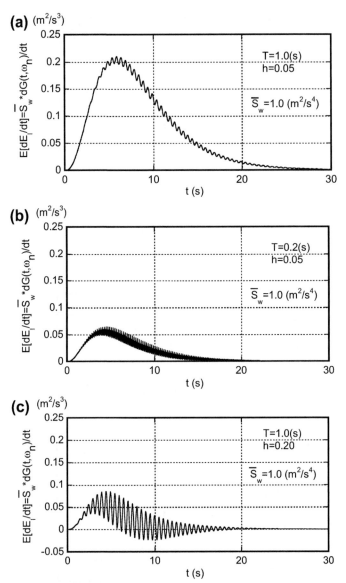

Figure 12.12 *Time history of* $E[dE_I/dt] = \bar{S}_w \times \partial G(t, \omega_n)/\partial t$ *for infinite PSD amplitude* $\bar{s}_w$*: (a)* T = 1.0(s), h = 0.05, *(b)* T = 0.2(s), h = 0.05, *(c)* T = 1.0(s), h = 0.20.

$T = 0.2(\text{s})$, $h = 0.05$ and Fig. 12.12(c) illustrates that for the model with $T = 1.0(\text{s})$, $h = 0.20$.

Consider the second example. The amplitude $\bar{s}_w$ of the PSD function $S_w(\omega)$ has been specified so as to be finite. The algorithm for the solution

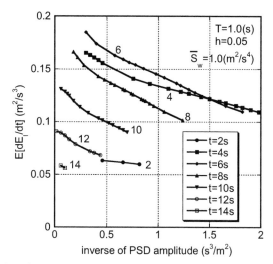

Figure 12.13 *Plot of mean energy input rate with respect to inverse of power spectral density amplitude for various times.*

explained in Fig. 12.7(b) has been used for the model with $T = 1.0(\text{s})$ and $h = 0.05$. Fig. 12.13 shows the mean earthquake energy input rate per unit mass $E[dE_I(t)/dt]$ with respect to the inverse of the PSD amplitude, $1/\bar{s}_w$, for every two seconds. This figure has been drawn by specifying the level of the dotted line in Fig. 12.7(b) and finding the intervals $\tilde{\Omega}_i$. The PSD amplitude $\bar{s}_w$ is obtained from $\bar{s}_w = \bar{S}_w/(2\sum_i \tilde{\Omega}_i)$ and the mean earthquake energy input rate $E[dE_I(t)/dt]$ is computed from Eqs. (12.18) and (12.25). The search of the maximum value of the function $E[dE_I(t)/dt]$ with respect to time for a given value of $\bar{s}_w$ corresponds to the procedure $\max_t \max_{S_w(\omega)} \{E[dE_I(t)/dt]\}$ instead of $\max_{S_w(\omega)} \max_t \{E[dE_I(t)/dt]\}$ as devised in this chapter. Fig. 12.13 implies that, while the maximum value of $E[dE_I(t)/dt]$ is attained around $t = 6(\text{s})$ for larger PSD amplitude $\bar{s}_w$ (smaller value of the inverse of the PSD amplitude), the maximum value of $E[dE_I(t)/dt]$ is attained around $t = 4(\text{s})$ for smaller PSD amplitude $\bar{s}_w$ (larger value of the inverse of the PSD amplitude).

An example of the critical PSD function for the model with $T = 1.0(\text{s})$ and $h = 0.05$ is shown in Fig. 12.14. The specified power of $S_w(\omega)$ is $\bar{S}_w = 1.0(\text{m}^2/\text{s}^4)$ and the specified PSD amplitude is $\bar{s}_w = 0.92(\text{m}^2/\text{s}^3)$. It can be seen that three rectangles with different widths constitute the critical PSD function in this case.

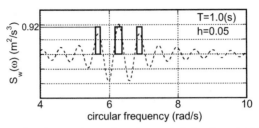

Figure 12.14 *Example of critical PSD function* $S_w(\omega)$ *for* $T = 1.0(s), h = 0.05$ *and* $\bar{S}_w = 1.0(m^2/s^4), \bar{s}_w = 0.92(m^2/s^3)$.

12.10. DETERMINISTIC EARTHQUAKE ENERGY INPUT RATE

In this chapter, the ensemble mean of earthquake energy input rate to an SDOF model has been treated in a probabilistic framework. In this section, a deterministic expression of earthquake energy input rate to an MDOF model is introduced (Yamamoto et al. 2011).

Consider a base-isolated building as shown in Fig. 12.15. Let $\mathbf{u}(t), \mathbf{M}, \mathbf{1}$ denote the horizontal floor displacement vector, the mass matrix and the vector consisting of unity only. The earthquake input energy to this model until time t may be expressed by

$$E_I(t) = -\int_0^t \dot{\mathbf{u}}(\tau)^T \mathbf{M}\mathbf{1}\ddot{u}_g(\tau)d\tau \tag{12.36}$$

Let us define a modified ground motion $\hat{\ddot{u}}_g(\tau; t)$ at time τ which has the same component until time t and a null component after time t (see Fig. 12.16, Ohi et al. 1991; Takewaki 2005b). This quantity is called "the truncated ground motion." The response velocity at time τ corresponding to $\hat{\ddot{u}}_g(\tau; t)$ is denoted by $\hat{\dot{u}}(\tau; t)$. The Fourier transforms of $\hat{\ddot{u}}_g(\tau; t)$ and $\hat{\dot{u}}(\tau; t)$ are expressed by $\hat{\ddot{U}}_g(\omega; t)$ and $\hat{\dot{U}}(\omega; t)$, respectively. From its definition, it is found that $\hat{\ddot{U}}_g(\omega; t_0) = \ddot{U}_g(\omega)$.

By introducing the truncated ground motion $\hat{\ddot{u}}_g(\tau; t)$ and the corresponding velocity response $\hat{\dot{u}}(\tau; t)$ and extending the integration limits, Eq. (12.36) can be rewritten as

$$E_I(t) = -\int_{-\infty}^{\infty} \hat{\dot{\mathbf{u}}}(\tau; t)^T \mathbf{M}\mathbf{1}\hat{\ddot{u}}_g(\tau; t)d\tau \tag{12.37}$$

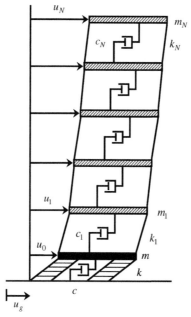

Figure 12.15 *N-story shear building model supported by base-isolation system.*

Application of the Fourier and inverse Fourier transformations (Lyon 1975; Ohi et al. 1985; Kuwamura et al. 1994; Ordaz et al. 2003; Takewaki 2004a, b) to Eq. (12.37) leads to

$$E_I(t) = -\int_{-\infty}^{\infty} \frac{1}{2\pi} \int_{-\infty}^{\infty} \hat{\mathbf{U}}(\omega;t)^T e^{i\omega\tau} \mathbf{M1} \hat{\ddot{u}}_g(\tau;t) d\omega d\tau$$

$$= \frac{1}{2\pi} \int_{-\infty}^{\infty} i\omega \mathbf{1}^T \mathbf{M}^T \mathbf{A}^{-1} \hat{\ddot{U}}_g(\omega;t) \mathbf{M1} \hat{\ddot{U}}_g(-\omega;t) d\omega$$

$$= \int_{0}^{\infty} |\hat{\ddot{U}}_g(\omega;t)|^2 F(\omega) d\omega \qquad (12.38)$$

In Eq. (12.38) $|\hat{\ddot{U}}_g(\omega;t)|^2$ is referred to as the *squared Fourier amplitude spectrum* **(SFAS)** and plays a principal role in the present formulation. In the derivation of Eq. (12.38), the relation $\hat{\mathbf{U}}(\omega;t) = -i\omega \mathbf{A}^{-1} \mathbf{M1} \hat{\ddot{U}}_g(\omega;t)$ is used and the function $F(\omega)$ is defined by

$$F(\omega) \equiv \mathrm{Re}[i\omega \mathbf{1}^T \mathbf{M}^T \mathbf{A}^{-1} \mathbf{M1}]/\pi \qquad (12.39)$$

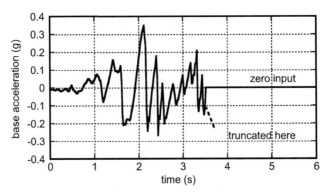

Figure 12.16 *Truncated ground motion.*

Since $\hat{\ddot{U}}_g(\omega; t_0) = \ddot{U}_g(\omega)$ from its definition, it follows that

$$E_I(t_0) = \int_0^\infty |\ddot{U}_g(\omega)|^2 F(\omega)d\omega \qquad (12.40)$$

From Eq. (12.38), the earthquake energy input rate $dE_I(t)/dt$ may be expressed by

$$\frac{dE_I(t)}{dt} = \int_0^\infty (d|\hat{\ddot{U}}_g(\omega; t)|^2/dt)F(\omega)d\omega \qquad (12.41)$$

This expression was derived first by Ohi et al. (1991) and it was pointed out that the expression $d|\hat{\ddot{U}}_g(\omega; t)|^2/dt$ is equivalent to "the instantaneous power spectrum" introduced and discussed by Page (1952).

Numerical examples of the deterministic earthquake energy input rate for a base-isolated 10-story shear building model (BI building or BI system) are presented. The story stiffnesses of the building are determined so that the 10-story shear building model with fixed-base has the fundamental natural period of 1.0(s) and the lowest eigenmode of the model with fixed-base is straight (inverted triangle). Then the lowest eigenmode (super-structural part) of the BI building is not straight. This procedure is based on the inverse problem approach (Nakamura and Yamane 1986). It is also assumed that the damping matrix of the super-structure with fixed base is proportional to the stiffness matrix of the super-structure and the damping ratio in the lowest mode of the super-structure with fixed base is 0.02. The stiffness k and damping coefficient c of the BI system have been determined so that the fundamental natural period of the BI building is 5.3(s) (see Takewaki 1998 for hybrid inverse problems for rigid building stiffnesses) and the damping ratio of the BI rigid building model is 0.2. The parameters of the building and the BI system are summarized in Table 12.1.

12.11. EXAMPLE 1 (MODEL OF SFAS OF TRUNCATED GROUND MOTION)

Since the concept of SFAS is somewhat complicated and it seems beneficial to consider example models, its model is constructed. An example of SFAS $|\hat{\ddot{U}}_g(\omega;t)|^2$ of truncated ground motions can be expressed by

$$|\hat{\ddot{U}}_g(\omega;t)|^2 = \begin{cases} 2\pi t \cdot S_u(\omega,t) = 2\pi t \cdot a(e^{-b_1 t} - e^{-b_2 t})S(\omega) & (0 \leq t \leq t_0) \\ 2\pi t_0 \cdot S_u(\omega,t_0) = 2\pi t_0 \cdot a(e^{-b_1 t_0} - e^{-b_2 t_0})S(\omega) & (t \geq t_0) \end{cases}$$

(12.42)

where t_0 is the terminal time (duration) of the input ground motion and the frequency function $S(\omega)$ denotes

$$S(\omega) = \left\{ \frac{\Omega^4 + 4\zeta^2\Omega^2\omega^2}{[\omega^2 - \Omega^2]^2 + 4\zeta^2\Omega^2\omega^2} \right\} S_0$$

(12.43)

In Eq. (12.43), Ω and ζ are the predominant circular frequency and damping parameter. Furthermore S_0 is a constant value.

The parameter a in Eq. (12.42) is introduced to characterize the intensity of the ground motion and is determined from the constraint on acceleration power expressed by

$$\overline{C}_A = \int_{-\infty}^{\infty} \ddot{u}_g(t)^2 dt = \frac{1}{\pi} \int_0^{\infty} |\ddot{U}_g(\omega)|^2 d\omega$$

$$= \frac{1}{\pi} \int_0^{\infty} |\hat{\ddot{U}}_g(\omega;t_0)|^2 d\omega = 2\pi t_0 \cdot a(e^{-b_1 t_0} - e^{-b_2 t_0}) \int_0^{\infty} S(\omega) d\omega$$

(12.44)

In this section, the short period ground motion models and long period ground motion models are taken into account. Table 12.2 shows the parameters for these artificial ground motion models. The long period ground motion models simulate a pulse wave due to near-field ground

Table 12.1 Parameters of building and base-isolation system

	plan	mass [kg/m²]	mass [kg]	horizontal stiffness [N/m]	damping coefficient [Ns/m]
building (10-story)	40m × 40m	800 (every story)	1.28×10^7 (reduced SDOF model)	5.05×10^8 (reduced SDOF model)	lowest-mode damping ratio for fixed-base model = 0.02
base-isolation story		2400	3.84×10^6	2.42×10^7	damping ratio for rigid building model = 0.2

Table 12.2 Input motion models

		predominant period of ground or ground motion $2\pi/\Omega$ (s)	duration (s)	a	b_1	b_2
Short period model	short period model 1	0.6	40	13.2	0.05	0.051
	short period model 2	0.8	40	17.6	0.05	0.051
Long period model	long period model 1	3.0	180	2.9	0.01	0.011
	long period model 2	5.3	180	4.8	0.01	0.011
	long period model 3	8.0	180	7.5	0.01	0.011

motions (Hall et al. 1995; Heaton et al. 1995; Jangid and Kelly 2001) or a long-period ground motion resulting from surface waves (Irikura et al. 2004; Ariga et al. 2006). The short period ground motion model 1 (predominant period = 0.6s) is treated here as the standard ground motion model and the parameter a for this model has been determined so that the maximum value of the Fourier amplitude spectrum at the predominant circular frequency $\Omega = 2\pi/0.6$[rad/s] attains the value 3 [m/s] computed for El Centro NS (Imperial Valley 1940). The damping parameter ζ is set to 0.4 and $S_0 = 0.2$(m^2/s^3).

Fig. 12.17 shows the time history of SFAS $|\hat{U}_g(\omega; t)|^2$ at 18(rad/s) of truncated ground motions extracted from El Centro NS 1940 and the corresponding time history of short period ground motion model 1. It can be seen that the short period ground motion model 1 can simulate very well the time history of SFAS $|\hat{U}_g(\omega; t)|^2$ of truncated motions extracted from El Centro NS 1940.

Fig. 12.18 illustrates the time histories of input energy for various introduced artificial ground motion models. It can be observed that the total earthquake input energy by the long period ground motion model 2 (predominant period = 5.3(s)) is the largest and those by the long period ground motion models 1 and 3 are the next. On the other hand, the total earthquake input energies by the short period ground motion models are rather small. It should be noted that the long period ground motion model 2 has a predominant period resonant to the fundamental natural period of the BI building. This means that the resonance of the fundamental natural period of the BI building with the predominant period of ground motions is one of the key issues in the evaluation of input energy to BI buildings (Ariga et al. 2006).

Fig. 12.19 shows the time histories of energy input rate for various ground motion models. It is found that the maximum energy input rate by the long period ground motion model 2 is the largest and those by the long period

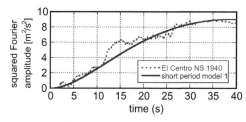

Figure 12.17 *Time history of SFAS at 18(rad/s) of truncated ground motions extracted from El Centro NS 1940 and the corresponding time history of short period model 1.*

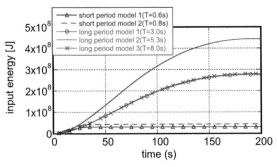

Figure 12.18 *Time histories of input energy for various ground motion models.*

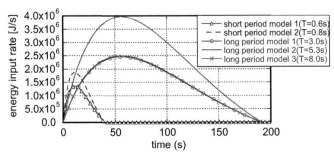

Figure 12.19 *Time histories of energy input rate for various ground motion models.*

ground motion models 1 and 3 are the next. Although the maximum energy input rates by the short period ground motion models are small, that by short period ground motion model 2 is comparable to that by the long period ground motion models 1 and 3. It should also be noted that the large energy input rate continues for a long time (almost 150s) in the long period ground motion models and large energy input can be predicted in the long period ground motion models from this figure. As in the case of input energy, the resonance of the fundamental natural period of the BI building with the predominant period of ground motions seems to be one of the key issues in the evaluation of energy input rate to BI buildings (Ariga et al. 2006).

12.12. EXAMPLE 2 (DETAILED ANALYSIS OF SFAS OF TRUNCATED GROUND MOTION AND COMPARISON OF TIME-DOMAIN AND FREQUENCY-DOMAIN ANALYSES)

It seems meaningful to investigate the frequency properties of SFAS of truncated ground motions with various truncated times. Fig. 12.20 presents

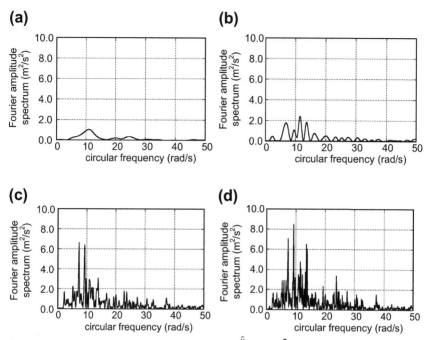

Figure 12.20 *Frequency distributions of SFAS* $|\hat{U}_g(\omega;t)|^2$ *of El Centro NS 1940 truncated at (a) 2.5s, (b) 5.0s, (c) 20.0s, (d) 40.0s.*

the frequency distributions of SFAS $|\hat{U}_g(\omega;t)|^2$ of El Centro NS (Imperial Valley 1940) truncated at (a) 2.5s, (b) 5.0s, (c) 20.0s, (d) 40.0s. It can be observed that, while a blunted spectrum is seen in the motions truncated at early stages, a sharp spectrum can be found in the motions truncated at later stages. It can also be understood that the intensity becomes larger as the truncated time passes (moves later).

Fig. 12.21 illustrates the frequency distributions of the time-rates $\partial|\hat{U}_g(\omega;t)|^2/\partial t$ of SFAS of El Centro NS 1940 truncated at (a) 2.5s, (b) 5.0s, (c) 20.0s, (d) 40.0s). As in Fig. 12.20, it can be observed that, while a blunted time-rate spectrum is seen in the motions truncated at early stages, a sharp spectrum can be found in the motions truncated at later stages. On the other hand, the intensity is largest at the intermediate truncated time different from the case for SFAS $|\hat{U}_g(\omega;t)|^2$.

Fig. 12.22 shows sample accelerations of a simulated nonstationary ground motion and a stationary ground motion. The stationary motion has also been generated to clarify the property of $|\hat{U}_g(\omega;t)|^2$ through

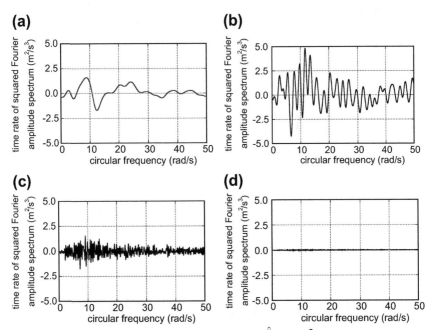

Figure 12.21 *Frequency distributions of time-rates $\partial|\hat{U}_g(\omega;t)|^2/\partial t$ of SFAS of El Centro NS 1940 truncated at (a) 2.5s, (b) 5.0s, (c) 20.0s, (d) 40.0s.*

the comparison with that for the nonstationary motion. This simulated stationary ground motion has been generated by the sum of sine waves with a band limited PSD function of $0.03(\text{m}^2/\text{s}^3)$ in the frequency range $0.1{-}5.1(\text{rad/s})$ instead of Eq. (12.43). On the other hand, the nonstationary motion has been multiplied by an envelope function

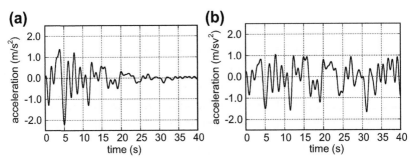

Figure 12.22 *A sample of simulated ground acceleration: (a) nonstationary input, (b) stationary input.*

given by Eq. (12.42) with parameters $a = 2.32$, $b_1 = 0.09$, $b_2 = 1.49$. These simulated ground motions enable one to compare (1) the time histories of earthquake input energy computed by the time-domain method (Eq. (12.36)) and the frequency-domain method (Eq. (12.38)) and (2) the time histories of energy input rate computed by the time-domain method $(dE_I(t)/dt = -\dot{\mathbf{u}}(t)^T\mathbf{M}1\ddot{u}_g(t))$ and the frequency-domain method (Eq. (12.41)). This comparison may be difficult in the model described by Eqs. (12.42) and (12.43) because it does not seem easy to obtain the corresponding time history of ground motion acceleration.

Fig. 12.23 illustrates the 3-D view of average SFAS of 100 ground motions truncated at continuously increasing time ((a) nonstationary, (b) stationary)). A characteristic similar to Fig. 12.21 (blunted spectrum at early stages and sharp spectrum at later stages) can be observed in both figures. It can also be seen that, since the nonstationary motion with high intensity almost ends at 20(s), the averaged frequency distributions of SFAS do not change much after the truncated time of around 20(s).

Fig. 12.24 shows the comparison of the time histories of earthquake input energy computed by the time-domain method (Eq. (12.36)) and the frequency-domain method (Eq. (12.38)) for a nonstationary ground motion. In Fig. 12.24, the plots by the time-domain method with different computational time increments ($dt = 0.005$, 0.01, 0.02s) are also illustrated. It can be observed that, while the accuracy of the time-domain method depends largely on the computational time increment, the frequency-domain method exhibits a stable result close to the result for $dt = 0.005$. This clearly shows the reliability of the frequency-domain method. Furthermore, it should be kept in mind that, owing to the introduction of Fourier amplitude spectra of truncated ground motions, the computational error does not accumulate even at later times in the frequency-domain formulation.

Fig. 12.25 illustrates the comparison of the time histories of energy input rate computed by the time-domain method $(dE_I(t)/dt = -\dot{\mathbf{u}}(t)^T\mathbf{M}1\ddot{u}_g(t))$ and the frequency-domain method (Eq. (12.41)) for a nonstationary ground motion. It can be observed that the frequency-domain method has almost an equivalent accuracy to the time-domain method. It should be noted that, because the analytical functions of ground motion accelerations with arbitrary truncated time and its Fourier transform exist in this case, an accurate comparison becomes possible.

(a) $\left|\hat{\ddot{U}}_g\left(\omega;t\right)\right|^2$

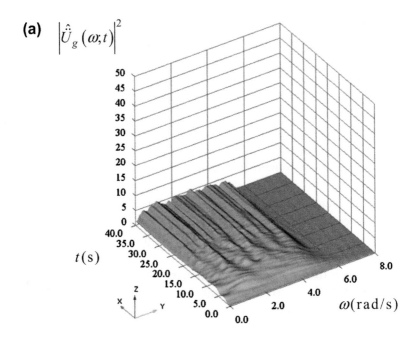

(b) $\left|\hat{\ddot{U}}_g\left(\omega;t\right)\right|^2$

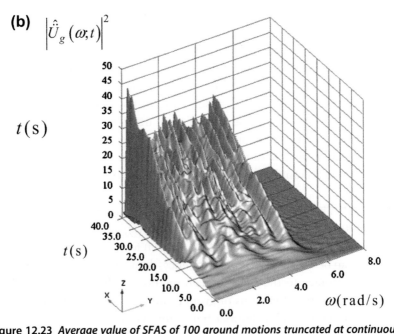

Figure 12.23 *Average value of SFAS of 100 ground motions truncated at continuously increasing time: (a) nonstationary input, (b) stationary input.*

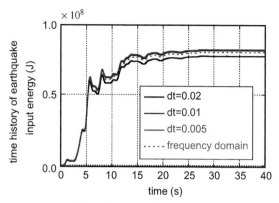

Figure 12.24 *Comparison of time histories of earthquake input energy computed by the time-domain method (Eq. (12.36)) and the frequency-domain method (Eq. (12.38)) for nonstationary input.*

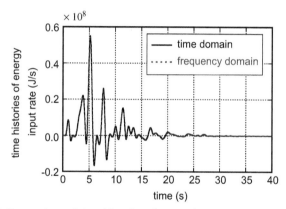

Figure 12.25 *Comparison of time histories of energy input rate computed by the time-domain method $(dE_I(t)/dt = -\dot{u}(t)^T M1\ddot{u}_g(t))$ and the frequency-domain method (Eq. (12.41)) for nonstationary input.*

12.13. CONCLUSIONS

The conclusions may be stated as follows:

(1) A probabilistic critical excitation method can be developed for the problem of maximizing the mean earthquake energy input rate to damped linear elastic SDOF models.

(2) An expression has been derived on the probabilistic earthquake input energy and its time derivative subjected to uniformly modulated stochastic ground motions. The integrand of the objective function to be maximized in the critical excitation problem can be expressed as the product of a time-frequency function representing the transfer energy input rate and the power spectral density function of a stochastic part in the input motion.

(3) An idea similar to the previously explained one can be used partially in finding and identifying the most unfavorable frequency content of the critical excitation which is described by a uniformly modulated excitation model. The key is the order interchange in the double maximization procedure of the objective function with respect to time and to the shape of the power spectral density function.

(4) An expression has also been derived on the probabilistic earthquake input energy and its time derivative subjected to complex *nonuniformly modulated* stochastic ground motions. The explained solution procedure may be extended to a more general problem for such nonuniformly modulated ground motion models.

(5) Numerical examples revealed the peculiar time-varying characteristics of the generalized nonstationary energy transfer function multiplied by the envelope function of the input motion model.

(6) The validity of the present solution algorithm for the infinite power spectral density amplitude and the finite power spectral density amplitude has been demonstrated numerically.

REFERENCES

Akiyama, H., 1985. Earthquake Resistant Limit-state Design for Buildings. University of Tokyo Press, Tokyo, Japan.

Ariga, T., Kanno, Y., Takewaki, I., 2006. Resonant behaviour of base-isolated high-rise buildings under long-period ground motions. Struct. Design Tall Spec. Build. 15 (3), 325–338.

Austin, M., Lin, W., 2004. Energy balance assessment of base-isolated structures. J. Eng. Mech. 130 (3), 347–358.

Bozorgnia, Y., Bertero, V., 2003. Damage spectra: characteristics and applications to seismic risk reduction. J. Struct. Eng. 129 (10), 1330–1340.

Conte, J.P., Peng, B.F., 1997. Fully non-stationary analytical earthquake ground motion model. J. Engrg. Mech. 123 (1), 15–24.

Crandall, S.H., 1958, 1963. Random Vibration, Vol.I.. II. MIT Press, Cambridge, MA.

Der Kiureghian, A., 1980. Structural response to stationary excitation. J. Eng. Mech. Div. 106 (6), 1195–1213.

Drenick, R.F., 1970. Model-free design of aseismic structures. J. Engrg. Mech. Div. 96 (4), 483–493.

Fang, T., Sun, M., 1997. A unified approach to two types of evolutionary random response problems in engineering. Archive of Applied Mechanics 67 (7), 496–506.

Hall, J.F., Heaton, T.H., Halling, M.W., Wald, D.J., 1995. Near-source ground motion and its effects on flexible buildings. Earthquake Spectra 11 (4), 569–605.

Heaton, T.H., Hall, J.F., Wald, D.J., Halling, M.W., 1995. Response of high-rise and base-isolated buildings to a hypothetical MW 7·0 blind thrust earthquake. Science 267 (5195), 206–211.

Housner, G.W., 1956. Limit design of structures to resist earthquakes. Proc. of the First World Conference on Earthquake Engineering. University of California, Berkeley, CA. Berkeley, 5:1–5:11.

Housner, G.W., 1959. Behavior of structures during earthquakes. J. Engrg. Mech. Div. 85 (4), 109–129.

Housner, G.W., Jennings, P.C., 1975. The capacity of extreme earthquake motions to damage structures. In: Hall, W.J. (Ed.), Structural and Geotechnical Mechanics: A volume honoring N.M.Newmark. Prentice-Hall Englewood Cliff, NJ, pp. 102–116.

Irikura, K., Kamae, K., Kawabe, H., 2004. Importance of prediction of long-period ground motion during large earthquakes. In: Annual Conference of the Seismological Society of Japan, Poster session (in Japanese).

Iyama, J., Kuwamura, H., 1999. Application of wavelets to analysis and simulation of earthquake motions. Earthquake Engrg. and Struct. Dyn. 28 (3), 255–272.

Jangid, R.S., Kelly, J.M., 2001. Base isolation for near-fault motions. Earthquake Engng Struct Dyn 30 (5), 691–707.

Kuwamura, H., Kirino, Y., Akiyama, H., 1994. Prediction of earthquake energy input from smoothed Fourier amplitude spectrum. Earthquake Engng. Struct. Dyn. 23 (10), 1125–1137.

Kuwamura, H., Iyama, J., Takeda, T., 1997a. Energy input rate of earthquake ground motion–matching of displacement theory and energy theory. J. Structural and Construction Engrg., Archi. Inst. of Japan 498, 37–42 (in Japanese).

Kuwamura, H., Takeda, T., Sato, Y., 1997b. Energy input rate in earthquake destructiveness–comparison between epicentral and oceanic earthquakes. J. Structural and Construction Eng., Archi. Inst. of Japan 491, 29–36 (in Japanese).

Lin, Y.K., 1967. Probabilistic theory of structural dynamics. McGraw-Hill, New York.

Lyon, R.H., 1975. Statistical energy analysis of dynamical systems. The MIT Press, Cambridge, MA.

Nakamura, T., Yamane, T., 1986. Optimum design and earthquake-response constrained design of elastic shear buildings. Earthquake Engineering and Structural Dynamics 14 (5), 797–815.

Nigam, N.C., 1983. Introduction to Random Vibrations. MIT Press, Cambridge, MA.

Ohi, K., Takanashi, K., Tanaka, H., 1985. A simple method to estimate the statistical parameters of energy input to structures during earthquakes. J Struct Construct Engng, AIJ 347, 47–55 (in Japanese).

Ohi, K., Takanashi, K., Honma, Y., 1991. Energy input rate spectra of earthquake ground motions. J. Structural and Construction Engrg., Archi. Inst. of Japan 420, 1–7 (in Japanese).

Ordaz, M., Huerta, B., Reinoso, E., 2003. Exact computation of input-energy spectra from Fourier amplitude spectra. Earthquake Engrg. and Struct. Dyn. 32 (4), 597–605.

Page, C.H., 1952. Instantaneous power spectra. Journal of Applied Physics 23 (1), 103–106.

Shinozuka, M., 1970. Maximum structural response to seismic excitations. J. Engrg. Mech. Div. 96 (5), 729–738.

Takewaki, I., 1998. Hybrid inverse eigenmode problem for a shear building supporting a finite element subassemblage. Journal of Vibration and Control 4 (4), 347–360.

Takewaki, I., 2001. Nonstationary random critical excitation for nonproportionally damped structural systems. Computer Methods in Applied Mechanics and Engineering 190 (31), 3927–3943.

Takewaki, I., 2002. Seismic critical excitation method for robust design: A review. J. Struct. Eng. 128 (5), 665–672.

Takewaki, I., 2004a. Frequency domain modal analysis of earthquake input energy to highly damped passive control structures. Earthquake Engrg. and Struct. Dyn. 33 (5), 575–590.

Takewaki, I., 2004b. Bound of earthquake input energy to soil-structure interaction systems. Proc. of the 11th Int. Conf. on Soil Dynamics and Earthquake Engrg. (ICS-DEE) and the 3rd Int. Conf. on Earthquake Geotechnical Engrg. (ICEGE). Berkeley, January 7–9, 2: 734–741.

Takewaki, I., 2004c. Bound of earthquake input energy. J. Struct. Eng. 130 (9), 1289–1297.

Takewaki, I., 2004d. Critical envelope functions for non-stationary random earthquake input. Computers & Structures 82 (20–21), 1671–1683.

Takewaki, I., 2005a. Bound of earthquake input energy to soil–structure interaction systems. Soil Dynamics and Earthquake Engineering 25 (7–10), 741–752.

Takewaki, I., 2005b. Closure to the discussion by Ali Bakhshi and Hooman Tavallali [to I. Takewaki's (2004) Bound of Earthquake Input Energy. J Struct. Eng. 130 (9), 1289–1297, 131(10), 1643–1644.

Takewaki, I., Fujimoto, H., 2004. Earthquake input energy to soil-structure interaction systems: A frequency-domain approach. Advances in Structural Engineering 7 (5), 399–414.

Trifunac, M.D., Hao, T.Y., Todorovska, M.I., 2001. On energy flow in earthquake response. Report CE 01–03, July. University of Southern California.

Uang, C-M., Bertero, V.V., 1990. Evaluation of seismic energy in structures. Earthquake Engrg. and Struct. Dyn. 19 (1), 77–90.

Vanmarcke, E.H., 1977. Structural response to earthquakes. In: Lomnitz, C., Rosenblueth, E. (Eds.), Seismic Risk and Engineering Decisions. Elsevier, Amsterdam, pp. 287–337.

Yamamoto, K., Fujita, K., Takewaki, I., 2011. Instantaneous earthquake input energy and sensitivity in base-isolated building. Struct. Design Tall Spec. Build. 20 (6), 631–648.

CHAPTER THIRTEEN

Critical Excitation for Multi-Component Inputs

Contents

13.1. INTRODUCTION

The ground motion is an outcome of ground shaking induced by seismic waves that are transmitted from an epicenter through the ground. Therefore, simultaneous consideration of multiple components of ground motion is realistic, reasonable and inevitable in the reliable design of structures. It is sometimes assumed practically that there exist principal axes in the ground motions (Penzien and Watabe 1975; Clough and Penzien 1993). Penzien and Watabe (1975) investigated many recorded ground motions and found a tendency of existence of those axes. It is also recognized in the literature that the principal axes change their directions in time during ground shaking. In recent structural design practice, the effect of the multi-component ground motions is often included by using the square root of the sum of the squares (SRSS) method, or the extended Complete Quadratic Combination rule (CQC3) method (Smeby and Der Kiureghian 1985). It may be beneficial to provide an overview of these methods.

In the SRSS method, the maximum responses to respective ground motions are combined based on the rule of SRSS. The SRSS method is the first-step method for combination and assumes statistical independence among the respective ground motions. However, it is well understood that the multi-component ground motions have some statistical dependence and some revisions may be desired.

Critical Excitation Methods in Earthquake Engineering
© 2013 Elsevier Ltd.
All rights reserved.
335

On the other hand, the CQC3 rule is well known as a response spectrum method that can take into account the effect of correlation between the components of ground motions, and this method is frequently used in the USA. Although an absolute value of a cross power spectral density (PSD) function has been described by the correlation coefficient, the CQC3 rule cannot treat directly, in the sense of direct treatment of both real and imaginary parts, the cross PSD functions of multi-component ground motions.

After some preliminary investigations on actual response data for actual inputs, useful proposals were made. Menun and Der Kiureghian (1998) and Lopez et al. (2000) employed the CQC3 method as the response evaluation method and discussed the critical states, e.g. a critical loading combination or a critical incident angle. It is clear that the derived results depend on the accuracy of the CQC3 method. Athanatopoulou (2005) and Rigato and Medina (2007) investigated the effect of incident angle of ground motions on structural response without use of the Penzien-Watabe (1975) model and pointed out the significance of considering multiple inputs in practical seismic design. The approach is applicable only to a set of recorded motions. In the references of Fujita et al. (2008a) and Fujita and Takewaki (2010), the cross PSD function in terms of both real and imaginary parts was discussed in more detail from the viewpoint of critical excitation. They concluded that the critical set of co-spectrum and quad-spectrum, the real and imaginary parts of the cross PSD function, can be characterized by the maximization of the inner product between (co-spectrum, quad-spectrum) and the characteristic functions ($f_1(t; \omega)$, $f_2(t; \omega)$) consisting of structural modal properties and envelope functions of inputs.

Critical excitation methods have been developed extensively since the pioneering work by Drenick in 1970 (see Chapter 1). In particular, in recent work, after 1995, there has been remarkable advancement (Manohar and Sarkar 1995; Sarkar and Manohar 1996, 1998; Takewaki 2001, 2002, 2004a, 2004b, 2006b; Abbas and Manohar 2002a, 2007; Takewaki et al. 2012). However, there has been little research on critical excitation methods under multiple-component inputs.

In the following section, the critical excitation for horizontal and vertical simultaneous inputs is discussed.

13.2. HORIZONTAL AND VERTICAL SIMULTANEOUS INPUTS

Consider a rigid block, as shown in Fig. 13.1, of mass m and mass moment I_g of inertia around the centroid C subjected to horizontal and vertical

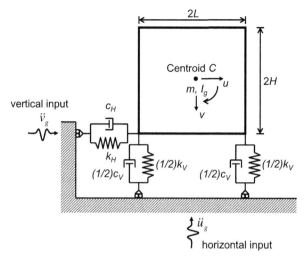

Figure 13.1 *Elastically supported rigid block subjected to horizontal and vertical simultaneous inputs.*

simultaneous base inputs. Let $\ddot{u}_g$ and $\ddot{v}_g$ denote the horizontal and vertical ground accelerations, respectively. The rigid block is supported by a set of a horizontal spring k_H and a dashpot c_H and another set of two vertical springs $(1/2)k_V$ and dashpots $(1/2)c_V$. These springs and dashpots represent the stiffness and damping of the ground on which the rigid block rests. The width and the height of the rigid block are denoted by $2L$ and $2H$, respectively. The motion of the block consists of a horizontal-rotational motion under a horizontal base input and a vertical motion under a vertical base input.

Let u and θ denote the horizontal displacement of the centroid C relative to the horizontal input and the angle of rotation of the rigid body, respectively, and let v denote the vertical displacement of the centroid C relative to the vertical input. The equations of motion of the rigid block in the swaying-rocking motion can be expressed as

$$
\begin{bmatrix} m & 0 \\ 0 & I_g \end{bmatrix} \begin{Bmatrix} \ddot{u} \\ \ddot{\theta} \end{Bmatrix} + \begin{bmatrix} c_H & -c_H H \\ -c_H H & c_H H^2 + c_V L^2 \end{bmatrix} \begin{Bmatrix} \dot{u} \\ \dot{\theta} \end{Bmatrix}
$$

$$
+ \begin{bmatrix} k_H & -k_H H \\ -k_H H & k_H H^2 + k_V L^2 \end{bmatrix} \begin{Bmatrix} u \\ \theta \end{Bmatrix} = - \begin{bmatrix} m & 0 \\ 0 & I_g \end{bmatrix} \begin{Bmatrix} 1 \\ 0 \end{Bmatrix} \ddot{u}_g
$$

(13.1)

By using the complex modal combination method, the rotational response of the rigid block can be expressed as

$$\theta(t) = \sum_{i=1}^{2} \mathrm{Re}\left[\beta^{(i)}\Theta^{(i)}q^{(i)}(t)\right] \tag{13.2}$$

where $\beta^{(i)}$ and $\Theta^{(i)}$ denote the i-th complex participation factor and the rotational angle in the i-th complex eigenmode. $q^{(i)}(t)$ is the i-th complex modal coordinate. The detailed explanation of these parameters can be found in Fujita et al. (2008b).

On the other hand, the vertical response of this model as a single-degree-of-freedom (SDOF) model can be derived as

$$v(t) = \int_{0}^{t} \{-\ddot{v}_g(\tau)\}g_V(t - \tau)d\tau \tag{13.3}$$

where $g_V(t) = (1/\omega_v')e^{-h_v\omega_v t}\sin\omega_v't$ $(\omega_v = \sqrt{k_V/m})$ is the impulse response function in the vertical vibration and $\omega_v' = \omega_v\sqrt{1 - h_v^2}$ is the damped natural circular frequency of the model in the vertical vibration. The parameter h_v is the damping ratio in the vertical vibration.

The horizontal input acceleration and a vertical input acceleration are to be described as a nonstationary random process.

$$\ddot{u}_g(t) = c_u(t)w_u(t) \tag{13.4a}$$

$$\ddot{v}_g(t) = c_v(t)w_v(t) \tag{13.4b}$$

where $c_u(t)$, $c_v(t)$ are deterministic envelope functions and $w_u(t)$, $w_v(t)$ are zero mean stationary Gaussian random processes. It is assumed that $c_u(t)$ and $c_v(t)$ are given and the power spectra of $w_u(t)$ and $w_v(t)$ are prescribed. A more elaborate model (nonuniformly modulated model, (see Takewaki 2006a) can be used if desired.

The corresponding critical excitation problem is to find the worst cross-spectrum (see Nigam 1983) and the corresponding cross-correlation function of the horizontal and vertical inputs ($w_u(t)$ and $w_v(t)$) that produce the maximum mean squares response of the uplift. The real part of the cross-spectrum is denoted by $C_{w_u w_v}$ and the imaginary part is indicated by $Q_{w_u w_v}$. This problem includes a difficulty of maximizing the objective function both to time and cross-spectrum between $w_u(t)$ and $w_v(t)$. To overcome this difficulty, an algorithm is devised such that the order of the

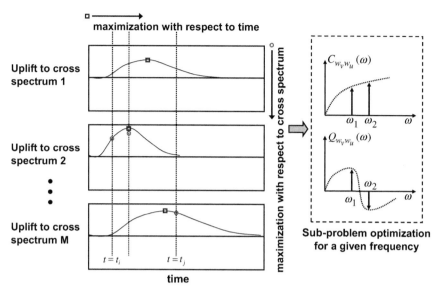

Figure 13.2 *Schematic diagram of the proposed procedure (order interchange of double maximization procedure including subproblem optimization).*

maximizations with respect to time and the cross–spectrum is interchanged. The schematic diagram of this procedure is shown in Fig. 13.2.

Consider next a moment-resisting frame, as shown in Fig. 13.3, subjected to horizontal and vertical simultaneous inputs (Fujita et al. 2008a; Fujita and Takewaki 2009). The input ground motions are modeled by Eqs. (13.4a, b). The corresponding critical excitation problem is to find the cross-spectrum $\left(C_{w_u w_v}, Q_{w_u w_v} \right)$ and the corresponding cross-correlation function of the worst horizontal and vertical inputs ($w_u(t)$ and $w_v(t)$) that produce the maximum root mean squares response of sum of bending moments due to horizontal and vertical inputs with respect to span length.

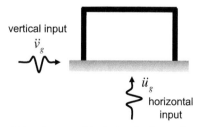

Figure 13.3 *Moment-resisting frame subjected to horizontal and vertical simultaneous inputs.*

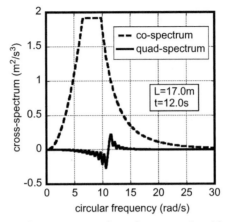

Figure 13.4 *Co-spectrum and quad-spectrum in critical input.*

Fig. 13.4 shows the co-spectrum and quad-spectrum of the critical input. It can be observed that the quad spectrum has a rather small value compared to the co-spectrum. Fig. 13.5 presents the phase angle of the cross spectrum of the critical input. This figure just corresponds to the quad spectrum in Fig. 13.4. Fig. 13.6 illustrates a sample of horizontal and vertical motion accelerations in the critical input. The vertical motion was constructed so as to exhibit a critical relation with a sample horizontal motion. Fig. 13.7 shows the amplitude ratio of vertical to horizontal motions in the critical input and the perfectly correlated input. It can be seen that an average of the ratio is almost 0.5. Fig. 13.8 presents the maximum value of root mean squares of the sum of bending moments due to horizontal and vertical inputs with

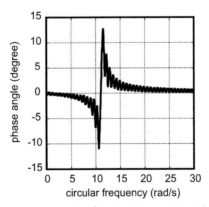

Figure 13.5 *Phase angle of cross spectrum in critical input.*

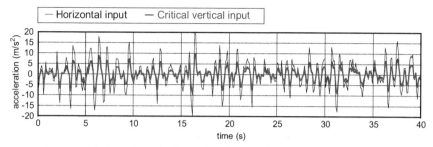

Figure 13.6 *Sample of horizontal and vertical motions in critical input.*

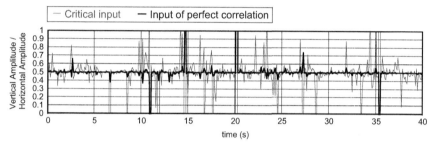

Figure 13.7 *Amplitude ratio of vertical motion to horizontal motion in critical input and perfectly correlated input.*

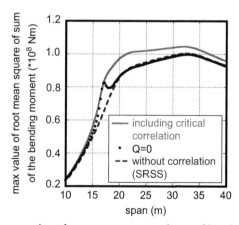

Figure 13.8 *Maximum value of root mean square of sum of bending moments due to horizontal and vertical inputs with respect to span length (critical correlation, perfect correlation, without correlation (SRSS)).*

respect to span length (critical correlation, perfect correlation, without correlation (SRSS)). It should be noted that the model with the span length of 17(m) has the same fundamental natural frequencies both in horizontal and vertical vibrations.

13.3. BI-DIRECTIONAL HORIZONTAL INPUTS

While the critical excitation problem under horizontal and vertical ground motion inputs is to find the worst cross-spectrum and the corresponding cross-correlation function of the horizontal and vertical inputs that produce the maximum mean squares response of the uplift, the critical excitation problem under two horizontal ground motion inputs is to find the worst cross-spectrum and the corresponding cross-correlation function of the two horizontal inputs that produce the maximum mean squares response. In the problem under horizontal and vertical ground motion inputs, the input directions coincide with the building principal axes in most cases. On the other hand, in the problem under two horizontal inputs, the input directions do not necessarily coincide with the building principal axes. This situation brings a further problem of great significance. The coherence function between two horizontal ground motion inputs is a function of the ratio of power spectra of two horizontal ground motion inputs and the angle of input to the building. For each angle of input to the building and the ratio of power spectra of two horizontal ground motion inputs, the coherence function between two horizontal ground motion inputs is computed. Then a technique similar to that for the problem under horizontal and vertical ground motion inputs can be used.

Consider a one-story one-span three-dimensional (3-D) frame as shown in Fig. 13.9. It is assumed that two axes X_1 and X_2 are perpendicular to each other and along the building structural axes. Let $S_{Z_1}(\omega)$ and $S_{Z_2}(\omega)$ denote the auto PSD functions along the principal axes Z_1, Z_2 of ground motions respectively. According to the Penzien and Watabe (P-W) model, two-dimensional ground motions (2DGM) along Z_1, Z_2 are regarded to be completely uncorrelated. The auto PSD functions of ground motions along X_1, X_2 are determined from the auto PSD functions of 2DGM along Z_1, Z_2. The auto PSD functions along X_1, X_2 are described by $S_{11}(\omega)$ and $S_{22}(\omega)$, respectively.

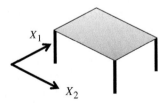

Figure 13.9 *Space frame structure subjected to bi-directional horizontal inputs.*

It can be shown that the sum of $S_{Z_1}(\omega)$ and $S_{Z_2}(\omega)$ is to be equal to the sum of $S_{11}(\omega)$ and $S_{22}(\omega)$. Furthermore, the coherence function between 2DGM along X_1 and X_2 is also denoted as

$$\rho_{12}\left(\gamma_{\text{org}}, \theta\right) = \frac{\left(1 - \gamma_{\text{org}}\right)\sin 2\theta}{\sqrt{\left(1 + \gamma_{\text{org}}\right)^2 - \left(1 - \gamma_{\text{org}}\right)^2 \cos^2 2\theta}} \tag{13.5}$$

where $\gamma_{\text{org}} = S_{Z_2}(\omega)/S_{Z_1}(\omega)$. θ is the angle of rotation (incident angle) between the two horizontal axes Z_1, X_1. Fig. 13.10 shows the coherence function expressed by Eq. (13.1) with various values of γ_{org} for varied rotation (incident) angle. In Fig. 13.10, when γ_{org} is zero, the coherence function ρ_{12} is reduced to 1 at any θ except $\theta = 0$ and $\theta = \pi/2$. This means that the components along X_1 and X_2 have perfect correlation under unidirectional ground motion along the major principal axis of ground motion.

The P-W model is often used in the modeling of multi-component ground motions. Although the coherence function of 2DGM along X_1 and X_2 can be given in terms of γ_{org} and θ as shown in Eq. (13.5), the cross PSD function cannot be treated directly in the CQC3 rule. For that reason, it is supposed in this chapter that the cross PSD function between 2DGM along X_1 and X_2 can take any value in the feasible complex plane.

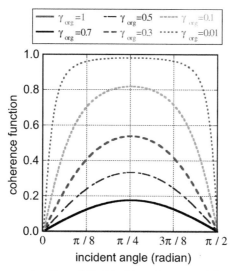

Figure 13.10 *Coherence function of 2DGM with various auto PSD ratios with respect to various incident angles in the Penzien-Watabe model (Fujita and Takewaki 2010).*

From the definition of the coherence function, the co-spectrum (real part of cross PSD) $C_{12}(\omega)$ and quad-spectrum (imaginary part of cross PSD) $Q_{12}(\omega)$ must satisfy the following relation.

$$C_{12}(\omega)^2 + Q_{12}(\omega)^2 \leq \left\{ \rho_{12}\left(\gamma_{\mathrm{org}}, \theta \right) \right\}^2 S_{11}(\omega) S_{22}(\omega) \qquad (13.6)$$

This model is called the extended P–W model hereafter. It may be possible to incorporate the extended P–W model into the stochastic response evaluation method. In that case, a new critical excitation problem can be constructed

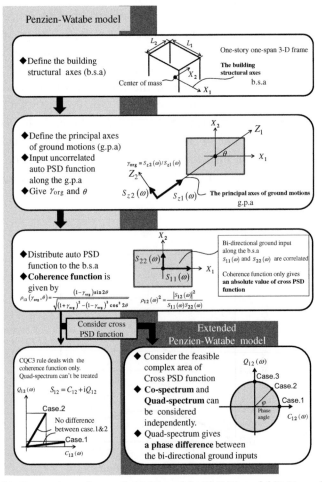

Figure 13.11 *Comparison of extended P-W model with P-W model (Fujita and Takewaki 2010).*

in which the critical cross PSD function is searched in the feasible complex plane represented by Eq. (11.2). This method can be regarded as an extended method of the CQC3 rule based on the P–W model.

Fig. 13.11 shows the comparison of the extended P–W model with the P–W model. In the extended P–W model, a feasible complex area of the cross PSD function is considered. In this framework, both the co-spectrum and quad-spectrum can be treated independently. The quad-spectrum gives a phase difference between the bi-directional ground inputs.

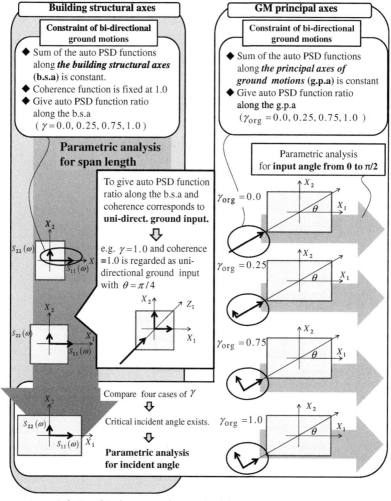

Figure 13.12 *Relationship between the method for building structural axes and the method for ground motion principal axes (Fujita and Takewaki 2010).*

Fig. 13.12 presents the relationship between the method for building structural axes and the method for ground motion principal axes. In the method for building structural axes, the sum of the auto PSD functions along the building structural axes is kept constant. On the other hand, in the method for ground motion principal axes, the sum of the auto PSD functions along the ground motion principal axes is kept constant.

Fig. 13.13 shows the schematic diagram of the proposed procedure. It is interesting to note that this optimization procedure utilizes the order interchange of the double maximization procedure including subproblem optimization with respect to cross-spectrum and time.

Fig. 13.14 is the schematic illustration of the present critical excitation problem. It is useful to understand the problem structure graphically.

In order to understand the property of the critically correlated ground motions more deeply, comparison with the perfectly correlated ground motions without time delay has been made. The structural plan is given as $L_1 = 25(m)$, $L_2 = 15(m)$. Fig. 13.15 shows a sample of the critical inputs (2-D horizontal ground motions) and the corresponding time history of the root-mean-square of column-end fiber stress.

Another comparison with the perfectly correlated ground motions without time delay has been made for different structural spans. The structural plan is given as $L_1 = 15(m)$, $L_2 = 25(m)$. Fig. 13.16(a) shows

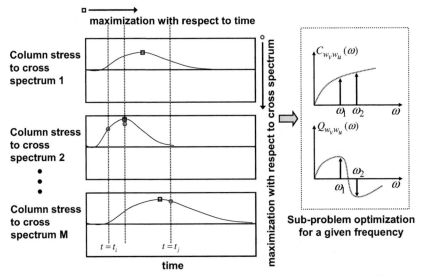

Figure 13.13 *Schematic diagram of the proposed procedure (order interchange of double maximization procedure including subproblem optimization).*

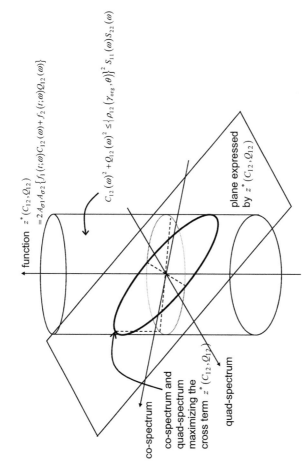

Figure 13.14 Schematic illustration of the present critical excitation problem.

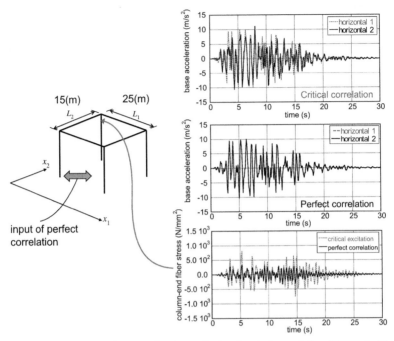

Figure 13.15 *One sample set of Monte Carlo simulation of the 2DGM (critically correlated and perfectly correlated) and corresponding root-mean-square of column-end fiber stress.*

two horizontal ground accelerations with the critical correlation for the input model of $\gamma_{org} = 0$ and $\theta = 0.106\pi[\text{rad}](= 19.0°)$(critical incident angle). This set has been generated by using random numbers. On the other hand, Fig. 13.16(b) indicates two horizontal ground accelerations with the perfect correlation without time delay for $\gamma_{org} = 0$ and $\theta = 0.106\pi[\text{rad}](= 19.0°)$. Fig. 13.17 illustrates the time history of the root-mean-square of column-end extreme-fiber stress to these two sets of horizontal ground accelerations. It can be observed that the response to the critically correlated ground accelerations could become about 1.5 times larger than that to the perfectly correlated ground accelerations without time delay.

13.4. INTERPRETATION USING INNER PRODUCT

The critical set of co-spectrum and quad–spectrum can be characterized by the maximization of the inner product between (co-spectrum,

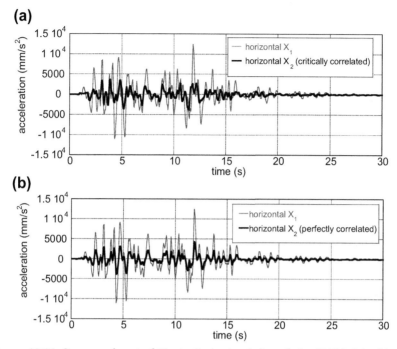

Figure 13.16 *One sample set of Monte Carlo simulation of the 2DGM: (a) critically correlated, (b) perfectly correlated (Fujita and Takewaki 2010).*

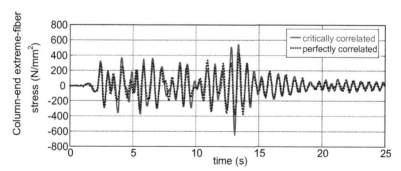

Figure 13.17 *Comparison of the column-end extreme-fiber stress to the critically correlated 2DGM with that to the perfectly correlated ones (Fujita and Takewaki 2010).*

quad-spectrum) and the characteristic functions $(f_1(t; \omega), f_2(t; \omega))$ consisting of structural modal properties and envelope functions of inputs. This interpretation was made by Abbas and Manohar (2002b) although they dealt with a multi-input problem.

13.5. CONCLUSIONS

The conclusions may be stated as follows:

(1) Critical excitation for multi-component inputs is an interesting and important problem in earthquake-resistant design of buildings. In the problem under horizontal and vertical ground motion inputs, the input directions coincide with the building principal axes in most cases. On the other hand, in the problem under two horizontal inputs, the input directions do not necessarily coincide with the building principal axes. This situation brings a further problem of great significance.

(2) The critical excitation problem under horizontal and vertical ground motion inputs is to find the worst cross-spectrum and the corresponding cross-correlation function of the horizontal and vertical inputs that produce the maximum mean squares response of the uplift. This problem includes a difficulty of maximizing the objective function both to time and cross-spectrum between horizontal and vertical ground motion inputs. An algorithm can be devised such that the order of the maximizations with respect to time and the cross-spectrum is interchanged.

(3) The critical excitation problem under two horizontal ground motion inputs is to find the worst cross-spectrum and the corresponding cross-correlation function of the two horizontal inputs that produce the maximum mean squares response. It should be noted that the coherence function between two horizontal ground motion inputs is a function of the ratio of power spectra of two horizontal ground motion inputs and the angle of input to the building. For each angle of input to the building and the ratio of power spectra of two horizontal ground motion inputs, the coherence function between two horizontal ground motion inputs is computed. Then a technique similar to that for the problem under horizontal and vertical ground motion inputs can be used.

(4) The solution algorithm both for the problem under horizontal and vertical ground motion inputs and that under two horizontal ground motion inputs can be interpreted and characterized by an inner product of the response quantities and the cross spectrum of multi-component inputs.

REFERENCES

Abbas, A.M., Manohar, C.S., 2002a. Investigations into critical earthquake load models within deterministic and probabilistic frameworks. Earthq. Eng. Struct. Dyn. 31 (4), 813–832.

Abbas, A.M., Manohar, C.S., 2002b. Critical spatially varying earthquake load models for extended structures. J. Struct. Engrg. (JoSE, India) 29 (1), 39–52.

Abbas, A.M., Manohar, C.S., 2007. Reliability-based vector nonstationary random critical earthquake excitations for parametrically excited systems. Struct. Safety 29 (1), 32–48.

Athanatopoulou, A.M., 2005. Critical orientation of three correlated seismic components. Eng. Struct. 27 (2), 301–312.

Clough, R.W., Penzien, J., 1993. Dynamics of Structures, second ed. Prentice Hall, New York.

Fujita, K., Takewaki, I., 2009. Property of critical excitation for moment-resisting frames subjected to horizontal and vertical simultaneous ground motions. Journal of Zhejiang University SCIENCE A 10 (11), 1561–1572.

Fujita, K., Takewaki, I., 2010. Critical correlation of bi-directional horizontal ground motions. Engineering Structures 32 (1), 261–272.

Fujita, K., Takewaki, I., Nakamura, N., 2008a. Critical disturbance for stress resultant in long-span moment-resisting frames subjected to horizontal and vertical simultaneous ground inputs. J. Structural and Construction Engineering, No. 626, 551–558 (in Japanese).

Fujita, K., Yoshitomi, S., Tsuji, M., Takewaki, I., 2008b. Critical cross-correlation function of horizontal and vertical ground motions for uplift of rigid block. Engineering Structures 30 (5), 1199–1213.

Lopez, O.A., Chopra, A.K., Hernandez, J.J., 2000. Critical response of structures to multicomponent earthquake excitation. Earthq. Eng. Struct. Dyn. 29 (12), 1759–1778.

Manohar, C.S., Sarkar, A., 1995. Critical earthquake input power spectral density function models for engineering structures. Earthq. Eng. Struct. Dyn. 24 (12), 1549–1566.

Menun, C., Der Kiureghian, A., 1998. A replacement for the 30%, 40%, and SRSS rules for multicomponent seismic analysis. Earthq. Spectra 14 (1), 153–163.

Nigam, N.C., 1983. Introduction to Random Vibrations. MIT Press, London.

Penzien, J., Watabe, M., 1975. Characteristics of 3-dimensional earthquake ground motions. Earthq. Eng. Struct. Dyn. 3 (4), 365–373.

Rigato, A.B., Medina, R.A., 2007. Influence of angle of incidence on seismic demands for inelastic single-storey structures subjected to bi-directional ground motions. Eng. Struct. 29 (10), 2593–2601.

Sarkar, A., Manohar, C.S., 1996. Critical cross power spectral density functions and the highest response of multi-supported structures subjected to multi-component earthquake excitations. Earthq. Eng. Struct. Dyn. 25, 303–315.

Sarkar, A., Manohar, C.S., 1998. Critical seismic vector random excitations for multiply supported structures. J. Sound and Vibration 212 (3), 525–546.

Smeby, W., Der Kiureghian, A., 1985. Modal combination rules for multicomponent earthquake excitation. Earthq. Eng. Struct. Dyn. 13 (1), 1–12.

Takewaki, I., 2001. A new method for non-stationary random critical excitation. Earthq. Engrg. Struct. Dyn. 30 (4), 519–535.

Takewaki, I., 2002. Seismic critical excitation method for robust design: A review. J. Struct. Eng. 128 (5), 665–672.

Takewaki, I., 2004a. Critical envelope functions for non-stationary random earthquake input. Computers & Structures 82 (20–21), 1671–1683.

Takewaki, I., 2004b. Bound of earthquake input energy. J. Struct. Eng. 130 (9), 1289–1297.

Takewaki, I., 2006a. Probabilistic critical excitation method for earthquake energy input rate. J Eng. Mech. 132 (9), 990–1000.

Takewaki, I., 2006b. Critical Excitation Methods in Earthquake Engineering. Elsevier Science, Oxford.

Takewaki, I., Moustafa, A., Fujita, K., 2012. Improving the Earthquake Resilience of Buildings: The Worst Case Approach. Springer, London.

CHAPTER FOURTEEN

Critical Excitation for Elastic-Plastic Response Using Deterministic Approach

Contents

14.1. INTRODUCTION

Critical excitation for elastic-plastic structural response was discussed in Chapter 5 of this book where the concept of statistical equivalent linearization was introduced. Although statistical equivalent linearization is useful because of its ability to include probabilistic properties of inputs and responses, it is also desired to treat an earthquake ground motion as a deterministic time series. This is because the representation of strong ground motions in terms of deterministic time series enables one to capture more directly the properties of critical excitation for elastic-plastic structures. The advantage of this approach lies in excluding the linearization technique and the associated approximation. In this direction, Moustafa (2002), Abbas and Manohar (2005), Abbas (2006), Moustafa (2002, 2009, 2011), Moustafa and Takewaki (2010b, c, 2011), Moustafa et al. (2010), Ueno et al. (2010, 2011) have developed some specific methods.

Moustafa (2002) and Abbas and Manohar (2005) formulated the critical excitation problem for nonlinear single-degree-of-freedom (SDOF) structures with simple elastic-plastic or nonlinear springs (e.g. Duffing oscillator). They minimized the Hasofer-Lind reliability index to find the critical input using the Response Surface Method (RSM) and First-Order Reliability Method (FORM).

Critical Excitation Methods in Earthquake Engineering
© 2013 Elsevier Ltd.
All rights reserved.
353

Abbas (2006) considered another critical excitation problem for nonlinear SDOF structures with simple elastic-plastic springs. The earthquake load was modeled as a deterministic time history that is expressed in terms of a deterministic Fourier series that is modulated by a predefined enveloping function. The resulting nonlinear optimization problem was tackled by using the sequential quadratic optimization method. This work was extended to multi-degree-of-freedom (MDOF) systems by Moustafa (2009).

Subsequently, Moustafa and Takewaki (2009, 2010a–c, 2011, 2012), Moustafa et al. (2010), Ueno et al. (2010, 2011), Moustafa (2011) developed several critical excitation methods based on the work of Moustafa (2002, 2009), Abbas and Manohar (2002, 2005) and Abbas (2006) by introducing damage indices into the definition of the seismic performance of elastic-plastic and bilinear structures under earthquake loads.

These studies concluded that the resonant ground motion has its energy in a narrow frequency range, close to the fundamental natural frequency of the linear structure, and can produce larger damage in the structure compared to ordinary records.

14.2. ABBAS AND MANOHAR'S APPROACH

Abbas and Manohar developed the critical excitation method within the deterministic and the probabilistic frameworks. In the deterministic framework, Abbas and Manohar (2002, 2005), Abbas (2006), Moustafa (2009) used the following Fourier series-type expression as the stationary part of the input ground acceleration.

$$\ddot{u}_{g0}(t) = \sum_{n=1}^{N_f} (A_n \cos \omega_n t + B_n \sin \omega_n t) \tag{14.1a}$$

Then the input ground acceleration is expressed as follows using an envelope function $e(t)$.

$$\ddot{u}_g(t) = e(t)\ddot{u}_{g0}(t) \tag{14.1b}$$

As constraints, they employed the following conditions.

$$\left[\int_0^\infty \ddot{u}_g^2(t)\, dt \right]^{\frac{1}{2}} \leq E$$

$$\max_{0<t<\infty} \left| \ddot{u}_g(t) \right| \le M_1$$

$$\max_{0<t<\infty} \left| \dot{u}_g(t) \right| \le M_2 \qquad\qquad (14.2)$$

$$\max_{0<t<\infty} \left| u_g(t) \right| \le M_3$$

$$M_5(\omega) \le \left| \ddot{U}_g(\omega) \right| \le M_4(\omega)$$

In Eq. (14.2), $\ddot{U}_g(\omega)$ is the Fourier transform of the ground motion acceleration $\ddot{u}_g(t)$. These constraints involve upper limits on the energy, peak ground acceleration, peak ground velocity, peak ground displacement and upper and lower limits on the Fourier transform of the ground motion. The quantities $E, M_1, M_2, M_3, M_4(\omega)$, $M_5(\omega)$ are estimated based on analysis of a set of past records that were measured at the site under consideration or at other sites with similar soil conditions. They introduced the reliability index, the damage index or the Hasofer-Lind reliability index as the objective function to define performance of the structure.

In the probabilistic approach, the constraints were imposed on the ground motion energy, variance and the entropy that represents the amount of disorder in the ground acceleration (see, e.g., Abbas and Manohar 2002 and Moustafa 2009).

This critical excitation problem has a property such that multiple local maxima exist and the solution depends strongly on the initial guess. However, this approach is general enough to extend to various types of critical excitation problems. This approach has been used in Moustafa and Takewaki's approach that is explained in the next section.

14.3. MOUSTAFA AND TAKEWAKI'S APPROACH

Moustafa and Takewaki extended Abbas and Manohar's approach to various critical excitation problems (Moustafa and Takewaki 2009, 2010a–c, 2011, 2012; Moustafa et al. 2010; Ueno et al. 2010, 2011).

As an example, consider a 2-story shear building model with braces, as shown in Fig. 14.1, as a fail-safe model subjected to a strong ground motion that can be decomposed into a body wave and a surface wave. The frame and braces are placed in parallel and this system has a fail-safe mechanism. The floor masses are $3.0 \times 10^6 \text{(kg)}$ and the story stiffnesses are $1.0 \times 10^5 \text{(N/m)}$.

Figure 14.1 *2-story shear building model with braces as a fail-safe model.*

The story shear force–deformation relation of the fail-safe model is shown in Fig. 14.2. The shear force sustained by the brace in the elastic range is half the total story shear force. The braces are modeled by an elastic-perfectly plastic relationship. The brace yields at the deformation u_{y1} and fails at the deformation u_{y2}. After the failure of the brace, only the frame resists the acceleration input. Fig. 14.3 shows the equilibrium of forces in the frame and brace before and after the brace failure. When a brace fails, the brace force is removed and the corresponding floor acceleration changes suddenly to satisfy

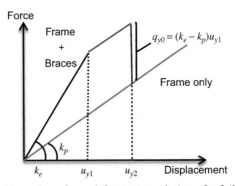

Figure 14.2 *Story shear force-deformation relation of a fail-safe model.*

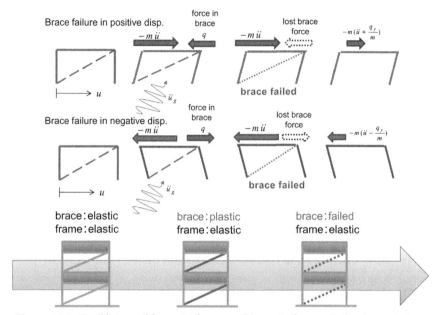

Figure 14.3 *Equilibrium of forces in frame and brace before and after brace failure.*

the equilibrium condition. The fundamental and second natural frequencies of the model with elastic braces are $f_1 = 0.98(\text{Hz}), f_2 = 2.58(\text{Hz})$ and those of the model with only the frame are $f_1 = 0.75(\text{Hz}), f_2 = 2.40(\text{Hz})$. The structural damping ratio of the model is assumed by 0.02, i.e. the damping matrix is proportional to the initial stiffness matrix. The relation of frequency contents of body wave and surface wave with fundamental and second natural frequencies is shown in Fig. 14.4.

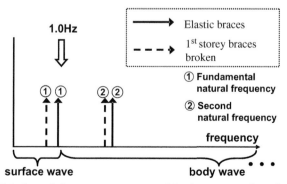

Figure 14.4 *Relation of frequency contents of body wave and surface wave with fundamental and second natural frequencies.*

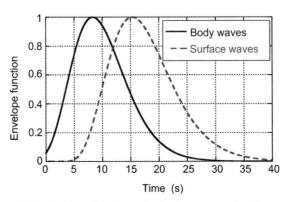

Figure 14.5 *Envelope functions for body wave and surface wave.*

The envelope function of the body wave and the surface wave are shown in Fig. 14.5. This was constructed based on the investigation on the past recorded ground motions. A sample of the body wave is shown in Fig. 14.6(a) and a sample of the surface wave is illustrated in Fig. 14.6(b). A specific frequency content is assumed here. Then the synthesized wave is presented in Fig. 14.6(c).

A worst-case scenario is considered here that the brace yields and fails under the body wave in an early stage and the frame without the brace is resonant to the surface wave in the later stage. The illustrative explanation of this worst-case scenario is shown in Fig. 14.7. The time history of the obtained critical excitation acceleration is illustrated in Fig. 14.8 and its frequency contents are shown in Fig. 14.9. The corresponding story shear force-deformation relations in the two stories are presented in Fig. 14.10. It can be understood that the braces in the first story fail and only the frame resists after that. Fig. 14.11 shows the energy time history for the fail–safe model subjected to the critical input.

14.4. CONCLUSIONS

The conclusions may be stated as follows:

(1) A critical excitation problem of elastic–plastic structures is an interesting problem of significant difficulty. Although statistical equivalent linearization is useful because of its nature to be able to include probabilistic properties of inputs and responses, it is also desired to treat an earthquake ground motion as a deterministic time series since it simulates actual recorded seismographs.

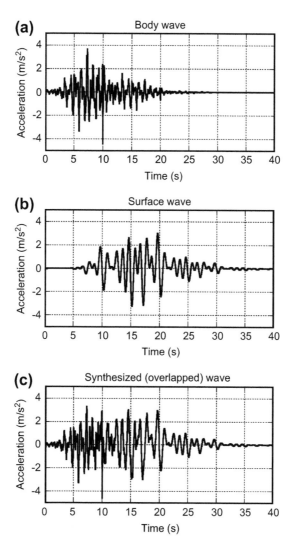

Figure 14.6 *A sample of critical body wave (a), surface wave (b) and synthesized wave (c).*

(2) Abbas (2006) and Moustafa (2009) considered a critical excitation problem for elastic-plastic SDOF and MDOF structures with simple elastic-plastic springs. The earthquake load was modeled as a deterministic time history that is expressed in terms of a Fourier series that is modulated by an enveloping function. The resulting nonlinear optimization problem was tackled by using the sequential quadratic optimization method.

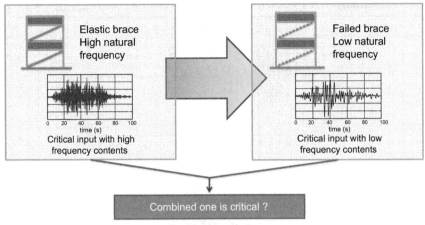

Figure 14.7 *Illustrative explanation of assumed worst-case scenario.*

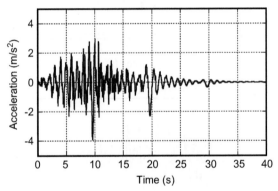

Figure 14.8 *Critical excitation for fail-safe model including body wave and surface wave.*

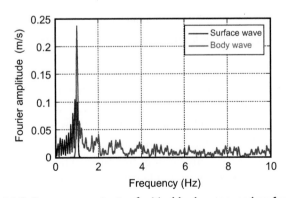

Figure 14.9 *Frequency contents of critical body wave and surface wave.*

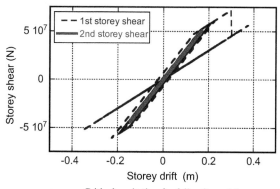

Critical excitation for fail-safe model

Figure 14.10 *Story shear force-deformation relations in two stories.*

(3) A critical excitation problem for elastic–plastic structures under a ground motion consisting of a body wave and a surface wave can be solved by the method due to Abbas (2006). The elastic–plastic structure damaged by the first attack of the body wave may experience further significant damage by the surface wave with longer frequency contents. This scenario has been confirmed by numerical examples exploring various types of seismic waves (body and surface waves) and their influence on damage of the structure under worst-case earthquake inputs.

(4) Future research in this direction needs to focus on extending the methods developed by this author and his co-workers to complicated engineering systems, such as piping systems in nuclear power plants and secondary systems.

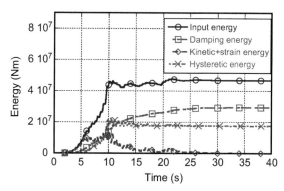

Figure 14.11 *Energy time history for fail-safe model subjected to critical input.*

REFERENCES

Abbas, A.M., 2006. Critical seismic load inputs for simple inelastic structures. Journal of Sound and Vibration 296 (4–5), 949–967.

Abbas, A.M., Manohar, C.S., 2002. Investigations into critical earthquake load models within deterministic and probabilistic frameworks. Earthq. Eng. Struc. Dyn. 31 (4), 813–832.

Abbas, A.M., Manohar, C.S., 2005. Reliability-based critical earthquake load models. Part 2: Nonlinear structures. Journal of Sound and Vibration 287 (4–5), 883–900.

Moustafa, A., 2002. Deterministic/Reliability-based critical earthquake load models for linear/nonlinear engineering structures. Doctoral dissertation, Department of Civil Engineering, Indian Institute of Science, Bangalore, India.

Moustafa, A., 2009. Critical earthquake load inputs for multi-degree-of-freedom inelastic structures. Journal of Sound and Vibration 325 (3), 532–544.

Moustafa, A., 2011. Damage-based design earthquake loads for Single-Degree-Of-Freedom inelastic structures. J. Struct. Eng. 137 (Special Issue), 456–467.

Moustafa, A., Takewaki, I., 2009. Use of probabilistic and deterministic measures to identify unfavorable earthquake records. Journal of Zhejiang University SCIENCE A 10 (5), 619–634.

Moustafa, A., Takewaki, I., 2010a. Deterministic and probabilistic representation of near-field pulse-like ground motion. Soil Dynamics and Earthquake Engineering 30 (5), 412–422.

Moustafa, A., Takewaki, I., 2010b. Critical characterization and modeling of pulse-like near-fault strong ground motion. Struct. Eng. & Mech. 34 (6), 755–778.

Moustafa, A., Takewaki, I., 2010c. Modeling critical ground-motion sequences for inelastic structures. Advances in Structural Engineering 13 (4), 665–679. Erratum in Advances in Structural Engineering, 14(1), 121, (2011).

Moustafa, A., Takewaki, I., 2011. Response of nonlinear single-degree-of-freedom structures to random acceleration sequences. Engineering Structures 33 (4), 1251–1258.

Moustafa, A., Takewaki, I., 2012. Earthquake ground motion of multiple sequences and associated structural response. Earthquakes and Structures 3 (5), 629–647.

Moustafa, A., Ueno, K., Takewaki, I., 2010. Critical earthquake loads for SDOF inelastic structures considering evolution of seismic waves. Earthquakes and Structures 1 (2), 147–162.

Ueno, K., Moustafa, A., Takewaki, I., 2010. Critical earthquake loads for passively controlled inelastic building structures considering evolution of seismic waves. Proc. of the 5th World Conference on Structural Control and Monitoring (5WCSCM). Tokyo.

Ueno, K., Fujita, K., Moustafa, A., Takewaki, I., 2011. Critical input for inelastic structures under evolving seismic waves. J. Structural and Construction Engineering 659, 79–87 (in Japanese).

CHAPTER FIFTEEN

Earthquake Resilience Evaluation of Building Structures with Critical Excitation Methods

Contents

15.1. INTRODUCTION

The term "earthquake resilience" is frequently used, especially after the Tohoku earthquake, which occurred off the Pacific coast of Japan on March 11, 2011. Earthquake resilience implies the ability to recover from some damaged states or the ability not to be damaged against various disturbances. When structural designers try to investigate earthquake resilience, they have to evaluate the earthquake performances of building structures with various uncertainties under a wide range of earthquake ground motions, preferably for critical excitation.

An efficient methodology is required to evaluate the robustness (variability of response) of a building with uncertain structural properties under uncertain ground motions. It is well known that base–isolated buildings and structural controlled buildings have large structural uncertainties due to wide variability of structural properties caused by temperature and frequency dependencies, manufacturing errors and the aging effect on earthquake resistant buildings. Furthermore, after the worst disaster of 2011, the Tohoku earthquake (Takewaki et al. 2011b; Takewaki et al. 2012a, b;

Takewaki and Fujita 2012), it is under discussion that base-isolated buildings are vulnerable to unexpected long-period ground motions.

Under these circumstances, it is desirable to evaluate the response variability caused by such structural variability and uncertain ground motions (Takewaki et al. 2011a, b). The method based on the convex model may be one possibility (Ben-Haim and Elishakoff 1990). Introduction of a bound on Fourier amplitude of input ground motions may be another approach (Takewaki and Fujita 2009). Independently, Kanno and Takewaki (2005, 2006a) proposed an efficient and reliable method for evaluating the robustness of structures under uncertainties based on the concept of the robustness function (Ben-Haim 2001; Takewaki and Ben-Haim 2005, 2008). However, it does not appear that an efficient and reliable method for evaluating the robustness of structures has been proposed.

An interval analysis is believed to be one of the most efficient and reliable methods to respond to this requirement. While a basic assumption of *"inclusion monotonic"* is introduced in usual interval analysis, the possibility should be taken into account of an occurrence of the extreme value of the objective function in an inner feasible domain of the interval parameters for more accurate and reliable evaluation of the objective function. It is shown that the critical combination of the structural parameters can be derived explicitly by maximizing the objective function by the use of the second-order Taylor series expansion. This method is called the Updated Reference-Point (URP) method (Fujita and Takewaki 2011a–c). This method is applicable to the problems for which the response functions are available for uncertain parameters. When we deal with nonlinear problems, it is necessary to combine the URP method (Fujita and Takewaki 2011a–c) with a kind of response surface method (Fujita and Takewaki 2012a).

15.2. ROBUSTNESS, REDUNDANCY AND RESILIENCE

The concepts of robustness, redundancy and resilience are closely interrelated. In general, robustness means insensitiveness of a system to parameter variation (Ben-Haim 2001; Takewaki and Ben-Haim 2005; Kanno and Takewaki 2005, 2006a–c, 2007; Takewaki 2008a). On the other hand, redundancy indicates the degree of safety, frequently expressed by a safety factor (Doorn and Hansson 2011) of a system against disturbances or the connectivity of components. In the latter meaning, a parallel system is regarded as a preferable system able to avoid overall system failure (the fail–safe system is a representative one). Resilience can be regarded as an ability of a system to recover from a damaged state or resist external disturbances and seems to be a more generic concept including robustness and redundancy (Takewaki et al. 2011a, 2012b). Recently the concept of resilience is getting much interest in broad fields of society (Ellingwood et al. 2006; Takewaki et al. 2011a; Committee on National Earthquake Resilience 2011; Poland 2012). In Committee on National Earthquake Resilience (2011), there are some explanations. The following are from this reference.

"The capability of an asset, system, or network to maintain its function or recover from a terrorist attack or any other incident" (DHS, 2006).

"The capacity of a system, community or society potentially exposed to hazards to adapt, by resisting or changing in order to reach and maintain an acceptable level of functioning and structure. This is determined by the degree to which the social system is capable of organizing itself to increase this capacity for learning from past disasters for better future protection and to improve risk reduction measures" (UN ISDR, 2006).

"The ability of social units (e.g. organizations, communities) to mitigate risk and contain the effects of disasters, and carry out recovery activities in ways that minimize social disruption while also minimizing the effects of future disasters. Disaster Resilience may be characterized by reduced likelihood of damage to and failure of critical infrastructure, systems, and components; reduced injuries, lives lost, damage, and negative economic and social impacts; and reduced time required to restore a specific system or set of systems to normal or pre-disaster levels of functionality" (MCEER, 2008).

As stated in Committee on National Earthquake Resilience (2011), the term resilience is often used loosely and inconsistently. After some discussions, the following definition is summarized in Committee on National Earthquake Resilience (2011).

A disaster-resilient nation is one in which its communities, through mitigation and predisaster preparation, develop the adaptive capacity to maintain important community functions and recover quickly when major disasters occur.

To investigate "earthquake resilience" in more detail, the following 18 tasks are considered in Committee on National Earthquake Resilience (2011):

Task 1: Physics of Earthquake Processes
Task 2: Advanced National Seismic System
Task 3: Earthquake Early Warning

Task 4: National Seismic Hazard Model

Task 5: Operational Earthquake Forecasting

Task 6: Earthquake Scenarios

Task 7: Earthquake Risk Assessments and Applications

Task 8: Post-earthquake Social Science Response and Recovery Research

Task 9: Post-earthquake Information Management

Task 10: Socioeconomic Research on Hazard Mitigation and Recovery

Task 11: Observatory Network on Community Resilience and Vulnerability

Task 12: Physics-based Simulations of Earthquake Damage and Loss

Task 13: Techniques for Evaluation and Retrofit of Existing Buildings

Task 14: Performance-based Earthquake Engineering for Buildings

Task 15: Guidelines for Earthquake-Resilient Lifeline Systems

Task 16: Next Generation Sustainable Materials, Components, and Systems

Task 17: Knowledge, Tools, and Technology Transfer to Public and Private Practice

Task18: Earthquake-Resilient Communities and Regional Demonstration Projects

15.3. REPRESENTATION OF UNCERTAINTY IN SELECTING DESIGN GROUND MOTIONS

The properties of earthquake ground motions are highly uncertain both in an epistemic and aleatory sense and it is believed to be a hard task to predict forthcoming events precisely (Geller et al. 1997; Stein 2003; Aster 2012). It has been made clear through numerous investigations that near-field ground motions (Northridge 1994, Kobe 1995, Turkey 1999 and Chi-Chi, Taiwan 1999) and the far-field motions (Mexico 1985, Tohoku 2011) have some peculiar, unpredictable characteristics.

In the history of earthquake resistant design of building structures, we have learned a lot of lessons from actual earthquake disasters such as the Nobi earthquake 1891 (Japan) and San Francisco earthquake 1906 (USA). After we encountered a major earthquake disaster, we upgraded the earthquake resistant design codes many times. However, the repetition of this revision never resolves the essential problem. To overcome this problem, the concept of critical excitation was introduced. Although the concept of active

structural control has been developed since around 1980, the actual installation of those devices has further difficulties. Based on these observations,
approaches based on the concept of "critical excitation" seem to be
promising. Drenick (1970) formulated this problem in a mathematical
framework and many researchers followed him. The detailed history can be
found in Takewaki 2007.

After the Tokachi-oki earthquake in 2003, long-period ground motions
have been receiving much interest in the field of earthquake resistant design.
The most difficult problem is that these long-period ground motions are
highly uncertain in nature and the existence of these ground motions was
not known during the construction of high-rise and super high-rise buildings. In order to take into account such highly uncertain long-period ground
motions, a new paradigm is desired. There are various buildings in a city
(Fig. 15.1). Each building has its own natural period of amplitude-
dependency and its original structural properties. When an earthquake
occurs, a variety of ground motions are induced in the city, e.g. a combination of body waves (including pulse wave) and surface waves, long-period

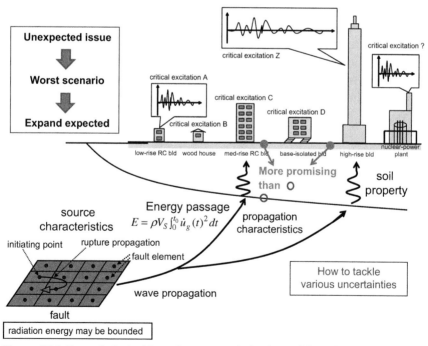

Figure 15.1 *Scenario to tackle various uncertainties in modeling design earthquake
ground motions. RC, reinforced concrete. From Takewaki (2008b).*

ground motions. The relation of the building natural period with the predominant period of the induced ground motion may lead to disastrous phenomena in the city (see Fig. 15.1). In other words, the most critical issue in seismic resistant design is resonance. Many past earthquake observations demonstrated such phenomena repeatedly, e.g. Mexico 1985, Northridge 1994, Kobe 1995. One of the promising approaches to this is to shift the natural period of the building through structural control (Takewaki 2009) and to add damping in the building. However, it is also true that structural control is developing now and more time is necessary to respond properly to uncertain ground motions.

It is discussed and believed that an earthquake has a bound on its magnitude and the earthquake energy radiated from the fault has a bound (Trifunac 2008). The problem is to find the most unfavorable ground motion for a building or a group of buildings (see Fig. 15.1). There are two possibilities in the specification of such bounds. One is to define a velocity power at the bottom of the basin based on the fault rupture mechanism and wave propagation characteristics. The other is to set the velocity power at the ground surface level (Takewaki and Tsujimoto 2011). In the case of definition at the bottom of the basin, the surface ground wave propagation has to be considered properly. However, this procedure may include another uncertainty. In this sense, the setting of the velocity power at the ground surface level may be preferable.

It may be interesting to present a case study on the setting of such a bound. The Fourier spectrum of a ground motion acceleration has been proposed at the rock surface depending on the seismic moment M_0, distance R from the fault, etc. (e.g. Boore 1983).

$$|A(\omega)| = CM_0 S(\omega, \omega_C) P(\omega, \omega_{\max}) \exp(-\omega R/(2\beta Q_\beta))/R \qquad (15.1)$$

Such a spectrum may contain uncertainties. One possibility or approach is to specify the acceleration or velocity power (Takewaki 2007) as a global measure and allow the variability of the spectrum. As for the Great East Japan Earthquake, $|A(\omega)|$ is reported to be about 0.5(m/s) near the fault region. However, this treatment has difficulty in confirming the reliability of the theory and of specification of the fault site. The change of ground motion by surface soil conditions is another difficulty. Based on this observation, a concept of critical excitation is introduced.

A significance of critical excitation methods can be explained by investigating the role of buildings in a city. In general there are two classes of buildings in a city. One is the important building, which plays an important role during and after disastrous earthquakes. The other is the ordinary building. The former should not be damaged during an earthquake and the latter may be damaged to some extent especially for critical excitation larger than code-specified design earthquakes. Just as the investigation on limit states of structures plays an important role in the specification of response limits and performance levels of structures during disturbances, the clarification of critical excitations for a given structure or a group of structures appears to provide structural designers with useful information in determining excitation parameters in a risk-based reasonable way. It is expected that the concept of critical excitation enables structural designers to make ordinary buildings more seismic-resistant and seismic-resilient (Takewaki et al. 2012b).

15.4. UNCERTAINTY EXPRESSION IN TERMS OF INFO-GAP MODEL

In this section, let us introduce and explain a new concept of structural design that combines load and structural uncertainties. For this purpose, it is absolutely necessary to identify the critical excitation and the corresponding critical set of structural model parameters. It is well recognized that the critical excitation depends on the structural model parameters and it is quite difficult to deal with load uncertainties and structural model parameter uncertainties simultaneously. In order to tackle these difficult problems, info–gap models of uncertainty (nonprobabilistic uncertainty models) by Ben-Haim (2001) are used. This concept enables the representation of uncertainties that exist in the load (input ground acceleration) and in parameters of the vibration model of the structure.

As a simple example, let us consider a vibration model, as shown in Fig. 15.2, with viscous dampers in addition to masses and springs. It is well

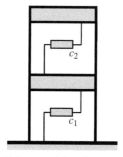

Figure 15.2 *Shear building model with uncertain viscous dampers.*

recognized in the field of structural control and health monitoring that viscous damping coefficients c_i of dampers in a vibration model are quite uncertain compared to masses and stiffnesses. By using a specific method for describing such uncertainty, the uncertain viscous damping coefficient of a damper can be expressed in terms of the nominal value $\tilde{c}_i$ and the unknown uncertainty level α as shown in Fig. 15.3 (Takewaki and Ben-Haim 2005).

$$C(\alpha, \tilde{\mathbf{c}}) = \left\{ \mathbf{c} : \left| \frac{c_i - \tilde{c}_i}{\tilde{c}_i} \right| \leq \alpha, \quad i = 1, \cdots, N \right\}, \alpha \geq 0 \tag{15.2a}$$

The inequality in Eq. (15.2a) can be rewritten as

$$(1 - \alpha)\tilde{c}_i \leq c_i \leq (1 + \alpha)\tilde{c}_i. \tag{15.2b}$$

This description is the same one used in the interval analysis (Moore 1966; Mullen and Muhanna 1999; Koyluoglu and Elishakoff 1998).

[Info-gap robustness function]

An uncertainty analysis called "the info-gap uncertainty analysis" was introduced by Ben-Haim (Ben-Haim 2001) for measuring the robustness (the degree of response insensitiveness to uncertain parameters) of a structure subjected to external loads. Simply speaking, the info-gap robustness is the greatest horizon of uncertainty, α, up to which the performance function $f(\mathbf{c}, \mathbf{k})$ does not exceed a critical value, f_C. The performance function may be a peak displacement, peak stress or earthquake input energy, and so on.

Let us define the following info-gap robustness function corresponding to the info-gap uncertainty model represented by Eq. (15.1a).

$$\hat{\alpha}(\mathbf{k}, f_C) = \max\{\alpha : \{ \max_{\mathbf{c} \in C(\alpha, \tilde{\mathbf{c}})} f(\mathbf{c}, \mathbf{k})\} \leq f_C\} \tag{15.3}$$

An illustrative explanation can be seen in Fig. 15.4.

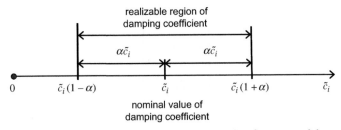

Figure 15.3 *Description of uncertainty with info-gap model.*

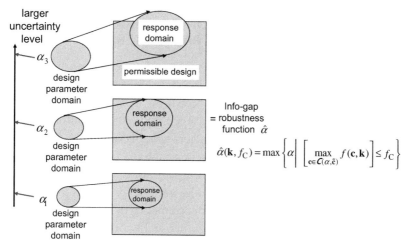

Figure 15.4 *Illustrative representation of concept of info-gap robustness function.*

Let us put $f_{C0} = f(\tilde{\mathbf{c}}, \mathbf{k})$ for the nominal damping coefficients. Then one can show that $\hat{\alpha}(\mathbf{k}, f_{C0}) = 0$ for the specific value f_{C0}, as shown in Fig. 15.5. Furthermore, let us define $\hat{\alpha}(\mathbf{k}, f_C) = 0$ if $f_C \leq f_{C0}$ (see Fig. 15.5). This means that when the performance requirement is too small, we cannot satisfy the performance requirement for any admissible damping coefficients. The definition in Eq. (15.3) also implies that the robustness is the maximum level of the structural model parameter uncertainty, α, satisfying the performance requirement $f(\mathbf{c}, \mathbf{k}) \leq f_C$ for all admissible variation of the structural model parameter represented by Eq. (15.1a).

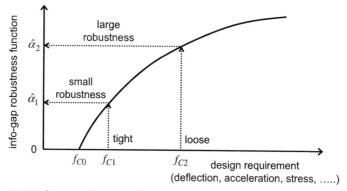

Figure 15.5 *Info-gap robustness function for tight and loose design requirements.*

15.5. WORST COMBINATION OF STRUCTURAL PARAMETERS AND INPUT PARAMETERS

Consider a general problem, as shown in Fig. 15.6, of finding a worst case under uncertainties of structural parameters and input ground motion parameters. The problem without uncertainty in input ground motion parameters was considered in Fujita and Takewaki (2012b) and the problem without uncertainty in structural parameters was investigated by Fujita et al. (2010). While the domain satisfying the constraints is referred to as the feasible domain, the domain defined by the info-gap model is called the info-gap domain. The case is meaningful where the info-gap domain is just included in the feasible domain. The edge point corresponds to the worst case.

The most challenging part is how to find such worst case in which both uncertainties of structural parameters and input ground motion parameters are taken into account. The worst case of input ground motion parameters is the function of structural parameters and their uncertainty levels. This relationship is extremely complicated and this problem can be a principal subject in the field of critical excitation.

As a promising method for investigating this subject, interval analysis and related methods have been developed (see e.g. Moore 1966; Alefeld and Herzberger 1983; Qiu et al. 1996; Mullen and Muhanna 1999; Koyluoglu and Elishakoff 1998; Qiu 2003; Chen and Wu 2004; Chen et al. 2009; Fujita and Takewaki 2011a–c).

Fig. 15.7 shows the objective functions in the cases of monotonic inclusion and nonmonotonic inclusion. In order to solve this problem of interval analysis, Fujita and Takewaki (2011a–c) developed two new methods. One is the fixed reference-point method (FRP method) shown in Fig. 15.8 and the other is the updated reference-point method (URP method) shown in Fig. 15.9.

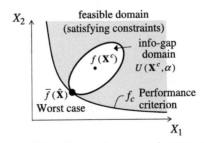

Figure 15.6 *Info-gap domain and worst case.*

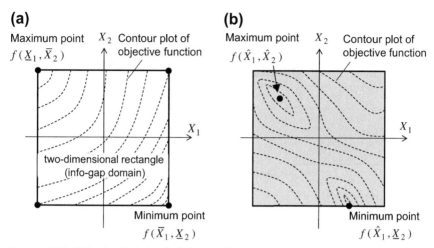

Figure 15.7 *Objective function in info-gap domain: (a) monotonic inclusion, (b) non-monotonic inclusion.*

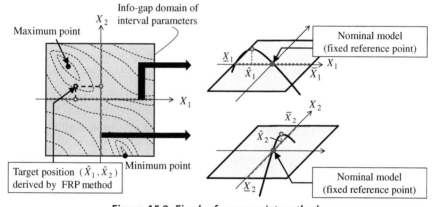

Figure 15.8 *Fixed reference-point method.*

15.6. REALITY OF RESONANCE AND ITS INVESTIGATION

The resonance of buildings with input ground motions has been an issue of great interest in earthquake structural engineering for a long time. Some actual examples come from Mexico 1985, Northridge 1994, Kobe 1995, and Tokachi-oki 2003. Ariga et al. (2006) discussed the issue for base-isolated buildings subjected to long-period ground motion after the experience of the Tomakomai ground motion during Tokachi-oki 2003. They drew attention to the large deformation of base-isolated high-rise buildings

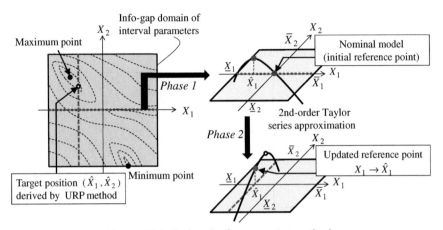

Figure 15.9 *Updated reference-point method.*

with rather long fundamental natural period. Takewaki et al. (2011b) investigated the response of super high-rise buildings in Tokyo and Osaka during Tohoku 2011. Some clear observations will be explained in the following.

The response of a 55-story super high-rise building in Osaka (height = 256m: T_1 = 5.8s (long-span direction), 5.3s (short-span direction)) is very famous because this building is owned by a public community and the data are openly available to many people. The building was shaken intensively regardless of the fact that Osaka is located about 800km from the epicenter (about 600km from the boundary of the fault region) and the JMA instrumental intensity was 3 in Osaka (Takewaki et al. 2011b; Takewaki et al. 2012a, b). Through the post-earthquake investigation, the natural periods of the building were found to be longer than the design values mentioned above reflecting the flexibility of pile-ground systems, the increase of a mass at the top and the damage to nonstructural partition walls and so on. It should be pointed out that the level of velocity response spectra of ground motions observed here (first floor) is almost the same as that at the Shinjuku station (K–NET) in Tokyo, and the top-story displacements are about 1.4m (short-span direction) and 0.9m (long-span direction). Most of the data for buildings at Shinjuku, Tokyo are not openly available because of the data release problem. Once these data are made available, the vibration equivalent to this Osaka's building may be reported.

Fig. 15.10 shows the ground acceleration, ground velocity and top-story displacement recorded or numerically integrated in this building. It can be observed that a clear resonant phenomenon occurs

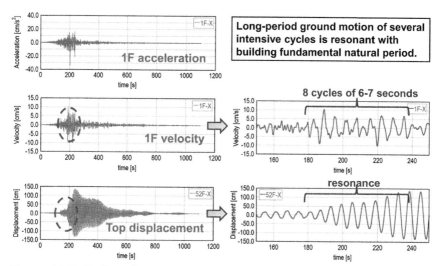

Figure 15.10 *Reality of resonance: ground acceleration, velocity and top-story displacement of a 55-story building in Osaka.*

during about eight cycles (ground fundamental natural period can be evaluated by $4H/V_s = 4 \times 1.6/1.0 = 6.4s$). It seems that such a clear observation has never been reported in super high-rise buildings around the world. This implies the need for consideration and code-specification of long-period ground motions in the seismic resistant design of super high-rise buildings in mega cities even though the site is far from the epicenter. It is also being discussed that the expected Tokai, Tonankai and Nankai event is closer to this building (about 160km from the boundary of the fault region) and several times the size of the ground motion may be induced during that event based on the assumption that body waves are predominant outside of the Osaka basin. However, the nonlinearity of surface ground and other uncertain factors may influence the amplification. Further investigation will be necessary. Seismic retrofitting using hysteretic steel dampers, oil dampers and friction dampers is being planned.

Fig. 15.11 shows the influence of resonance, duration of ground motion and damping deterioration on input energy response. The mechanism is investigated in detail on the increase of credible bound of input energy for the velocity power constraint due to uncertainties in input excitation duration (lengthening) and in structural damping ratio (decrease). As for uncertainties in the excitation predominant period and in the natural period of a structure, the resonant case is critical and corresponds to the worst case

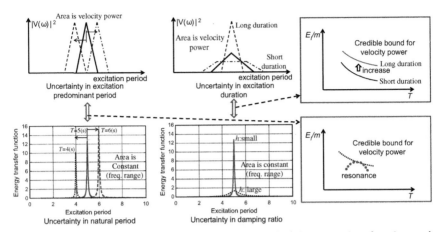

Figure 15.11 *Effect of input motion predominant period, input motion duration and structural damping ratio uncertainties on bound of input energy: Resonance between input and structure and increase of credible bound of input energy for velocity power constraint due to lengthening of input excitation duration and decrease of damping ratio of structure.*

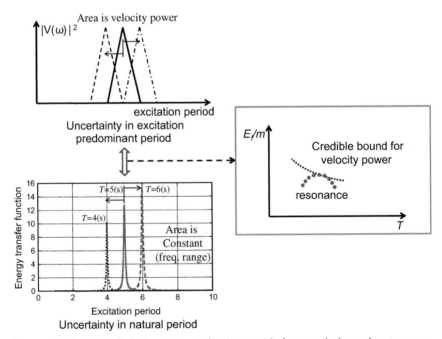

Figure 15.12 *Effect of input motion predominant period uncertainties on input energy: resonance between input and structure.*

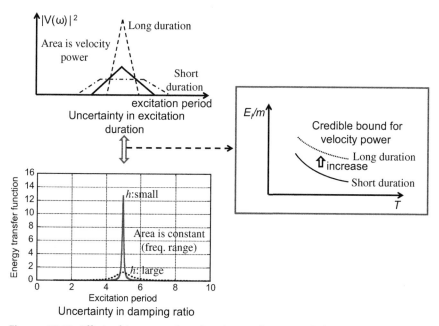

Figure 15.13 *Effect of input motion duration and structural damping ratio uncertainties on bound of input energy: Increase of credible bound of input energy for velocity power constraint due to lengthening of input excitation duration and decrease of damping ratio of structure.*

(Fig. 15.12). It can be understood from Fig. 15.13 that the lengthening of input excitation duration and decrease of structural damping ratio due to damping mechanism deterioration may have caused large input in the super high-rise building in Osaka bay area mentioned above. In particular, the decrease of the structural damping ratio induces unsmoothing of the energy spectrum and the period region to increase the energy spectrum happened to coincide with the fundamental natural period of the super high-rise building (see Takewaki et al. 2012a, b).

15.7. CONCLUSIONS

The conclusions may be stated as follows:

(1) The concepts of robustness, redundancy and resilience are closely interrelated. In general, robustness means insensitiveness of a system to parameter variation. On the other hand, redundancy indicates the degree of safety of a system against disturbances or the parallel system avoiding overall system failure (the fail-safe system is a representative one). Resilience can be regarded as an ability of a system to recover from a damaged state or resist external disturbances and seems to be a more generic concept including robustness and redundancy.

(2) Several uncertainties in earthquake ground motions can be explained by a model of Boore. By introducing the input energy bound, structural designers can restrict such uncertainties to a limited level.

(3) The info-gap model can be an approach to describe uncertainty in earthquake ground motions and structural parameters. The method to obtain the worst combination of structural parameters and input parameters is desired. The fixed reference-point method and updated reference-point method based on interval analysis can be promising methods.

(4) The influence of resonance, duration of ground motion and damping deterioration on input energy response can be investigated with the credible bound theory.

REFERENCES

Alefeld, G., Herzberger, J., 1983. Introduction to Interval Computations. Academic Press, New York.

Ariga, T., Kanno, Y., Takewaki, I., 2006. Resonant behaviour of base-isolated high-rise buildings under long-period ground motions. Struct. Design Tall Spec. Build. 15 (3), 325–338.

Aster, R., 2012. Expecting the unexpected: black swans and seismology. Seismological Research Letters 83 (1), 5–6.

Ben-Haim, Y., 2001. Information-Gap Decision Theory: Decisions under Severe Uncertainty. Academic Press, San Diego.

Ben-Haim, Y., Elishakoff, I., 1990. Convex Models of Uncertainty in Applied Mechanics. Elsevier, Amsterdam.

Boore, D.M., 1983. Stochastic simulation of high-frequency ground motions based on seismological models of the radiated spectra. Bulletin of the Seismological Society of America 73 (6A), 1865–1894.

Chen, S.H., Wu, J., 2004. Interval optimization of dynamic response for structures with interval parameters. Comp. Struct. 82 (1), 1–11.

Chen, S.H., Ma, L., Meng, G.W., Guo, R., 2009. An efficient method for evaluating the natural frequencies of structures with uncertain-but-bounded parameters. Comp. Struct. 87 (9–10), 582–590.

Committee on National Earthquake Resilience–Research, Implementation, and Outreach; Committee on Seismology and Geodynamics; National Research Council, 2011. National Earthquake Resilience: Research, Implementation, and Outreach. The National Academies Press, Washington, D.C.

DHS (U.S. Department of Homeland Security), 2006. National Infrastructure Protection Plan. Available at. www.fas.org/irp/agency/dhs/nipp.pdf (accessed August 11, 2012).

Doorn, N., Hansson, S.O., 2011. Should probabilistic design replace safety factors? Philos. Technol. 24 (2), 151–168.

Drenick, R.F., 1970. Model-free design of aseismic structures. J. Engrg. Mech. Div. 96 (4), 483–493.

Ellingwood, B.R., et al., 2006. Best practices for reducing the potential for progressive collapse in buildings. USA: National Institute of Standards and Technology, Gaithersburg, MD, USA.

Fujita, K., Takewaki, I., 2011a. Earthquake response bound analysis of uncertain base-isolated buildings for robustness evaluation. J. Structural and Construction Engineering 666, 1453–1460 (in Japanese).

Fujita, K., Takewaki, I., 2011b. An efficient methodology for robustness evaluation by advanced interval analysis using updated second-order Taylor series expansion. Eng. Struct. 33 (12), 3299–3310.

Fujita, K., Takewaki, I., 2011c. Sustainable building design under uncertain structural-parameter environment in seismic-prone countries. Sustainable Cities and Society 1 (3), 142–151.

Fujita, K., Takewaki, I., 2012a. Robustness evaluation on earthquake response of base-isolated buildings with uncertain structural properties under long-period ground motions. Architectoni.ca Journal 1 (1), 46–59.

Fujita, K., Takewaki, I., 2012b. Robust passive damper design for building structures under uncertain structural parameter environments. Earthquakes and Structures 3 (6), 805–820.

Fujita, K., Moustafa, A., Takewaki, I., 2010. Optimal placement of viscoelastic dampers and supporting members under variable critical excitations. Earthquakes and Structures 1 (1), 43–67.

Geller, R.J., Jackson, D.D., Kagan, Y.Y., Mulargia, F., 1997. Earthquakes cannot be predicted. Science 275 (5306), 1616.

Kanno, Y., Takewaki, I., 2005. Approximation algorithm for robustness functions of trusses with uncertain stiffness under uncertain forces, J. Structural and Construction Engineering. No. 591, 53–60 (in Japanese).

Kanno, Y., Takewaki, I., 2006a. Robustness analysis of trusses with separable load and structural uncertainties. Int. J. Solids and Structures 43 (9), 2646–2669.

Kanno, Y., Takewaki, I., 2006b. Sequential semidefinite program for maximum robustness design of structures under load uncertainty. Journal of Optimization Theory and Application 130 (2), 265–287.

Kanno, Y., Takewaki, I., 2006c. Confidence Ellipsoids for Static Response of Trusses with Load and Structural Uncertainties. Computer Methods in Applied Mechanics and Engineering 196 (1–3), 393–403.

Kanno, Y., Takewaki, I., 2007. Worst case plastic limit analysis of trusses under uncertain loads via mixed 0-1 programming. Journal of Mechanics of Materials and Structures 2 (2), 247–273.

Koyluoglu, H.U., Elishakoff, I., 1998. A comparison of stochastic and interval finite elements applied to shear frames with uncertain stiffness properties. Comp. and. Struct. 67 (1–3), 91–98.

MCEER (Multidisciplinary Center for Earthquake Engineering Research), 2008. MCEER research: Enabling disaster-resilient communities. Available at. Seismic Waves (November), 1–2. www.nehrp.gov/pdf/SeismicWavesNov08.pdf (accessed August 11, 2012).

Moore, R.E., 1966. Interval Analysis. Prentice-Hall, Englewood Cliffs, New Jersey.

Mullen, R., Muhanna, R., 1999. Bounds of structural response for all possible loading combinations. J. Struct. Eng. 125 (1), 98–106.

Poland, C., 2012. Creating disaster resilient cities. In: Bouquet, Alan (Ed.), Sustainable Cities. Adam Nethersole Publisher, London, pp. 136–138.

Qiu, Z., 2003. Comparison of static response of structures using convex models and interval analysis method. Int. J. Numer. Meth. Engng. 56 (12), 1735–1753.

Qiu, Z., Chen, S., Song, D., 1996. The displacement bound estimation for structures with an interval description of uncertain parameters. C. Numer. Meth. Engng. 12 (1), 1–11.

Stein, R.S., 2003. Earthquake conversations. Scientific American 288 (1), 72–79.

Takewaki, I., 2007. Critical Excitation Methods in Earthquake Engineering. Elsevier, Oxford.

Takewaki, I., 2008a. Robustness of base-isolated high-rise buildings under code-specified ground motions. Struct. Design Tall Spec. Build. 17 (2), 257–271.

Takewaki, I., 2008b. Critical excitation methods for important structures. Invited as a Semi-Plenary Speaker, EURODYN 2008, July 7–9, Southampton, England.

Takewaki, I., 2009. Building Control with Passive Dampers: Optimal Performance-based Design for Earthquakes. John Wiley & Sons (Asia), Singapore.

Takewaki, I., Ben-Haim, Y., 2005. Info-gap robust design with load and model uncertainties. J. Sound & Vibration, Special Issue: Uncertainty in Structural Dynamics 288 (3), 551–570.

Takewaki, I., Ben-Haim, Y., 2008. Info-gap Robust Design of Passively Controlled Structures with Load and Model Uncertainties, Chapter 19. In: Tsompanakis, Yiannis, Nikos, D., Lagaros, Papadrakakis, Manolis, Taylor, Francis (Eds.), Structural Design Optimization Considering Uncertainties, pp. 531–548.

Takewaki, I., Fujita, K., 2009. Earthquake Input Energy to Tall and Base-isolated Buildings in Time and Frequency Dual Domains. Struct. Design Tall Spec. Build. 18 (6), 589–606 (2008 Paper of the Year).

Takewaki, I., Fujita, K., 2012. Tohoku (Japan) earthquake and its impact on design of super high-rise buildings. Proc. of ICTAM2012 Congress (The 23rd International Congress on Theoretical and Applied Mechanics). August 19–24, 2012, Beijing, China.

Takewaki, I., Tsujimoto, H., 2011. Scaling of design earthquake ground motions for tall buildings based on drift and input energy demands. Earthquakes and Structures 2 (2), 171–187.

Takewaki, I., Fujita, K., Yamamoto, K., Takabatake, H., 2011a. Smart Passive Damper Control for Greater Building Earthquake Resilience in Sustainable Cities. Sustainable Cities and Society 1 (1), 3–15.

Takewaki, I., Murakami, S., Fujita, K., Yoshitomi, S., Tsuji, M., 2011b. The 2011 off the Pacific coast of Tohoku earthquake and response of high-rise buildings under long-period ground motions. Soil Dynamics and Earthquake Engineering 31 (11), 1511–1528.

Takewaki, I., Fujita, K., Yoshitomi, S., 2012a. Uncertainties of long-period ground motion and its impact on building structural design. Proc. of One Year after 2011 Great East Japan Earthquake, International Symposium on Engineering Lessons Learned from the Giant Earthquake. March 3–4, Tokyo, 1005–1016.

Takewaki, I., Moustafa, A., Fujita, K., 2012b. Improving the Earthquake Resilience of Buildings: The Worst Case Approach. Springer, London.

Trifunac, M., 2008. Energy of strong motion at earthquake source. Soil Dyn. Earthq. Engrg. 28 (1), 1–6.

UN ISDR (United Nations International Strategy for Disaster Reduction), 2006. Hyogo Framework for Action 2005–2015: Building the Resilience of Nations and Communities to Disasters. Extract from the final report of the World Conference on Disaster Reduction (A/CONF.206/6). March 16, 2005.

INDEX

Note: Page numbers with "f" denote figures.

Printed and bound by CPI Group (UK) Ltd, Croydon, CR0 4YY

08/05/2025

01864854-0001